PSAM 12
Probabilistic Safety Assessment and Management
22–27 June 2014 • Sheraton Waikiki, Honolulu, Hawaii, USA

CONFERENCE PROCEEDINGS
Volume 5 - Tuesday PM I

PSAM 12

Probabilistic Safety Assessment and Management

22 - 27 June, 2014

Sheraton Waikiki, Honolulu, Hawaii USA

CONFERENCE PROCEEDINGS

Volume 5

Tuesday PM I

Foreword

It is was our honor to welcome you to Honolulu, Hawaii, for the twelfth rendition of the Probabilistic Safety Assessment and Management (PSAM) Conference. The planning for PSAM Honolulu began back in 2007 (before PSAM 9 in Hong Kong), when we looked at several locations around the United States, included Arizona, California, Boston, and even considered locations in Oceania. Based upon the feedback both during and after the conference, PSAM 12 proved to be a great success.

We would like to thank all of the volunteers, those that served before, during, and after the Conference. Members of the Technical Program Committee, the Organizing Committee, the session chairs, and the presenters have our gratitude for making PSAM 12 the most memorable PSAM yet.

This publication represents the technical proceedings for the Conference. Due to the large number of published papers (a total of 391), we have subdivided the technical content (papers) into five volumes, one for each day of the conference.

On behalf of the International Association for Probabilistic Safety Assessment and Management Board of Directors, we hope that this publication will provide a valuable technical resource in addition to a reminder of the memorable stay in the Hawaiian Islands.

Dr. Curtis Smith
Technical Program Chairs

Dr. Todd Paulos
General Chair

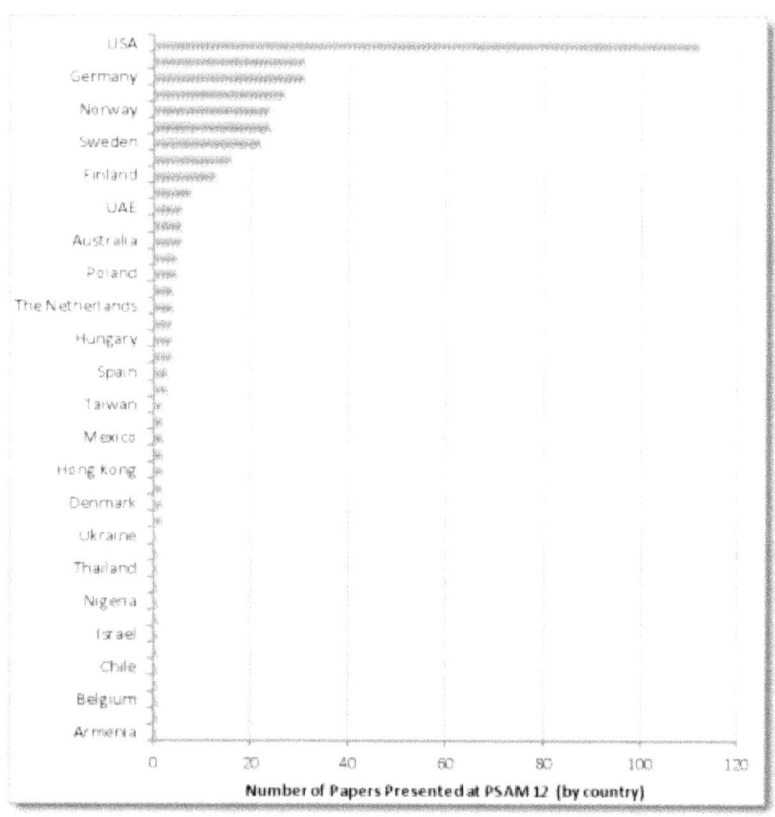

Number of Papers Presented at PSAM 12 (by country)

Sponsors

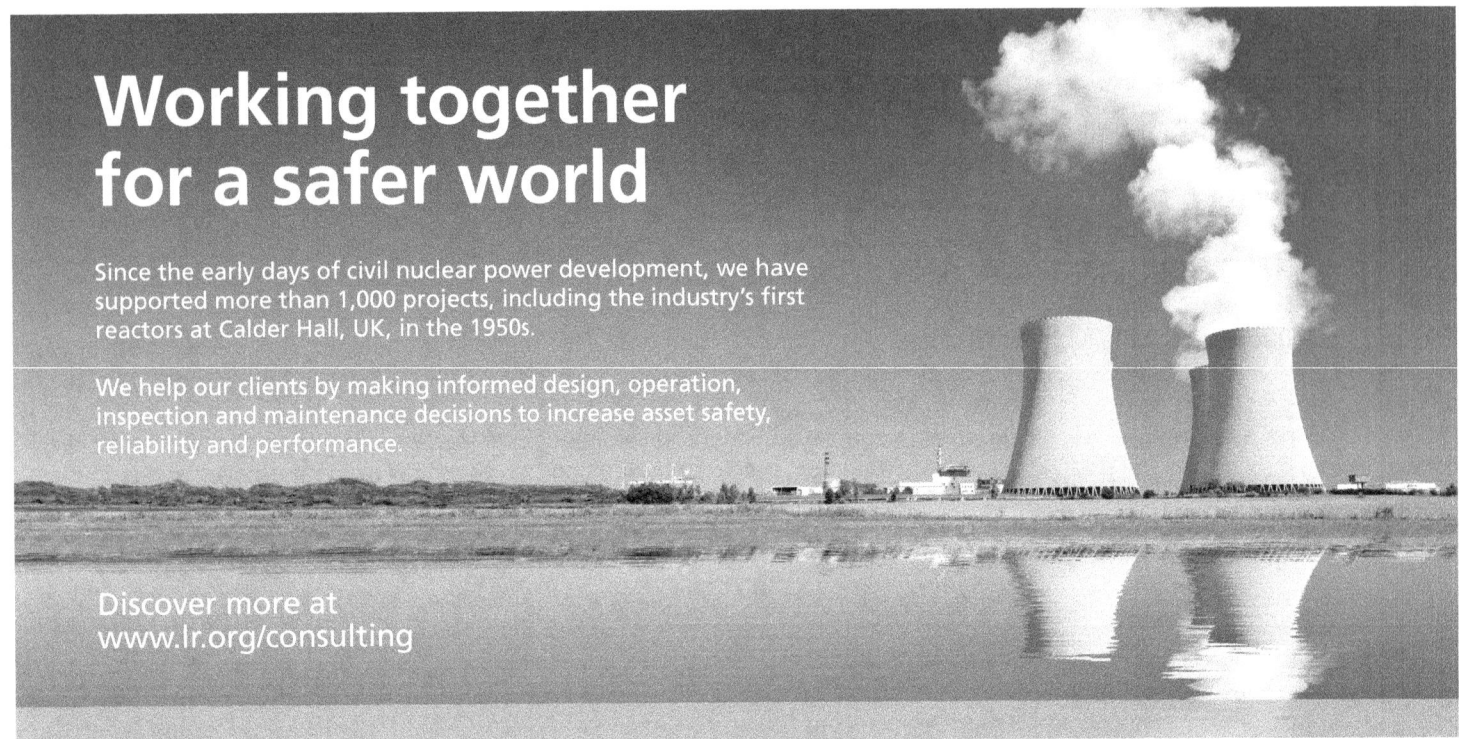

Sponsors

Technical Program Committee

Technical Program Chair: Curtis Smith, INL USA
Assistant Technical Program Chairs: Steve Epstein, Lloyd's Register Japan
Vinh Dang, PSI Switzerland
Ted Steinberg, QUT Australia

We would like to thank the members of the PSAM 12 Technical Program Committee. These individuals helped to make PSAM 12 a success by reviewing abstracts, technical papers, organizing sessions, and providing technical leadership for the conference.

Technical Committee Members:

- Roland Akselsson
- S. Massoud (Mike) Azizi
- Tito Bonano
- Ronald Boring
- Roger Boyer
- Mario Brito
- Kaushik Chatterjee
- Vinh Dang
- Claver Diallo
- Nsimah Ekanem
- Steve Epstein
- Fernando Ferrante
- Federico Gabriele
- Ray Gallucci
- S. Tina Ghosh
- David Grabaskas
- Katrina Groth
- Seth Guikema
- Steve Hess
- Christopher J. Jablonowski
- Moosung Jae
- Jeffrey Joe
- Vyacheslav S. Kharchenko
- James Knudsen
- Zoltan Kovacs
- Ping Li
- Harry Liao
- Francois van Loggerenberg
- Jerome Lonchampt
- Soliman A. Mahmoud
- Diego Mandelli
- Donoval Mathias
- Zahra Mohaghegh
- Thor Myklebust
- Cen Nan
- Mohammad Pourgolmohammad
- Marina Roewekamp
- Clayton Smith
- Shawn St. Germain
- Ted Steinberg
- Kurt Vedros
- Smain Yalaoui
- Robert Youngblood
- Enrico Zio

Organizing Committee

General Chair: Dr. Todd Paulos
General Vice Chair: Prof. Stephen Hora, USC
Technical Program Chair: Curtis Smith, INL USA
Webmaster, Registration, Support for Papers/Abstracts Submission and Review: Hanna Shapira, TICS

PSAM12 - Probabilistic Safety Assessment and Management

Table of Content

Page Paper

9 102 **A PRA Application to Support Outage Schedule Planning at OL1 and OL2 Units**
Hannu Tuulensuu
Teollisuuden Voima Oyj, Eurajoki, Finland

16 120 **Loss Of Offsite Power Frequency Calculation II**
Zhiping Li
Callaway Energy Center-Ameren Missouri, Fulton, USA

27 320 **Mean Fault Time for Estimation of Average Probability of Failure on Demand PFDavg**
Isshi KOYATA (a), Koichi SUYAMA (b), and Yoshinobu SATO (c)
a) The University of Marine Science and Technology Doctoral Course, Course of Applied Marine Environmental Studies, Tokyo, Japan and Japan Automobile Research Institute, Tokyo, Japan,
b) The University of Marine Science and Technology Doctoral Course, Professor, Tokyo, Japan, c) Japan Audit and Certification Organization for Environment and Quality, Tokyo, Japan

35 164 **Reliability Analysis Including External Failures for Low Demand Marine Systems**
HyungJu Kim, Stein Haugen (a), and Ingrid Bouwer Utne (b)
a) Department of Production and Quality Engineering NTNU, Trondheim, Norway, b) Department Marine Technology NTNU, Trondheim, Norway

44 339 **Heterogeneous Redundancy Analysis based on Component Dynamic Fault Trees**
Jose Ignacio Aizpurua, Eñaut Muxika (a), Ferdinando Chiacchio (b), and Gabriele Manno (c)
a) University of Mondragon, Mondragon, Spain, b) University of Catania, Catania, Italy, c) Strategic Research & Innovation DNV GL, Høvik, Norway

55 151 **Probabilistic Analysis of Asteroid Impact Risk Mitigation Programs**
Jason C. Reinhardt, Matthew Daniels, and M. Elisabeth Paté-Cornell
Stanford University, Stanford, United States of America

68 72 **Physics-based Entry, Descent and Landing Risk Model**
Ken Gee (a), Loc Huynh (b), and Ted Manning (a)
a) NASA Ames Research Center, Moffett Field, USA, b) Science and Technology Corporation, Moffett Field, USA

78 121 **Physics-Based Fragment Acceleration Modeling for Pressurized Tank Burst Risk Assessments**
Ted A. Manning, Scott L. Lawrence
NASA Ames Research Center, Moffett Field, CA, USA

90 192 **A Failure Propagation Modeling Method for Launch Vehicle Safety Assessment**
Scott Lawrence, Donovan Mathias, and Ken Gee
NASA Ames Research Center

102 191 **An Integrated Reliability and Physics-based Risk Modeling Approach for Assessing Human Spaceflight Systems**
Susie Go, Donovan Mathias (a), Chris Mattenberger (b), Scott Lawrence, and Ken Gee (a)
a) NASA Ames Research Center, Moffett Field, CA, USA, b) Science and Technology Corp., Moffett Field, CA, USA

114 19 **Apportioning Transient Combustible Fire Frequency via Areal Factors: More Complicated Than It May Seem**
Raymond H.V. Gallucci
U.S. Nuclear Regulatory Commission (USNRC), MS O-10C15, Washington, D.C. 20555

122 267 **Characterizing Fire PRA Quantitative Models: An Evaluation of the Implications of Fire PRA Conservatisms**
M.B. Saunders, E.T. Burns
ERIN Engineering and Research, Inc., Walnut Creek, California, USA

134 454 **Approach for Integration of Initiating Events into External Event Models**
Nicholas Lovelace, Matt Johnson (a), and Michael Lloyd (b)
a) Hughes Associates, Lincoln, NE, USA, b) Risk Informed Solutions Consulting Services, Ball Ground, GA, USA

140 44 **Development of Margin Assessment Methodology of Decay Heat Removal Function Against External Hazards ▫ Project Overview and Preliminary Risk Assessment Against Snow ▫**
Hidemasa Yamano, Hiroyuki Nishino, Kenichi Kurisaka, and Takaaki Sakai (a),
Takahiro Yamamoto, Yoshihiro Ishizuka, Nobuo Geshi, Ryuta Furukawa, and Futoshi Nanayama (b), and Takashi Takata (c)
a) Japan Atomic Energy Agency, Ibaraki, Japan, b) National Institute of Advanced Industrial Science and Technology, Ibaraki, Japan, c) Osaka University, Osaka, Japan

151 590 **Screening of Seismic-Induced Fires**
James C. Lin, Donald J. Wakefield (a), and John Reddington (b)
a) ABSG Consulting Inc., Irvine, California, United States, b) First Energy Nuclear Operating Company, Akron, Ohio, United States

Table of Content

Page   Paper

162   *28*   **Minimization of Vulnerability for a Network under Diverse Attacks**
Jose Emmanuel Ramirez-Marquez (a) and Claudio Rocco (b)
a) School of Systems and Enterprises, Stevens Institute of Technology, Hoboken, NJ, USA, b) Facultad de Ingeniería, Universidad Central de Venezuela, Caracas, Venezuela

174   *90*   **Applications of Bayesian Networks for Evaluating Nuclear I&C Systems**
Jinsoo Shin, Rahman Khalil Ur (a), Hanseong Son (b), and Gyunyoung Heo (a)
a) Kyung Hee University, Yongin-si, Gyeonggi-do, Korea, b) Joongbu University, Geumsan-gun, Chungnam, Korea

184   *367*   **Portfolio Analysis of Layered Security Measures**
Samrat Chatterjee, Stephen C. Hora, Heather Rosoff
CREATE, University of Southern California

195   *32*   **Cyber security: the Risk of Supply Chain Vulnerabilities in an Enterprise Firewall**
Marshall A. Kuypers, Greg Heon, Philip Martin, Jack Smith, Katie Ward, and Elisabeth Paté-Cornell
Stanford University, Stanford, CA

205   *489*   **Security Informed Safety Assessment of Industrial FPGA-Based Systems**
Vyacheslav Kharchenko (a,b), Oleg Illiashenko (a), Eugene Brezhnev (a,b), Artem Boyarchuk (a), Vladimir Golovanevskiy (c)
a) National Aerospace University KhAI, Kharkiv, Ukraine, b) Centre for Safety Infrastructure Oriented Research and Analysis, Kharkiv, Ukraine, c) Western Australian School of Mines, Curtin University, Australia

216   *47*   **Uncertainty of the Thermal-Hydraulic Model Analysis**
Yu YU, Yingqiu HU, Junchi CAI, Shengfei WANG, Fenglei NIU
School of Nuclear Science and Engineering, Beijing Key Laboratory of Passive Nuclear Safety Technology, North China Electric Power University, Beijing, China

224   *239*   **Sensitivity Analysis and Failure Damage Domain Identification of the Passive Containment Cooling System of an AP1000 Nuclear Reactor**
Francesco Di Maio, Giancarlo Nicola (a), Yu Yu (b) and Enrico Zio (a,c)
a) Energy Department, Politecnico di Milano, Milano, Italy, b) North China Electric Power University, Beijing, China, c) Chair on System Science and Energetic Challenge, European Foundation for New Energy, Electricite de France, Ecole Centrale, Paris, and Supelec, Paris, France

235   *374*   **The Development of a Demonstration Passive System Reliability Assessment**
Matthew Bucknor, David Grabaskas, and Acacia Brunett
Nuclear Engineering Division, Argonne National Laboratory, Argonne, IL U.S.

247   *153*   **Visual Monitoring Path Forecasting for Digital Human-Computer Interface in Nuclear Power Plant and its Application**
Hu Hong, Zhang Li (a), Jiang Jian-Jun (b), Yi Can-Nan (a), Dai Li-Cao (b), Chen Qin-Qin (a)
a) Ergonomics and safety management Institute, HuNan Institute of Technology, Hengyang, China, b) Human Factors Institute, University of South China, Hengyang, China

256   *13*   **Individual Differences in Human Reliability Analysis**
Jeffrey C. Joe and Ronald L. Boring
Idaho National Laboratory, Idaho Falls, ID, USA

266   *281*   **Cultural Profiles of Non-MCR Operators Working in Domestic NPPs**
Jinkyun Park, and Wondea Jung
Korea Atomic Energy Research Institute, Daejeon, Rep. of Korea

274   *219*   **Improving Scenario Analysis for HRA**
Claire Taylor
OECD Halden Reactor Project, Institute for Energy Technology (IFE), Halden, Norway

285   *308*   **Can we Quantify Human Reliability in Level 2 PSA?**
Lavinia Raganelli (a,b), Barry Kirwan (c)
a) Imperial College, London, United Kingdom, b) Corporate Risk Associate, London, United Kingdom, c) Eurocontrol, Brétigny-sur-Orge, France

A PRA application to support outage schedule planning at OL1 and OL2 units

Hannu Tuulensuu[a]
[a] Teollisuuden Voima Oyj, Eurajoki, Finland

Abstract: For Olkiluoto 1 (OL1) and Olkiluoto 2 (OL2) nuclear power plant units, planned outages are done annually. Every second year a refuelling outage (duration about 1 week) and every second year a maintenance outage (duration about 2-3 weeks) is performed. To ensure nuclear safety in such short outage times, well planned outage schedules are required. Because of this, a PRA application to support the outage schedule planning has been developed.

The PRA application for outage risk management has six goals: (1) to support outage schedule planning, (2) to assess plant modifications during outage, (3) to estimate core damage and radioactive release frequencies of the outage, (4) to identify "weak points" of the outage, (5) to teach risk-informed thinking to the outage schedule coordinators and (6) to develop the outage PRA models.

The PRA application is performed hour by hour throughout the whole outage. A risk profile for the outage as a function of time is the main result of the analysis. The assessment is updated when outage schedule is updated. Based on the results, the PRA application to support outage schedule planning is an efficient way to improve risk management during outages.

Keywords: PRA, application, outage, risk management

1. INTRODUCTION

Planned outages are normal part of the plant operation. Planned outages are done annually for Olkiluoto 1 (OL1) and Olkiluoto 2 (OL2). Normal practice is 12 months fuel cycle including every second year a refuelling outage (duration about 1 week) and every second year a maintenance outage (duration about 2-3 weeks). Because the duration of the outage is short, well planned outage schedule is required. The principal schedule of every planned outage is known several years in advance. The preparation of detailed outage schedule starts immediately after completion of the previous outage. The preparation continues whole fuel cycle including monthly organized outage schedule meetings with experts in different engineering areas like maintenance, testing and nuclear safety.

To ensure nuclear safety in such short outage times, a PRA application to support outage schedule planning has been developed. The PRA application for outage risk management is based on the outage PRA models. PRA modelling for "average" outages is explained in the chapter 2. The chapter 3 describes how those PRA models are used in the application for outage risk management including further development areas and the chapter 4 concludes and summaries the work.

2. PRA MODELLING FOR AVERAGE OUTAGE

Finnish Regulatory Body (STUK) requires that all operation modes and transitions between the modes shall be included in the PRA. For the units OL1 and OL2, following operation modes are covered in the PRA: power operation, refueling and maintenance outages, transition from power operation to hot shut-down state, planned shut-down from hot shut-down state to cold shut-down state and start-up from outage to power operation.

The first version of the outage PRA was developed in the early 1990s. After that, several updates have been done to maintain the outage PRA for living PRA purposes. The outage PRA represents kind of an average outage based on operating history since 1990s. Durations of different outages, initiating event frequencies and unavailability of systems due to maintenance are all determined using operating history.

In the PRA models, the outage is divided into six plant operational states (POS). Each POS has unique characteristics, e.g. water level and residual heat removal capacity. Physical behavior after initiating events is studied for each POS and based on those studies success criteria for the event trees are determined. The outage PRA is modelled using "large fault trees - small event trees"-method approach. Table 1 describes POS applied during outage.

Table 1. Description of plant operational states (POS)

POS	Description
T0	Starts from the shutdown state when the reactor pressure decreases below 1,2 MPa and ends when reactor water filling is started.
T1A	Starts when reactor water level is started to lift up and ends when water level reaches flange level.
T1B	Starts when water level reaches flange level and ends when the reactor lid is opened.
T2	Starts when the reactor lid is opened and water level is started to lift up and ends when residual heat removal is done using the systems 321 and 324 (reactor pools are filled up with water)
T3	Starts when residual heat removal is possible using only the system 324 (considering single failure criterion) and ends when water level is back to flange level.
T4	Starts when water level is in flange level and ends when the reactor lid is installed and ends when the control rods are started to pull out from the reactor (criticality).

The outage PRA is widely applied for risk-informed assessments e.g. in evaluation of plant modifications, Technical Specifications and plant disturbance events. The current PRA level 1 results for the OL1/OL2 units are shown in figure 1. The overall core damage frequency is 1,2E-5/a, for the power operation 9,4E-6/a, for the planned shutdown 1,1E-6/a, for the start-up 1,1E-6/a and for the outage 5,5E-7/a.

Figure 1. The distribution of the CDF for operating states

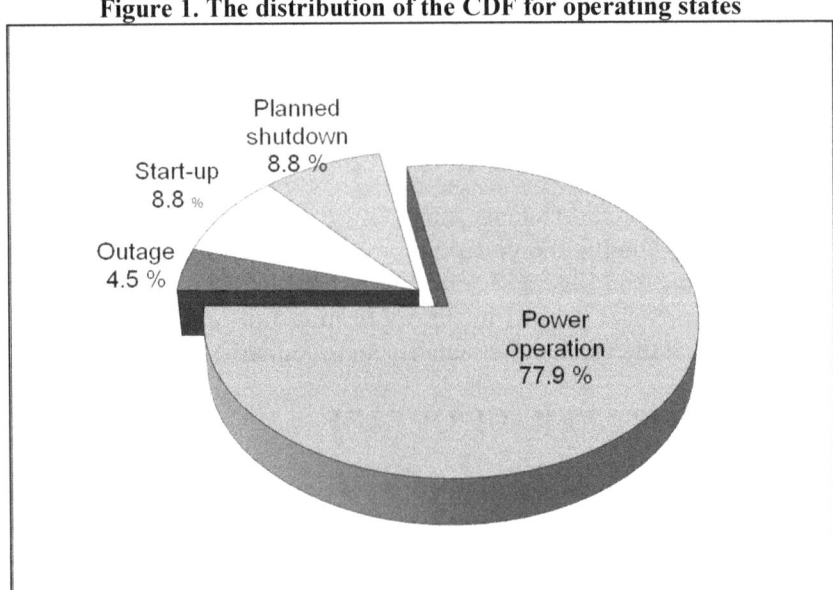

The initiating events groups for outages can be divided into three main categories: (1) initiating events causing loss of residual heat removal, (2) initiating events causing leakages from the reactor and (3)

internal fires. The relative contribution of the outage core damage frequency for different initiating event groups are shown in figure 2.

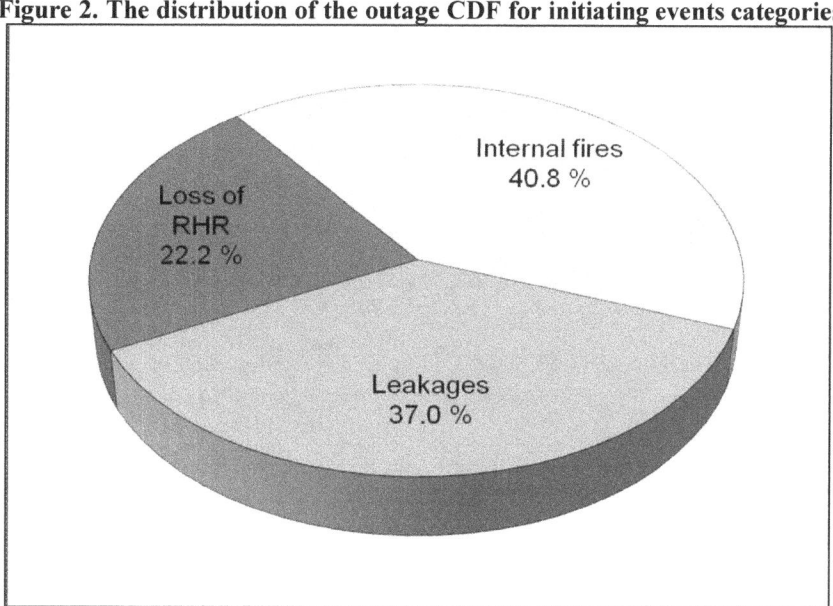

Figure 2. The distribution of the outage CDF for initiating events categories

3. APPLICATION FOR SPECIFIC OUTAGES

The main purpose of the application is to support outage schedule planning using the existing average outage PRA models to calculate momentary core damage and radioactive release frequency during the outage. The application was used for the first time in planning of the 2013 outages and several modifications to the outage schedule were done based on the results of the assessment. Good experiences and feedback were received from maintenance and operational personnel. Further development of the application was performed after the year 2013 outages by development of more automated data handling and batch calculations with a MS Excel tool.

3.1. The method description

The application for specific outages is based on the outage schedule printouts printed out in five different forms: (1) Main schedule, (2) Reactor schedule, (3) Residual heat removal schedule, (4) System maintenance schedule and (5) Criticality safety schedule. The main schedule shows the outage schedule in a simplified manner and only the most important information is included. The reactor schedule includes detailed time schedule for work packages in the reactor during outage. The residual heat removal schedule includes planned unavailability of the residual heat removal systems during the outage. The system maintenance schedule includes the time schedule for planned unavailability of the other systems and the criticality safety schedule shows the time schedule for the work packages affecting on criticality safety. An example of the outage schedule (main schedule) is shown in figure 3.

Figure 3. Main schedule

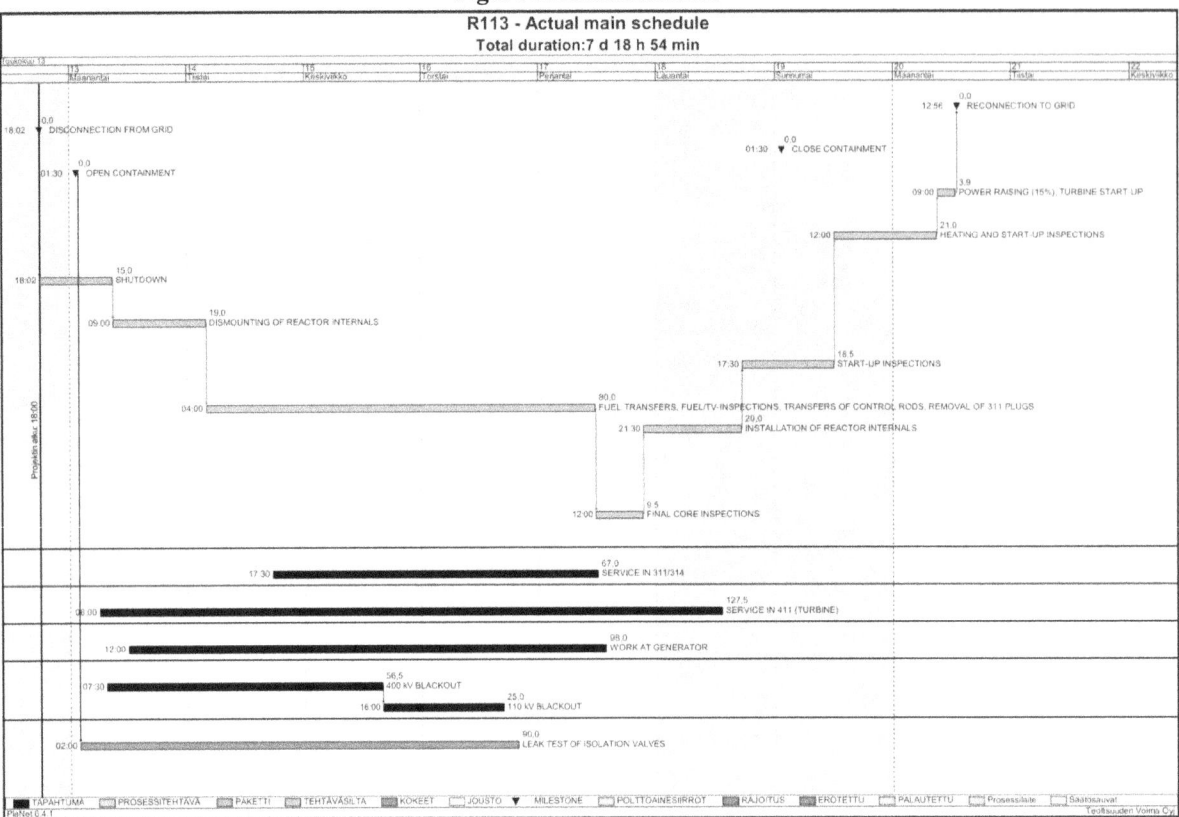

One hour accuracy is used for the outage schedule printouts and for that reason the application includes individual PRA assessment hour by hour throughout the whole outage. The PRA assessment includes five steps for each hour: (1) identification of planned unavailability of systems/subsystems, (2) identification of possible initiating events and estimation of initiating event frequencies (per hour), (3) identification of possible temporary plant modifications, (4) calculation of core damage frequency (and radioactive release frequency) and (5) verification that minimal cut sets are reasonable.

For the step (1), the residual heat removal schedule and the system maintenance schedule is used for identification of planned unavailability of the systems and subsystems. For the step (2), the reactor schedule identifies the work packages which have the possibility to cause initiating event "leakage", e.g. maintenance of the reactor coolant pumps. It is assumed that initiating events "loss of residual heat removal" and "internal fires" have constant initiating frequency during outages. In the step (3), if temporary plant modifications cause changes in the PRA-model, a separate PRA analysis is done. When steps 1, 2 and 3 are done for every hour of the outage, a minimal cut set calculation is executed to get the core damage and radioactive release frequencies for each hour (step 4). Step 5 is executed to ensure that the number of mistakes in the PRA assessment is minimal. The overall outage core damage and radioactive release frequency is a sum of each hour's core damage and radioactive release frequency. As a result, a risk profile for the specific outage can be drawn. An example of the risk profile (blue line) is presented in figure 4.

The probabilistic design objectives for Finnish nuclear power plants are explained in the YVL A.7. The numerical objective for the mean value of the probability of core damage is less than 1E-5/a and for the release exceeding the target value defined in the Government Resolution (716/2013) must be smaller than 5E-7/a. For short term operations, like during plant modifications, preventive maintenance and testing, the design goal can be exceeded. For short term operations, TVO's internal criterion is that average release frequency must not exceed 5E-5/a. Because most of the time during outage the containment is open, meaning that core damage leads directly to radioactive release, the short term objective for radioactive release frequency is used as an internal target value for the application of the specific outage. If the target value 5E-5/a is divided by 8760 hours (one year), the result is the target value for acceptable short-term release frequency, which is about 6E-9/h. The internal target value is shown red in figure 4.

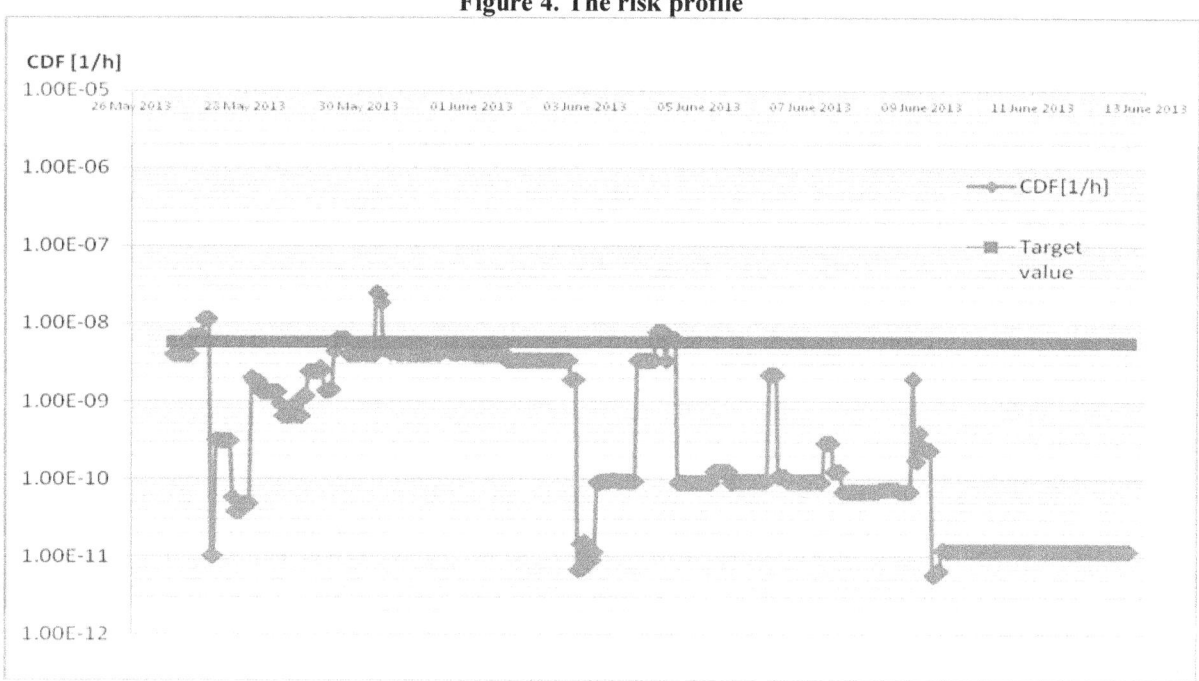

Figure 4. The risk profile

The risk peaks can be easily identified from the risk profile. If the target value is exceeded, actions to decrease the risk peak will be discussed with the outage schedule coordinators.

3.2. Data handling and batch calculations

The application for specific outage requires flexible data handling, because several changes to the schedules are normally implemented between the first version of the schedule and the official, final schedule before the outage. The risk profile is calculated based on the first version of the outage schedule and it is updated when changes to the outage schedule is implemented. A specific program suitable for Microsoft Excel is designed for input data handling. The purpose of the program is to easily feed information from the outage time schedules into a form usable for minimal cut set calculations.

The program includes several features for data handling. Each outage time period (chapter 2) is handled separately in the program. Initiating event frequencies and system unavailabilities can be feed to the program and they can be easily modified if changes to the outage schedule are implemented. Figure 5 shows an example of the PRA assessment for one specific hour.

Figure 5. An example of the PRA assessment for one specific hour

	SYSTEM/SUBSYSTEM UNAVAILABILITY				
	---------- Data records <transfer data> ----------				
	Name	Distribution	Status	Rel model	
30.5.2013 6:00	SIMO_A-SUB_T3	Beta	Failed	Operating	Note: Temporary power supply on
	327P003M1A	Beta	Failed	Standby	Note: Pump 327P002 unavailable
	621IRT3..H	Point Value	Failed	Operating	Note: 400kV grid unavailable
	653G101D1A	Beta	Failed	Operating	Note: Dieselgenerator 653G101 unavailable
	321IRT3..H-Z	Point Value	Failed	Operating	Note: System 321 unavailable
	664IRT3..E-B	Point Value	Failed	Operating	Note: Subsystem B of the system 664 unavailable
	INITIATING EVENTS				
	---------- Data records <transfer data> ----------				
	Name	Distribution	Status	Rel model	RelP1
	IRT-B3-T3	LogNormal	Normal	Frequency	9.60E-09 Note: Maintenance of the control rod drive unit
	IRT-A1LT3	LogNormal	Normal	Frequency	4.12E-06 Note: Opening of valve 312V5, V6, V7 or V8 would cause leakage
	IRT-A2HT3	LogNormal	Normal	Frequency	1.69E-04 Note: Opening of overpressure valve would cause leakage
	IRT-A1HT3	LogNormal	Normal	Frequency	3.95E-05 Note: Opening of 321 V1 or V2 would cause leakage

The application for the specific outage requires an efficient PRA-program to calculate hundreds of calculations in reasonable time. FinPSA is a PRA tool, developed by STUK and currently maintained by Technical Research Centre of Finland (VTT), containing parallel computation properties and it works with common Windows applications [1]. TVO has FinPSA cluster including three HP servers, each equipped with eight processors (four cores and four virtual cores). OL1 and OL2 PRA level 1 minimal cut set generation time is about two minutes using all 24 processors.

In addition, FinPSA includes a feature to add a task file for batch calculations. The task file enables to do several calculations automatically one after another. For each calculation, a number of unique PRA-model parameters can be set, e.g. fail basic events and set initiating event frequencies. After the steps 1-3 are executed (chapter 3.1) in the Microsoft Excel program, it is possible to do the step 4 by using the FinPSA task file. The Microsoft Excel program can create automatically a suitable task file for FinPSA from the Excel data. The result of the batch calculation is an output file containing results from the minimal cut set calculations. User can choose the output file's information, e.g. the amount of the most important minimal cut sets and the basic event importance measures. The step 5 can be done by using the results of the output file.

3.2. Future development

The PRA application to support outage schedules is in development phase. At the moment, it is used to support outage schedule planning before outage. The plan is to increase the range of usage to calculate core damage and radioactive release frequencies during outage and for verifying calculations after the outage is performed.

The average outage PRA model and documentation will be updated during year 2014. Update of the outage event trees, initiating event frequencies and system unavailabilities based on the operating history will be included in the next update. Weather phenomena's will be also included to the outage PRA model and initiating event frequencies for internal fires during outages will be updated.

4. CONCLUSION

The PRA application to support outage schedule planning is described above. The risk profile during the outage is the main result of the analysis and it shows the possible risk peaks that may occur based on the outage schedule.

The application requires a flexible PRA-model for modifications and an efficient PRA program to calculate hundreds of calculations in reasonable time. FinPSA includes the necessary features e.g. parallel calculation, integration to the Windows Applications and batch calculations, to make minimal cut set calculations possible. TVO has also developed a Microsoft Excel program to ease information transfer from the outage schedules to the PRA tool FinPSA.

Based on the results and feedback, the PRA application to support outage schedule planning is an efficient method to improve nuclear safety during outages.

References

[1] I.Niemelä, *"FinPSA: New features in PRA software"*, PSAM9 conference paper, 2008

Loss of Offsite Power Frequency Calculation II

Zhiping Li
Callaway Energy Center-Ameren Missouri, Fulton, USA

Abstract: The availability of alternating current (AC) power is essential for safe operation and accident recovery at commercial nuclear power plants. Normally, AC power is supplied by offsite sources via the electrical grid. Loss of this offsite power has significant contribution to the overall risk at nuclear power plants. Reliable offsite power is one key to minimizing the probability of severe accidents. The probability of losing all offsite power is an important input to nuclear power plant probabilistic safety assessments. Several studies have analyzed data on LOOP and/or offsite power restoration. However, significant differences in LOOP event description, category, duration, and applicability exist between the LOOP events used in NUREG/CR-6890 and the EPRI LOOP Reports. Different LOOP frequency calculation methods are used in NUREG/CR-6890 and in the EPRI's LOOP Reports. While the author was updating LOOP frequency for some nuclear power plants, it was found that there is a need to clarify how the LOOP frequency should be calculated. Loss of Offsite Power Frequency Calculation was presented to PSA2013, Columbia, SC in September 2013. A LOOP frequency calculation for an inland plant is performed. Insight about site specific LOOP frequency calculation and some discussion about applicability of LOOP events are presented. In addition, in Loss of Offsite Power Frequency Calculation II, LOOP frequencies for different categories will be calculated. Comparison and discussions about different LOOP frequency calculation methods will be presented.

Keywords: LOOP, LOSP, Frequency.

1. INTRODUCTION

The availability of alternating current (AC) power is essential for safe operation and accident recovery at commercial nuclear power plants. Normally, AC power is supplied by offsite sources via the electrical grid. Loss of this offsite power can have a major negative impact on a power plant's ability to achieve and maintain safe shutdown conditions. Risk analyses performed for U.S. commercial nuclear power plants indicate that the loss of all AC power contributes over 70% of the overall risk at some plants [1]. Clearly, loss of offsite power (LOOP, also referred to as LOSP) and subsequent restoration of offsite power are important inputs to plant probabilistic safety assessments (PSAs). These inputs must reflect current industry performance in order for PRAs to accurately estimate the risk from LOOP initiated scenarios.

Several studies have analyzed data on LOOP and/or offsite power restoration [2–6]; NUREG/CR-6890, *Reevaluation of Station Blackout Risk at Nuclear Power Plants* [1], extends the analysis to 2004. NUREG-1032, *Evaluation of Station Blackout Accidents at Nuclear Power Plants* [2], evaluated LOOP data from U.S. commercial nuclear reactors over the period 1968–1985. NUREG/CR-5496, *Evaluation of Loss of Offsite Power Events at Nuclear Power Plants: 1980–1996* [3], looked at data from 1980–1996. A more general report, NUREG/CR-5750, *Rates of Initiating Events at U.S. Nuclear Power Plants: 1987–1995* [4], covered a wide variety of initiating events, including LOOP for the period 1987–1995. NUREG/CR-6928, *Industry-Average Performance for Components and Initiating Events at U.S. Commercial Nuclear Power Plants* [5], covers LOOP frequencies which were based on NUREG/CR-6890. Electric Power Research Institute (EPRI) reports covering LOOP events have been issued periodically; the latest EPRI report covers LOOP events from 2003 to 2012 [6]. And NUREG-1784, *Operating Experience Assessment—Effects of Grid Events on Nuclear Power Plant Performance* [7], focuses on a subset of LOOP events (1985–2001) and the effects of deregulation on such events. That report contains more detailed engineering information concerning deregulation and its effects on the electrical grid and related LOOP events. NRC's *Analysis of Loss of Offsite Power Events 2010 Update* [8] collected and analyzed the LOOP data from calendar years 1986-2010. The data covered both critical (at power) and shutdown operations at these plants.

Reference [9] identified significant differences in LOOP event description, category, duration, and applicability between the LOOP events used in NUREG/CR-6890 and Entergy Nuclear South (ENS) plants' LOOP packages, which were based on EPRI LOOP reports with plant specific applicability analysis. It listed the differences between the data in NUREG/CR-6890 and EPRI reports and evaluated the applicability of the LOOP events to ENS plant specific PSA model.

While the author was updating LOOP frequency for some plants, it was found that there is a need to clarify how the LOOP frequency should be calculated. This paper provides the authors insight about site-specific LOOP frequency calculation and some discussion about applicability of LOOP events.

2. DEFINITION

LOOP or LOSP event - the simultaneous loss of electrical power to all unit safety buses (also referred to as emergency buses, Class 1E buses, and vital buses) requiring all emergency power generators to start and supply power to the safety buses. The nonessential buses may also be deenergized as a result of this.

Loss of Preferred Offsite Power - the interruption of the preferred power supply to the essential and nonessential switchgear buses necessitating or resulting in the use of emergency AC power supplies.

Plant-Centered LOOP event - a LOOP event in which the design and operational characteristics of the nuclear power plant unit itself play the major role in the cause and duration of the loss of offsite power. Plant-centered failures typically involve hardware failures, design deficiencies, human errors, and localized weather-induced faults such as lightning. The line of demarcation between plant-centered and switchyard-centered events is the nuclear power plant main and station power transformers high-voltage terminals. Plant-centered LOOP events occur within the plant, up to but not including the auxiliary or station transformers.

Switchyard-Centered LOOP event - a LOOP event in which the equipment, or human-induced failures of equipment, in the switchyard play the major role in the loss of offsite power. Switchyard-centered failures typically involve hardware failures, design deficiencies, human errors, and localized weather-induced faults such as lightning. The line of demarcation between switchyard-related events and grid-related events is the output bus bar in the switchyard. Switchyard-centered events occur within the switchyard, up to and including the output bus bar.

Grid-Related LOOP event – a LOOP event in which the initial failure occurs in the interconnected transmission grid that is outside the direct control of plant personnel. Failures that involve transmission lines from the site switchyard are usually classified as switchyard-centered events if plant personnel can take actions to restore power when the fault is cleared. However, the event should be classified as grid related if the transmission lines fail from voltage or frequency instabilities, overload, or other causes that require restoration efforts or corrective action by the transmission operator.

Weather-Related LOOP event - a LOOP event caused by severe or extreme weather. Severe weather is defined to be weather with forceful and non-localized effects. A LOOP is classified as a severe-weather event if it was judged that the weather was widespread, not just centered at the power plant site, and capable of major disruption. An example is storm damage to transmission lines instead of just debris blown into a transformer. This does not mean that the event had to actually result in widespread damage, as long as the potential was there. Examples of severe weather include thunderstorms, snow, and ice storms. Lightning strikes, though forceful, are normally localized to one unit, and so are coded as plant centered or switchyard centered. LOOP events involving hurricanes, strong winds greater than 125 miles per hour, and tornadoes are included in a separate category—extreme-weather-related LOOPs. Weather-Related LOOP event may overlap with evaluation of external events.

Extreme-Weather-Related LOOP event - a LOOP event caused by extreme weather. Examples of extreme weather are hurricanes, strong winds greater than 125 miles per hour, and tornadoes. Extreme-weather-related LOOP events are also distinguished from severe weather-related LOOP events by their potential to cause significant damage to the electrical transmission system and long offsite power restoration times. Extreme-weather-related events are included in the weather-related events category in NUREG/CR-6890 and EPRI's LOOP reports.

3. LOOP EVENT CATEGORIES

In NUREG/CR-6890, the LOOP events are classified based on the operating state of the plant at the time of the LOOP events. The LOOP categories in NUREG/CR-6890 are refined to four categories: plant-centered, switchyard-centered, grid-related and weather-related. NUREG/CR-6890 uses three categorization schemes to classify LOOP events. The first classifies LOOP events according to whether the plant was shut down or operating when the LOOP occurred and the consequences of the LOOP. The three main categories of LOOPs are those that occur (1) while a plant is shut down (LOOP-SD), (2) during critical operation and involve a plant trip (LOOP-IE), and (3) during critical operation but the plant is able to continue critical operation without a plant trip (LOOP-NT). LOOP-IE events are further subdivided, following the initiating event nomenclature in NUREG/CR-5750, into those in which the LOOP event causes the reactor trip (initial plant fault event or LOOP-IE-I) and those in which the LOOP occurs after the reactor trip. These latter events are included in the functional impact initiating event classification in NUREG/CR-5750, and include those in which the reactor trip causes a LOOP to occur (consequential LOOP or LOOP-IE-C) and those in which the reactor trip and LOOP are unrelated but occur during the same transient (LOOP-IE-NC). Each LOOP event is placed into one of the LOOP categories: LOOP-SD, LOOP-NT, LOOP-IE-I, LOOP-IE-C, or LOOP-IE-NC. This classification scheme helps determine which LOOP events should be included when determining LOOP frequency estimates. [1]

The EPRI's LOOP reports assign categories as Ia, Ib, IIa, IIb, III and IV. The definitions of EPRI's LOOP events categories are as: [6]

Ia - no offsite power available for 30 minutes or longer to the safety buses.

Ib - no offsite power available for less than 30 minutes to the safety buses.

IIa - with the unit on-line, the startup/shutdown sources of offsite power for the safety buses become deenergized. The main generator remains on-line (connected to the offsite grid) and power for the safety buses is available from a unit auxiliary transformer.

IIb - with the unit on-line, the startup/shutdown sources of offsite power for the safety buses remain energized but in question. There is low or unstable grid voltage, or there might be if the unit trips, or trips along with a LOCA and emergency safety feature actuation. The main generator remains on-line (connected to the offsite grid) and power for the safety buses is available from a unit auxiliary transformer.

III - the unit auxiliary source of power for the safety buses becomes deenergized or unavailable, but offsite power for the safety buses remains available, or can be made available, from a startup/shutdown source. Utilization of this source may require a fast or slow automatic transfer, or manual switching from the control room. A loss of unit auxiliary power that is the result of a unit trip is not a Category III event. To be a Category III event the loss of power from the unit auxiliary source must be the initiating event and precede the unit trip. Most problems that trip the unit off-line are not Category III events. A Category III event is more properly associated with a failure of main electrical power hardware that makes near term availability of the unit auxiliary source of power for the safety buses unlikely.

IV - no offsite power available during cold shutdown because of special maintenance conditions that do not occur during or immediately following operations.

The Category I events (Ia and Ib) could occur at any time. The Category IV events can only occur during cold shutdown because of special maintenance conditions and these events are not applicable to power operation. The Category IIa, IIb, III events are partial loss of power events.

4. LOOP FREQUENCIES AT POWER

The proposed total LOOP initiator frequency for at-power model is calculated as follows:

$$f_{LOOP-At-Power} = \frac{\# \text{ of Applicable LOOP Events}}{Rx \text{ Calendar Years}} \quad (1)$$

The unit of LOOP initiator frequency for at-power model is per reactor-calendar-year or per year. The concepts of *Reactor Calendar Years* and *Reactor Critical Years* may cause confusion to PRA analysts. Reference [11] provides an alternate description of the same concepts as by Note 1 to supporting requirement IE-C5 of the AMSE/ANS PRA Standard [10].

The ASME/ANS PRA Standard [10] IE-C5 requires the initiating event frequency to be expressed per Reactor Calendar Year (also commonly expressed as per Reactor-Year, which is the terminology that will be used in the remainder of this document) in order to be consistent with the needs of Reg. Guide 1.174 (that is, for comparison to the quantitative acceptance guidelines). This represents the annual risk contribution to CDF/LERF from at power operations, and, therefore, reflects the time the plant is at power.

Some applications, however, require the analyst to consider the conditional probability of core damage/large early release given the plant is at power. One such example is a risk monitor, which uses PRA to assess the risk of the plant at a given configuration and operating state. For this application we consider initiating event frequencies in units of per Reactor Critical Year, or the annual frequency of the initiating event assuming the reactor is critical the entire year.

The LOOP events can occur any time, regardless of plant operating state. When LOOP frequency is used in the PRA model, the value obtained from equation (1) should normally be multiplied by the plant availability factor, $F_{at\text{-}power}$, where

$F_{at\text{-}power}$ = fraction of year that, on average, the plant is at power, for example, 90% [0].

The time needed to restore offsite power after a LOOP event varies on different event categories. NUREG/CR-6890 [0] concludes that Plant-centered and switchyard-centered LOOPs have the lowest mean duration, while weather-related LOOPs have the highest. Similarly, the plant-centered and switchyard-centered probability of exceedance versus duration curves lie below those for the grid-related LOOPs, while the weather-related curve lies above all the others. In order to perform power recovery analysis, four LOOP frequencies can be calculated based on the four applicable event categories as:

$$f_{LOOP-At-Power-Plant} = \frac{\#\ of\ Applicable\ Plant\ LOOP\ events}{Rx\ Calendar\ Years} \quad (2)$$

$$f_{LOOP-At-Power-Switchyard} = \frac{\#\ of\ Applicable\ Switchyard\ LOOP\ events}{Rx\ Calendar\ Years} \quad (3)$$

$$f_{LOOP-At-Power-Grid} = \frac{\#\ of\ Applicable\ Grid\ Related\ LOOP\ events}{Rx\ Calendar\ Years} \quad (4)$$

$$f_{LOOP-At-Power-Weather} = \frac{\#\ of\ Applicable\ Weather\ LOOP\ events}{Rx\ Calendar\ Years} \quad (5)$$

Based on EPRI's LOOP Report [0], for the past 10 years (2003–2012), on average, the frequency of losing all offsite power was approximately 0.03 per year per unit. This is slightly higher than the average for the period from 2002 through 2011 (0.02 per year). There has been a slight upward trend in both the frequency and duration of losses of offsite power, in part due to a higher rate of weather-related events in recent years. The findings confirm that natural phenomena (weather and seismic activity) are important contributors to the loss of offsite power experience.

The significant differences between NUREG and EPRI LOOP events cannot be simply resolved by combining the NUREG LOOP events with the EPRI LOOP events. In the LOOP frequency calculation, the site-specific LOOP initiators have to be examined to make sure all the applicable LOOP events are included appropriately. For example, for an inland plant, it is normally not vulnerable to the high wind effects and the flood surge effects of hurricanes. The LOOP events involving hurricanes should be excluded from the applicability to the site. For a site in an area that is not susceptible to severe snow/ice storms, LOOP events involving severe snow/ice storms should be excluded.

Table 1 provides the generating unit years from 2003 to 2012. Table 2 shows the applicable LOOP events at power to Callaway, an inland plant. Please note that one plant-centered LOOP event is exclude; four weather-related LOOP events are excluded; eight Switchyard-Centered LOOP events are also excluded because of their applicability. The excluded LOOP events are listed in Table 3. So the total applicable number of LOOP events for Callaway is 17 for the past 10 year period from 2003 to 2012. It should be noted that each reactor event is treated independently if two or more reactor plants were affected by a common cause LOOP event. These events are treated as they were multiple independent events. The LOOP frequencies at power for Callaway are calculated as:

$$f_{LOOP-At-Power} = \frac{\#\ of\ Applicable\ LOOP\ Events}{Rx\ Calendar\ Years} = \frac{17}{1035.66}$$

Probabilistic Safety Assessment and Management PSAM 12, June 2014, Honolulu, Hawaii

$$= 1.64\text{E-}2 \text{ (/rx-calendar-yr)}$$

$$f_{LOOP-At-Power-Plant} = \frac{\# of\ Applicable\ Plant\ LOOP\ events}{Rx\ Calendar\ Years} = \frac{1}{1035.66}$$
$$= 9.66\text{E-}4 \text{ (/rx-calendar-yr)}$$

$$f_{LOOP-At-Power-Switchyard} = \frac{\# of\ Applicable\ Switchyard\ LOOP\ events}{Rx\ Calendar\ Years}$$
$$= 3.86\text{E-}3 \text{ (/rx-calendar-yr)}$$

$$f_{LOOP-At-Power-Grid} = \frac{\# of\ Applicable\ Grid\ Related\ LOOP\ events}{Rx\ Calendar\ Years}$$
$$= 5.79\text{E-}3 \text{ (/rx-calendar-yr)}$$

$$f_{LOOP-At-Power-Weather} = \frac{\# of\ Applicable\ Weather\ LOOP\ events}{Rx\ Calendar\ Years}$$
$$= 5.79\text{E-}33 \text{ (/rx-calendar-yr)}$$

For at-power PRA model quantification, the LOOP frequency should be multiplied by plant availability factor.

Table 1: Generating Unit Years from 2003 to 2012

Year	Unit Capability (%)	EPRI Generating Years	Reactor Critical Years
2003	91.4	103	94.14
2004	91.4	103	94.14
2005	92.0	103	94.76
2006	91.4	103	94.14
2007	91.5	103.66	94.85
2008	91.0	104	94.64
2009	91.3	104	94.95
2010	91.3	104	94.95
2011	91.3	104	94.95
2012	91.2	104	94.85
10 Year Total	91.4	1035.66	946.38

Note: Unit Capabilities taken from the 2012 INPO Annual Report

Table 2: Applicable LOOP Events at Power to Callaway Plant (2003-2012)

Site Name	Unit Number	Date	Category	EPRI Category	Condition	Applicable to Site
Palisades	0	3/25/2003	PLANT	Ia	Refueling	YES
Palo Verde	1	6/14/2004	GRID	Ia	100% Power	YES
Palo Verde	2	6/14/2004	GRID	Ia	100% Power	YES
Palo Verde	3	6/14/2004	GRID	Ia	100% Power	YES
Catawba	1	5/20/2006	SWITCHYARD	Ia	100% Power	YES
Catawba	2	5/20/2006	SWITCHYARD	Ia	100% Power	YES
Surry	1	10/7/2006	SWITCHYARD	Ia	100% Power	YES
Duane Arnold	1	2/24/2007	WEATHER	Ia	Refueling	YES
Oyster Creek	0	7/12/2009	GRID	Ia	100% Power	YES
Surry	1	4/16/2011	WEATHER	Ia	100% Power	YES
Surry	2	4/16/2011	WEATHER	Ia	98.3% Power	YES
Browns Ferry	1	4/27/2011	WEATHER	Ia	75% Power	YES
Browns Ferry	2	4/27/2011	WEATHER	Ia	75% Power	YES
Browns Ferry	3	4/27/2011	WEATHER	Ia	100% Power	YES
North Anna	1	8/23/2011	GRID	Ia	100% Power	YES
North Anna	2	8/23/2011	GRID	Ia	100% Power	YES
Wolf Creek	0	1/13/2012	SWITCHYARD	Ia	100% Power	YES

Table 3: Excluded LOOP Events to Callaway Plant (2003-2012)

Nulcear Unit	Date	Category	Comment
Brunswick 1	8/14/2004	WEATHER	Hurricane Charley
St. Luice 1	9/25/2004	WEATHER	Hurricane Jeane, Salt Spray
St. Luice 2	9/25/2004	WEATHER	Hurricane Jeane, Salt Spray
Brunswick 2	11/1/2006	SWITCHYARD	Loss of preferred offsite power
Point Beach 1	1/15/2008	SWITCHYARD	Loss of preferred offsite power
Byron 2	3/25/2008	SWITCHYARD	Loss of preferred offsite power
Salem*	7/29/2003	SWITCHYARD	At Callaway one off-site power feeds one safety bus and the other off-site power source feeds the other safety bus as a normal configuration. This event does not apply to Callaway
Dresden*	5/5/2004	SWITCHYARD	At Callaway this would be a partial LOOP as Callaway's could have lost one off-site source on clearing for a breaker failure
Nine Mile Point*	5/13/2008	SWITCHYARD	This event is not applicable to Callaway because of its multiple transmission line sources and separation that one off-site power feeds one safety bus and the other off-site power source feeds the other safety bus as a normal configuration
Millstone*	5/24/2008	SWITCHYARD	Callaway has separation in the switchyard and for the off-site power sources. One off-site power feeds one safety bus and the other off-site power source feeds the other safety bus. A single relay failure could only cause a loss of one of the sources (partial LOOP)
Braidwood*	7/30/2009	SWITCHYARD	Callaway does not power the safety busses from the unit auxiliary transformer and uses a 2 out of 2 coincidence on its sudden pressure relays
Byron 2	1/30/2012	SWITCHYARD	At Callaway one off-site power feeds one safety bus and the other off-site power source feeds the other safety bus as a normal configuration. This event does not apply to Callaway
Catawba 1,2	4/4/2012	PLANT	Callaway does not use underfrequency trips in the off-site power sources and the off-site power sources are independent of the main generator
Oyster Creek	10/29/2012	WEATHER	Hurricane Sandy

Note: * excluded after detail analysis by site Electric/I&C engineer mainly because of switchyard design differences

5. LOOP FREQUENCIES FOR SHUTDOWN

It is assumed that the at-power events could just as easily occur with the plant at shutdown as with it at power, therefore, the at-power LOOP events are included in the estimation of the frequency of LOOP while at shutdown. The at power applicable LOOP events are considered to be category Ia or Ib events from EPRI Reports and the shutdown LOOP events are category IV. The LOOP frequency contributed by category IV events at shutdown can be calculated as:

$$f_{LOOP-IV} = \frac{\# of\ Applicable\ Category IV\ Events}{Rx\ Shutdown\ Years} \qquad (6)$$

The total LOOP frequency for shutdown model is as:

$$f_{LOOP-SD} = f_{LOOP-At-Power} + f_{LOOP-IV} \qquad (7)$$

If there are no special maintenance conditions that do not occur during or immediately following operations and that could cause LOOP during shutdown, then the LOOP frequency for shutdown at this configuration equals the LOOP frequency for at-power model and it is as:

$$f_{LOOP-SD} = f_{LOOP-At-Power} \qquad (8)$$

For Callaway, there are four category IV LOOP events that are applicable from 2003 to 2012 (EPRI category IV events only, given in Table VI). The LOOP frequency contributed by category IV events is as:

$$f_{LOOP-IV} = \frac{\# of\ Applicable\ Category IV\ Events}{Rx\ Shutdown\ Years}$$
$$= \frac{4}{(1035.66 - 946.38)} = \frac{4}{89.28} = 4.48E-2 \qquad (/\text{rx-shutdown-yr})$$

The total LOOP frequency for shutdown model is as:

$$f_{LOOP-SD} = f_{LOOP-At-Power} + f_{LOOP-IV}$$

$$= 1.64E\text{-}2 + 4.48E\text{-}2 = 6.12E\text{-}2\ (/\text{rx-shutdown-yr})$$

Converting to hours:

$$f_{LOOP-SD} = 6.12E\text{-}2/365/24 = 6.99E\text{-}6\ (/\text{shutdown-hour})$$

Table 4: Applicable LOOP IV Events to Callaway Plant (2003-2012)

Site Name	Unit Number	Date	Category	EPRI Category	Condition	Applicable to Site
Millstone	3	4/25/2007	SWITCHYARD	IV	Refueling	YES
Wolf Creek	1	4/7/2008	SWITCHYARD	IV	Refueling	YES
Diablo Canyon	1	5/12/2007	GRID	IV	Refueling	YES
Point Beach	1	11/27/2011	SWITCHYARD	IV	CSD	YES

6. DISCUSSION

While the author was reviewing LOOP frequency calculations for some plants, it was found that the following equation is used to calculate the LOOP frequency at power:

$$f_{LOOP-At-Power} = \frac{\# \, of \, Applicable \, LOOP \, Events}{Rx \, Critical \, Years} \tag{9}$$

It is obviously that equation (9) overestimates the LOOP frequency. For the past 10 year time period (2003-2012), the industry average unit capability is 91.4%. So the LOOP frequency at-power by equation (9) is overestimated by about 9.4%. This is similar as multiplying the result from equation (1) by an additional term $1/F_{at-power}$.

In at-power PSA, for initiating event frequencies in unit of 1/reactor-calendar-year, usually these frequencies are used by multiplying the plant availability factor. Per Reference [11], EPRI-3002000774, this does apply to the at-power LOOP frequency. This statement is different from Reference [12]. For online risk monitor tools, such as, EOOS, Safety Monitor, the LOOP frequency may not be multiplied by the plant availability factor.

Some LOOP events, particularly weather-related or grid-centered events, may not be applicable for some plants based on geographical and climatological conditions. For example, an inland plant is not susceptible to hurricane and salt spray events.

Detail analysis from engineering department is very helpful to determine the applicability of plant-centered LOOP events and switchyard-centered LOOP events.

The LOOP events at power, during which no plant trip was observed were not included in the frequency analyses in NUREG/CR-6890 [0] and NRC's analysis of loop events 2010 update [0]. In some other LOOP frequency calculations, these events are also excluded. This may underestimate the realistic site specific LOOP frequency. It is suggested to review the LOOP events that do not cause a scram to determine if the event would result in a scram at the associated site. If the event would result in a scram, then the event should be included in the LOOP frequency calculation for the site. For example, the LOOP event at Nine Mile Point 1 (05/13/2008) is an Ia event for most plants, which did not involve a reactor trip.

In NUREG/CR-6890, for shutdown operation, only the LOOP-SD events were used. In an unusual situation (such as the current situation in Japan with most units are shutdown), this could bring significant deviation, even unreasonable result, most likely, it may significantly underestimate the maintenance activities contribution to LOOP at shutdown. So it is suggested to use equation (3) and (4) to calculate the LOOP frequency for shutdown model.

References

[1] S.A. Eide, et al., *Reevaluation of Station Blackout Risk at Nuclear Power Plants (Analysis of Loss of Offsite Power Events: 1986-2004)*, U.S. Nuclear Regulatory Commission, NUREG/CR-6890, December 2005.

[2] P. W. Baranowsky, *Evaluation of Station Blackout Accidents at Nuclear Power Plants*, U.S. Nuclear Regulatory Commission, NUREG-1032, June 1988.

[3] C. L. Atwood, et al., *Evaluation of Loss of Offsite Power Events at Nuclear Power Plants: 1980–1996*, U.S. Nuclear Regulatory Commission, NUREG/CR-5496, November 1998.

[4] J. P. Poloski, et al., *Rates of Initiating Events at U.S. Nuclear Power Plants: 1987–1995*, U.S. Nuclear Regulatory Commission, NUREG/CR-5750, February 1999.

[5] S.A. Eide, et al., *Industry-Average Performance for Components and Initiating Events at U.S. Commercial Nuclear Power Plants*, U.S. Nuclear Regulatory Commission, NUREG/CR-6928, February 2007.

[6] J. Gisclon, et al., *Losses of Off-Site Power at U.S. Nuclear Power Plants—Summary of Experience Through 2012*, Electric Power Research Institute, July 2013.

[7] W. S. Raughley, et al., *Operating Experience Assessment—Effects of Grid Events on Nuclear Power Plant Performance*, U.S. Nuclear Regulatory Commission, NUREG-1784, December 2003.

[8] NRC, *Analysis of Loss of Offsite Power Events 2010 Update*, February 2012.

[9] Yunlong Li, et al., "Applicability of Loss of Offsite Power (LOSP) Events in NUREG/CR-6890 for Entergy Nuclear South (ENS) Plant's LOSP Calculations", *ICONE 14-89644*, July 2006.

[10] ASME/ANS RA-Sa-2009, *Standard for Level 1/Large Early Release Frequency Probabilistic Risk Assessment for Nuclear Power Plant Applications*, February 2009.

[11] EPRI 3002000774, *EPRI Guidelines for PRA Data Analysis*, December 2013.

[12] Zhiping Li, *Loss of Offsite Power Frequency Calculation*, ANS PSA 2013, September 2013.

Mean fault time for estimation of average probability of failure on demand PFD$_{avg}$.

Isshi KOYATA[a]*, Koichi SUYAMA[b], and Yoshinobu SATO[c]

[a] The University of Marine Science and Technology Doctoral Course, Course of Applied Marine Environmental Studies, Tokyo, Japan
   Japan Automobile Research Institute, Tokyo, Japan
[b] The University of Marine Science and Technology Doctoral Course, Professor, Tokyo, Japan
[c] Japan Audit and Certification Organization for Environment and Quality, Tokyo, Japan

Abstract: In functional safety standards, the safety integrity of safety-related system operated in the low demand-mode of operation is defined as its average probability of dangerous failure on demand, PFD$_{avg}$. In this paper, we firstly formulate the PFD$_{avg}$ resulting from the undetected failures being maintained by proof tests from the two-viewpoints of the mean fault time, the reliability, and the risk assessment of safety-related system. Based on the formulation, the mean fault time is derived using the proof test interval for 1-out-of-n redundant systems. The mean fault time is useful for the exact estimation of safety integrity using Markov-state transition diagrams.

Keywords: Functional safety standard, Safety integrity, Low demand mode of operation, PFD$_{avg}$

1. INTRODUCTION

Basic safety standard IEC61508 that defined the Functional safety of electrical/electronic/ programmable electronic safety-related system (as below safety-related system) is classified as a low demand mode and continuous or high demand mode of the safety-related system operation mode.
In standard safety integrity level at a low demand mode derived that the average probability of dangerous failure on demand, i.e., PFD$_{avg}$ multiplied by demand rate.

PFD$_{avg}$ is derived from using mean fault time between proof test intervals of safety-related system.

This paper introduces the average probability of dangerous failure on demand, i.e., PFD$_{avg}$ by two view points of reliability and hazardous or harmful event rate.
And so on, from the results of two viewpoints it provide the method of introduce the mean fault time at redundant system.

2. HAZARDOUS OR HARMFUL EVENT LOGIC

There are two states of hazardous or harmful event logic, namely:

- Event logic 1 in case of that safety related system is failure state at at first and then demand occurs; and

- Event logic 2 in case of that safety-related system is demand state at first and then failure occurs.

The relationship of two states of hazardous or harmful event logic is depicted in Figure 1.

---

* ikoyata@jari.or.jp

Event logic 1 is equivalent to low demand mode of safety related system, and event logic 2 is equivalent to high demand or continuous mode of safety related system.

In this paper the mean fault time is derived from the results of formulated PFD_{avg} at three case view points for derive the mean down time between proof tests in low demand mode.

It is recommended that safety integrity level, i.e., failure (risk) event rate is evaluated correctly not only to use PFD_{avg} but also to use the Markov graph modeling.

It is mandatory to use the mean down time at modeling of the repair about the Dangerous Undetected failure.

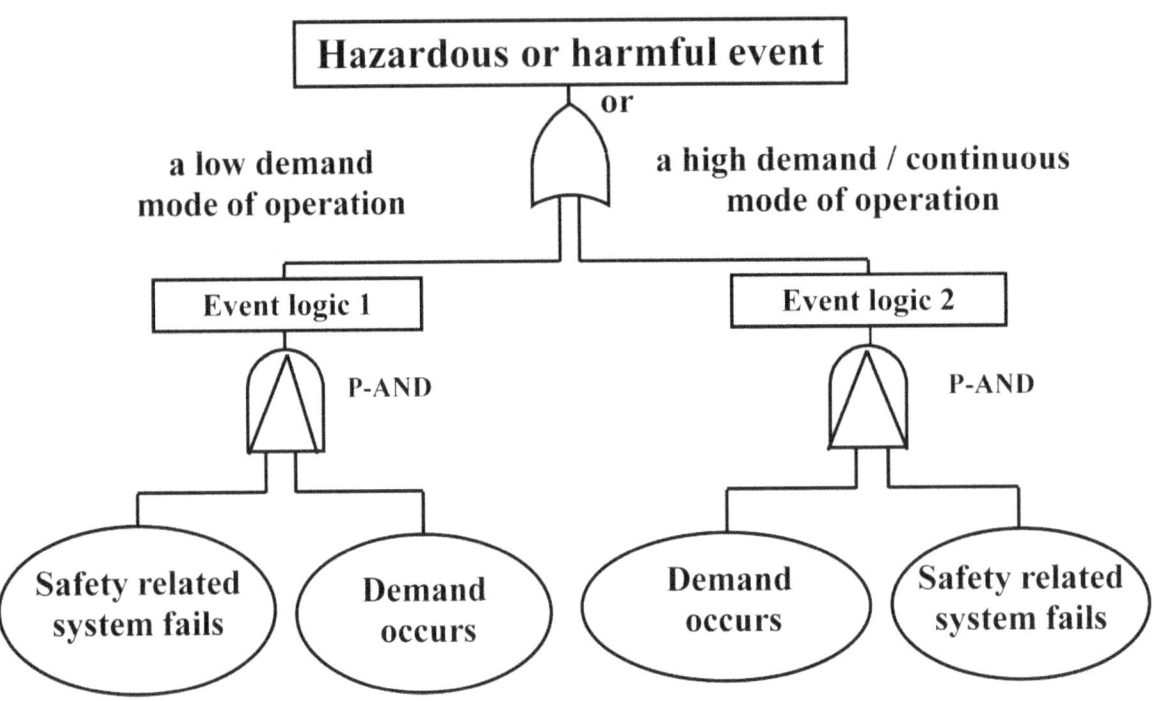

Figure 1 - The concept of hazardous or harmful event

3. BASIC FORMULATION

3.1. Average probability of dangerous failure on demand; PFD_{avg}

PFD_{avg} is expressed as "average probability of dangerous failure on demand" in Part-4 : "Definitions & abbreviations" of Functional safety standards IEC61508.
In standard NOTE 2; PFD_{avg} is expressed as "the dangerous undetected failures occurred since the last proof test and genuine on demand failures caused by the demands (proof tests and safety demands) themselves".

In functional safety standard; IEC61508 Part-1: General requirements, IEC61508 requires that safety integrity level is introduced from "the average probability of dangerous failure on demand of functional safety".

The dangerous failures on demand of safety function are occurred by dangerous detected failure by self test and dangerous undetected failure by repair at proof test not by self test.

In this report, later, it is focused on only dangerous undetected failure.
And the time of proof test and repair time are ignored small enough compared to the time interval of proof test : T.

The relationship between proof test and item status is shown in Figure 2.

Where
T is the time interval of proof test;
K (n) is the number of the proof test time meaning of K(1, 2, 3, ..., n, n+1);

T_a is the normal state time from restoration occurs at (K+1)-th proof test to failure for the 1-st time occurs during (n, n+1] proof test; and
T_b is the fault time from failure to restoration occurs during (n, n+1]-th proof test.

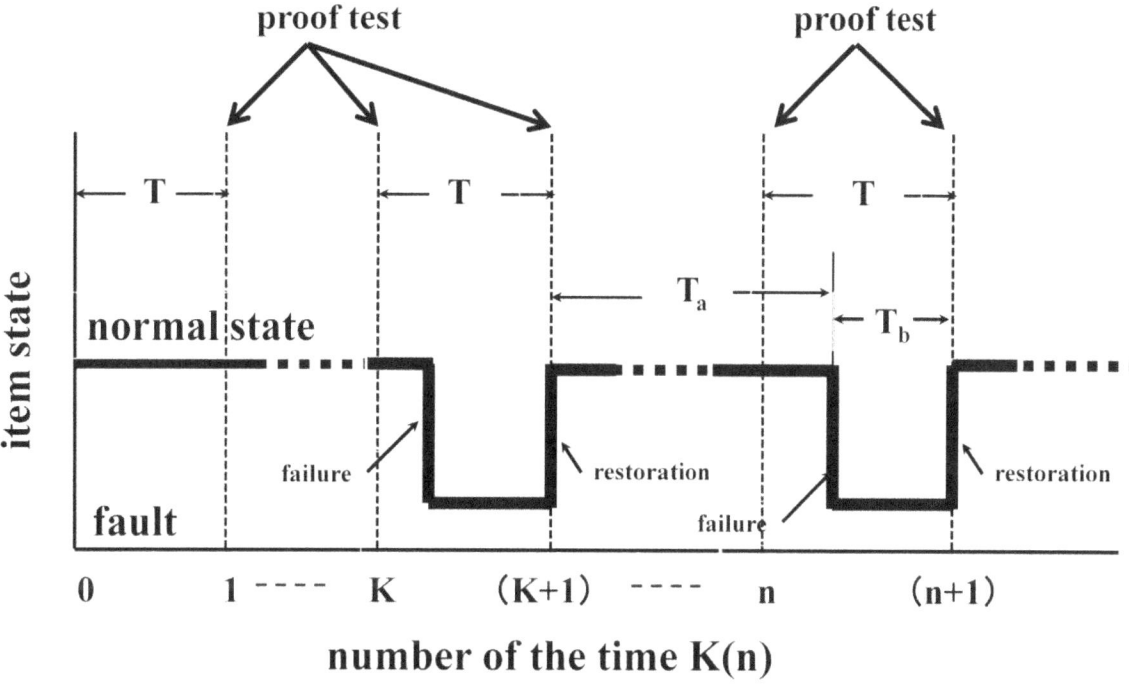

Figure 2 - Relationship between proof test and status of item

PFD_{avg} is derived by the following formula from the relationship $T_a^*+T_b^*$.

$$PFD_{avg} = T_b^* / (T_a^*+T_b^*) \tag{1}$$

Where
T_a^* is the mean time of T_a; and
T_b^* is the mean time of T_b.

T_a^* and T_b^* are derived by mean of T_a and T_b as below;

$$T_a^* = \frac{1}{l}\sum_{i=1}^{l} T_a\,i \tag{2}$$

$$T_b^* = \frac{1}{m}\sum_{j=1}^{m} T_b\,j \tag{3}$$

Further T_a^* is derived by failure rate of dangerous undetected failure.

$$T_a^* = 1 / \lambda_{DU} \tag{4}$$

PFD_{avg} is derived by the following formula from Figure 2 and the above relationship.

$$PFD_{avg} = T_b^* / (T_a^* + T_b^*) = T_b^* / (1/\lambda_{DU} + T_b^*) \fallingdotseq T_b^* \lambda_{DU} \tag{5}$$

$$(\text{because } 1 / \lambda_{DU} \gg T_b^*)$$

As above, "average probability of dangerous failure on demand", i.e., PFD_{avg} is described by the relationship of up state and down state of item in proof test.

3.2. Failure distribution function of 1-out-of-1 system

Failure distribution function of each system is described in order to derive the PFD_{avg} in each viewpoints.

Failure distribution function of series system, i.e., 1-out-of-1 system is described.

Failure distribution function is set to F(t) and F(t) is sufficiently smaller than 1.

$$F(t) \cong \lambda_{DU} t \ll 1 \tag{6}$$

From a series system of 1-out-of-1, the failure distribution function in proof test interval is mentioned to Figure 3.

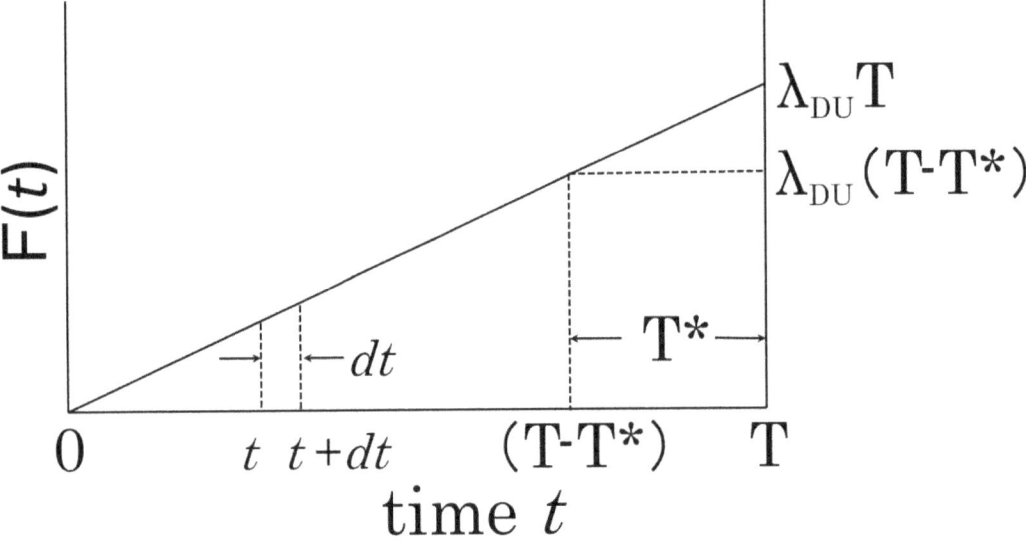

Figure 3 - a failure distribution function of 1-out-of-1 system

Figure 3 shows a failure distribution function of 1-out-of-1 system

Where
 T is proof test interval; and
 T* is the mean fault time.

3.3. Failure distribution function of 1-out-of-n system

It is derived from the relationship in 1-out-of-n system of redundant system.

When I set the failure distribution function (unreliability) equal F (t), from a redundant system,

$$F(t) \cong \lambda_{DU} t \ll 1 \qquad (7)$$

From a redundant system of 1-out-of-n, the failure distribution function in proof test interval is mentioned to Figure 4.

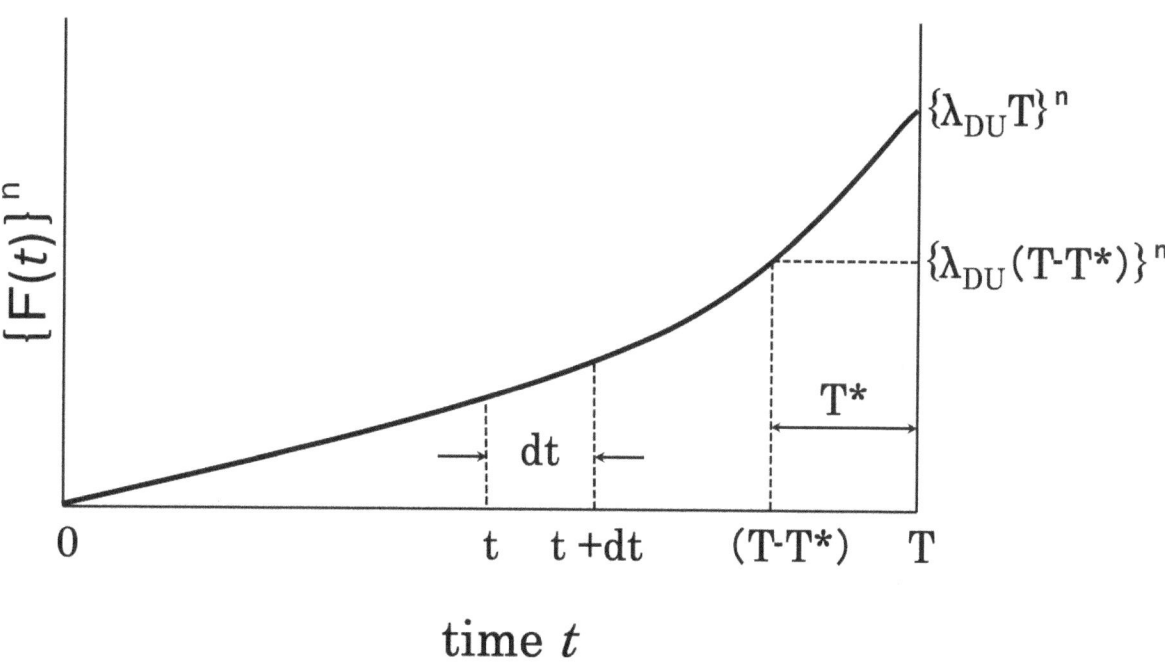

Figure 4 - a failure distribution function of 1-out-of-n system

Figure 4 shows a failure distribution function of 1-out-of-n system

Where
  T is proof test interval; and
  T* is the mean fault time.

3.4. Relationship between mean fault time and PFD_{avg}

It is derived from mean fault time of 1-out-of-n system to PFD_{avg}.

This means that the system fails at proof test interval during (0, T], and demand occurs at from during $((T-T_b^*)$ to T].

In this condition PFD_{avg} is

$$PFD_{avg} = \Pr \{\text{system fails at proof test during (0, T), and system fails from during } (T-T_b^*) \text{ to T at demand occurs during (0, T]}. \qquad (8)$$

That is

$$\text{PFD}_{avg} = \Pr\{\text{system fails at proof test during } (0, T)\} \Pr\{\text{demand occurs during } (T - T_b^*, T] \mid \text{demand occurs during } (0, T]\}$$

$$= \Pr\{\text{system fails at proof test during } (0, T)\} \Pr\{\text{demand occurs during } (T - T_b^*, T] / \Pr\{\text{demand occurs during } (0, T]\}$$

$$= \Pr\{\text{system fails at proof test during } (0, T)\} (T_b^* / T)$$

$$= \{F(T)\}^n (T_b^* / T)$$

$$= \{\lambda_{DU} T\}^n (T_b^* / T) \tag{9}$$

In condition of Figure 4, average probability of functional failure on demand is

$$\text{PFD}_{avg} = \{\lambda_{DU} T\}^n (T_b^* / T) \tag{10}$$

4. TWO METHODS FOR DERIVATION OF PFD$_{avg}$

4.1. Calculation of reliability

It is derived from reliability of 1-out-of-n system to PFD$_{avg}$.

$$\text{PFD}_{avg} =$$

$$\int_0^T \Pr\{\text{system fails during } (t, T + dt], \text{ and demand occurs during } (t, T] \text{ in proof test}\} dt \tag{11}$$

That is

$$\text{PFD}_{avg} =$$

$$\int_0^T \Pr\{\text{system fails during } (t, T + dt], \text{ and demand occurs during } (t, T] \text{ in proof test}\} dt$$

$$= \int_0^T n F(t)^{n-1} \frac{(T-t)}{T} \lambda_{DU} dt$$

$$= \int_0^T n \{\lambda_{DU} t\}^{n-1} \frac{(T-t)}{T} \lambda_{DU} dt$$

$$= \int_0^T n \{\lambda_{DU} t\}^{n-1} \frac{(T-t)}{T} \lambda_{DU} dt$$

$$= n \lambda_{DU}^n \int_0^T t^{n-1} dt - n \lambda_{DU}^n \frac{1}{T} \int_0^T t^n dt$$

$$= n \lambda_{DU}^n \left[\frac{t^n}{n}\right]_0^T - \lambda_{DU}^n \frac{n}{T} \left[\frac{t^{n+1}}{n+1}\right]_0^T$$

$$= \{\lambda_{DU} T\}^n - \{\lambda_{DU} T\}^n \frac{n}{n+1}$$

$$= \{\lambda_{DU}T\}^n \left(1 - \frac{n}{n+1}\right)$$

$$= \{\lambda_{DU}T\}^n \left(\frac{1}{n+1}\right)$$

$$= \frac{\{\lambda_{DU}T\}^n}{n+1}$$

Average probability of functional failure on demand

In condition of Figure 4, average probability of functional failure on demand is

$$\text{PFD}_{avg} = \frac{\{\lambda_{DU}T\}^n}{n+1} \qquad (12)$$

4.2. Calculation of hazardous or harmful event rate

It is derived from hazardous or harmful event rate of 1-out-of-n system to PFD_{avg}.

That is

an average probability $F(t)$ of system fault in proof test$(0, T]$, when demand occurs during $(0, T]$ by event rate λ_M.

$$\text{PFD}_{avg} = \int_0^T F(t)^n \lambda_M \, dt$$

$$= \lambda_M \int_0^T \{\lambda_{DU}t\}^n \, dt$$

$$= \lambda_M \{\lambda_{DU}\}^n \int_0^T t^n dt$$

$$= \lambda_M \{\lambda_{DU}\}^n \left[\frac{t^n}{n+1}\right]_0^T$$

$$= \lambda_M \{\lambda_{DU}\}^n \left(\frac{T^{n+1}}{n+1}\right)$$

$$= \lambda_M \{\lambda_{DU}T\}^n (T / n+1) \qquad (13)$$

On the other hands,

$$\text{PFD}_{avg} = \Pr\{\text{risk occurs during } (0, T]\} / \Pr\{\text{demand occurs during } (0, T]\}$$

$$= \left\{\lambda_M \{\lambda_{DU}T\}^n \frac{T}{n+1}\right\} / \{\lambda_M T\}$$

$$= \{\lambda_{DU}T\}^n / (n+1) \qquad (14)$$

4.3. Calculation of mean fault time

Mean fault time T_b^* is derived from each PFD_{avg} at clause 4.1. and clause 4.2. .

From formula (10),

$$PFD_{avg} = \{\lambda_{DU} T\}^n (T_b^*/T) \tag{15}$$

From formula (12) or formula (14),

$$PFD_{avg} = \frac{\{\lambda_{DU} T\}^n}{(n+1)} \tag{16}$$

From formula (15) and formula (16),

$$\{\lambda_{DU} T\}^n \frac{T_b^*}{T} = \frac{\{\lambda_{DU} T\}^n}{(n+1)}$$

In this,

$$T_b^* = T/(n+1) \tag{17}$$

5. SUMMARY

From the result of this paper, mean fault time T_b^* is

$$T_b^* = T/(n+1) \tag{18}$$

It is judged correctly because of same results are derived from two methods, i.e. ,reliability and hazardous or harmful event rate.

When we evaluate safety integrity level correctr, we use a Markov-state transition diagrams.

The mean fault time is useful to modelling of repair of dangerous un-detected failure by a Markov state-transition diagram.

References

[1] IEC 61508, *"Functional safety of electrical / electronic / programmable electronic safety-related systems, Part 1 "*, IEC, Geneva, Dec. 1998 - Feb. 2000.
[2] IEC 61508, *"Functional safety of electrical / electronic / programmable electronic safety-related systems, Part 4 "*, IEC, Geneva, Dec. 1998 - Feb. 2000.
[3] IEC 61508, *"Functional safety of electrical / electronic / programmable electronic safety-related systems, Part 6 "*, IEC, Geneva, Dec. 1998 - Feb. 2000.
[4] Yoshinobu SATO, *"Basis of Functional safety / Machinery Safety standard and Risk assessment - evaluation of SIL, PL, Automotive SIL "*, NIKKAN KOGYO SHIMBUN,LTD. , pp 20-27, Aug. 2011.

Reliability Analysis Including External Failures for Low Demand Marine Systems

KIM HyungJu[a*], HAUGEN Stein[a], and UTNE Ingrid Bouwer[b]
[a] Department of Production and Quality Engineering NTNU, Trondheim, Norway
[b] Department Marine Technology NTNU, Trondheim, Norway

Abstract: Marine systems fail not only due to equipment failure, but also because of external events like fire or flooding. Fire in the engine room, for example, can damage the main engine and it can lead to propulsion system failure regardless of the reliability of the main engine itself. Many redundancy requirements to vessels include these external events as a cause to system failure and they require physically separated redundancy in order to prevent system failure by a single fire or flooding. We need to consider external events, as well as equipment failure when analyzing the reliability of a marine system.

The main objective of this paper is to introduce a reliability analysis models for (i) equipment failure and (ii) external failure of low demand marine systems. A Markov model is suggested to calculate the hazardous event frequency (HEF) in this study. The paper also investigates the contribution of the two different types of failures (i & ii). The paper provides a case study of a fire pump in a passenger ship which analyses the contribution of each failure type.

Keywords: Reliability Analysis, Low Demand Marine System, Markov Model, External Failures, Redundancy, Safe Return to Port (SRtP), Dynamic Positioning (DP), Redundant Propulsion (RP), Hazardous Event Frequency (HEF).

1. INTRODUCTION

There are several redundancy requirements for vessels which have been issued by International Maritime Organization (IMO) and Classification Societies: IMO equipment class [1], Notations of Classification Societies for Dynamic Positioning (DP) system [2] and Redundant Propulsion (RP) [3], and Safe Return to Port (SRtP) regulation for passenger ships [4-6]. All redundancy requirements include external failures, as well as internal failures, because marine systems fail due to both types of failures. The former is a random hardware failure, which is relevant for reliability of equipment, while the latter is a failure caused by external events, which here are defined as fire and flooding [1-5]. For example, a fire in the vessel's engine room can damage the main engine and lead to loss of propulsion regardless of the reliability of the main engine itself. We therefore need to consider both types of failures when analyzing the reliability of a marine system. Even though reliability analysis methods and models for internal failures have evolved over several decades by various authors [7, 8], there are not much research on reliability analysis of external failures.

The main objective of this paper is to introduce a reliability analysis model and investigate the contribution of the two different types of failures of low demand marine systems: internal failure and external failure. If the effect of external failures is much greater than internal failures, then we need to pay much more attention to that kind of failures and study the feasibility of reliability analysis methods more in depth. On the other hand, if the effect of external failures is very small and negligible, then we may not need any further study on reliability analysis of external failures. The research in this paper investigates hazardous event frequency (HEF) of 1oo2 system to compare the impact of the two types of failures.

---

* Tel: +47 73 59 71 17; fax: +47 73 59 28 96
E-mail: hyung-ju.kim@ntnu.no

An external failure is some kind of a common cause failure (CCF) and therefore one may argue that it could be included in the analysis of an internal failure as a CCF. However, an external failure may be a single failure as well as a CCF. For example, two independent fires may damage a redundant system at the same time. Considering an external failure as a CCF only is therefore an imperfect model and external failures should be included in the model as a single failure too.

One challenge of analyzing external failures is that, for some systems, a demand itself is a cause of system failure. A system may be damaged by a demand, and therefore it may fail to respond to the demand. For instance, if a fire occurs in a vessel and it damages the fire pump or other essential components for the firefighting system, the firefighting system cannot extinguish the fire. In this case, the fire is a demand, as well as a cause of system failure. This study focuses on these kinds of systems in the following sections.

The rest of this paper is organized as follows: the research method is described in Section 2. Section 3 and 4 introduce the Markov models for internal failure and external failure, and case studies for each kind of failures are given in Section 5. Finally, discussion and concluding remarks are presented in Section 6.

2. RESEARCH METHOD

The impact of external failure and internal failure should be compared by using an identical method and unit. An external failure is closely related to a demand, and therefore the demand rate should be included in the model when analyzing external failures. Various authors have shown that a Markov approach is suitable for reliability modeling, including for the demand rate [9-14], and Jin, et al. [14] have suggested a Markov model for internal failure of a 1oo2 system. This paper is based on this model for reliability analysis of internal failures because 1oo2 system is the most common redundant concept in marine systems. Assessing the impact of external failures has not been attempted by previous research and therefore this study suggests a Markov model for external failures of a 1oo2 system.

With those two Markov models of each type of failures, this study calculates HEF [13, 14] and studies the impact and contribution of each kind of failures.

3. MARKOV MODEL FOR INTERNAL FAILURE OF 1oo2 SYSTEM

3.1 Assumptions

It is assumed that the system components are exposed to dangerous undetected (DU) failures only. A dangerous detected (DD) failure is assumed to be revealed and repaired immediately, and therefore it is negligible. A DU-failure is discovered during a functional test, and the functional test and repair actions are assumed perfect. All failure rates are constant even when components are responding to a demand, and the system satisfies the Markov property [7]. The standard β factor model is used to model CCFs. It is also assumed that the two components are identical and the system composes passive standby with perfect switching.

3.2 System States

The Markov model has six system states from state 0 to state 5, and they are defined in Table 1. System components have three kinds of states: "Available", "Functioning" and "Failure". "Available" means that the component is able to respond to a demand, and "Functioning" means that the component is already responding to a demand. "Failure" means that the component is unable to respond to a demand because of a DU-failure [14].

Table 1 System States of Internal Failure

System State	Component State	Demand
5	2 Available	Non-demand
4	1 Available, 1 Functioning	On-demand
3	1 Available, 1 Failure	Non-demand
2	1 Functioning, 1 Failure	On-demand
1	2 Failures	Non-demand
0	2 Failures	On-demand

3.3 System Transitions

System transitions for internal failure of 1oo2 system are illustrated in Figure 1. State 5 is the initial state, where all components are able to respond without any demand. State 0 is the hazardous state where both of redundant components are failed by DU-failure with an occurrence of a demand. State 4 represents that one of the components responds to a demand without any failures. State 3 is the state where one of the components has DU-failure without demand, and state 2 is one DU-failure with a demand. Even though both of the components are failed in state 1, it is not a hazardous state because there is no demand. The transition rates of this Markov model are defined in Table 2.

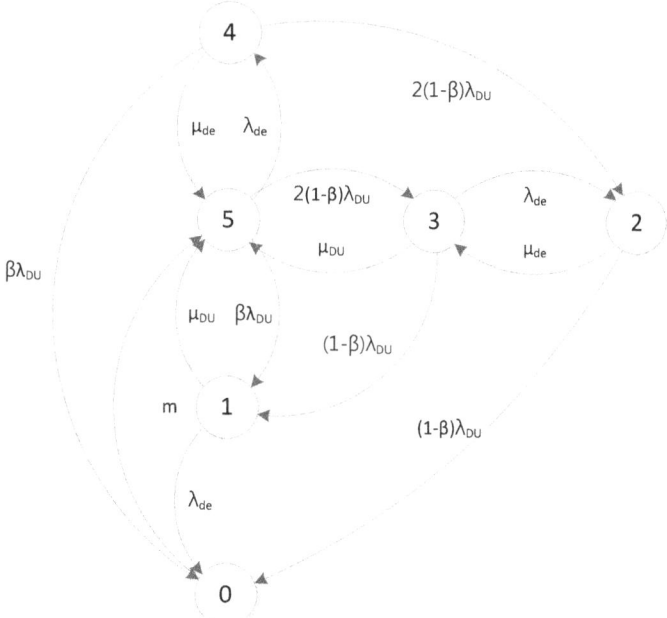

Figure 1 Markov Transition Diagram of Internal Failure

Table 2 Markov Transition Rates of Internal Failure

Transition Rates	Description
λ_{de}	Demand rate
μ_{de}	Demand duration rate
$(1-\beta)\lambda_{DU}$	Single DU-failure rate
$\beta\lambda_{DU}$	DU-CCF rate
μ_{DU}	DU repair rate
m	Renewal rate

4. MARKOV MODEL FOR EXTERNAL FAILURE OF 1oo2 SYSTEM

4.1 Assumptions

It is assumed that the system components are exposed to external failures only, and the external events are demands of the system, for instance a fire and a firefighting system. The fire is an external event which can lead to failure of the firefighting system and, at the same time, the fire is the demand of the firefighting system. If the system is not damaged by the external event (demand), the system successfully responds to the demand and the system is restored to an "as good as new" state after the demand duration. If the system is damaged by the external event, it is not possible to repair onboard, and the vessel should return to port to repair it. It is also assumed that the two components are identical and the system composes passive standby with perfect switching.

4.2 System States

The Markov model has five system states from state 0 to state 4, and they are given in Table 3. System components have three kinds of states, and they are same as in Section 3.2, except the "Failure" state. "Failure", in this model, means that the component is unable to respond to a demand because of an external failure.

Table 3 System States of External Failure

System State	Component State	Demand
4	2 Available	Non-demand
3	1 Available, 1 Functioning	On-demand
2	1 Functioning, 1 Failure	On-demand
1	1 Available, 1 Failure	Non-demand
0	2 Failures	On-demand

4.3 System Transitions

System transitions for external failures of 1oo2 system are illustrated in Figure 2. State 4 represents the initial state where all components are able to respond without any demand. State 0 is the hazardous state where both components are failed by external failure with an occurrence of a demand. In state 3, the external event (demand) occurs, but it does not damage the system. The system therefore responds to the demand, and the system state moves to the initial state 4 after the demand duration. State 2 is the state where one of the components is damaged by the demand and the other is responding to the demand. After demand duration, the system state moves to state 1 where one of the components is able to respond and the other is not. It is not possible to move directly from state 4 to state 1, because the demand is the cause of the system failure. Moreover, in this model, there is no state where both components are failed without demand, like state 1 in Section 3.2, because of the same reason. One principle of this model is "no demand, no failure".

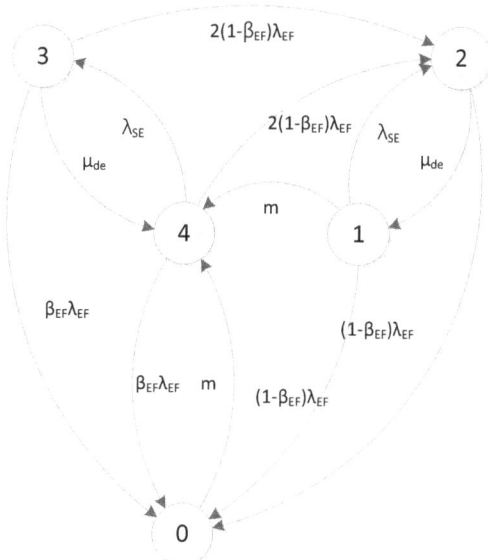

Figure 2 Markov Transition Diagram of External Failure

Table 4 Markov Transition Rates of External Failure

Transition Rates	Description
λ_{SE}	Survival rate by external event
μ_{de}	Demand duration rate
$(1-\beta_{EF})\lambda_{EF}$	Single external failure rate
$\beta_{EF}\lambda_{EF}$	External CCF rate
m	Renewal rate

λ_{EF} represents the failure rate of a component by external events (demand), while λ_{SE} is a survival rate. They may be calculated as

$$\lambda_{EF} = \lambda_{de} \cdot EF \qquad (1)$$

$$\lambda_{SE} = \lambda_{de} \cdot (1 - EF) \qquad (2)$$

where EF denotes the conditional probability that the demand damages system given a demand occurs. Beta factor for external failure (β_{EF}) is a value which represents whether redundant components are physically separated or not. If the two redundant components are perfectly separated and not damaged by a single external event at the same time, β_{EF} equals to 0.

5. CASE STUDY

To test the models from the previous sections two different types of fire water supply systems are analyzed. One complies with the redundancy requirement with physical separation and the other does not. The case study calculates HEF of internal failures and external failures for each of the two systems and investigates the impact and the contribution to the overall reliability of each type of failures.

5.1 System Description

The main function of the fire water supply system is to provide water, in case of an occurrence of a fire, to extinguish the fire. The system consists of two fire pumps, the power supply system, power cables, the control system, control cables, water pipes, valves, and many necessary accessories. For the sake of simplicity, this case study only considers the two fire pumps with a 1oo2 configuration.

Two different types of fire water supply systems are given in Figure 3; "System A" is a physically separated system, while "System B" is a physically not-separated system. The physical separation is based on marine redundancy requirements [1-4]. They assume that physically separated components are not damaged by a single fire at the same time, while physically not-separated components fail simultaneously if a fire occurs in the compartment where they are located. Applying this physical separation affects the reliability of the system related to external failures, and therefore the research in this case study includes both of the systems in the case studies.

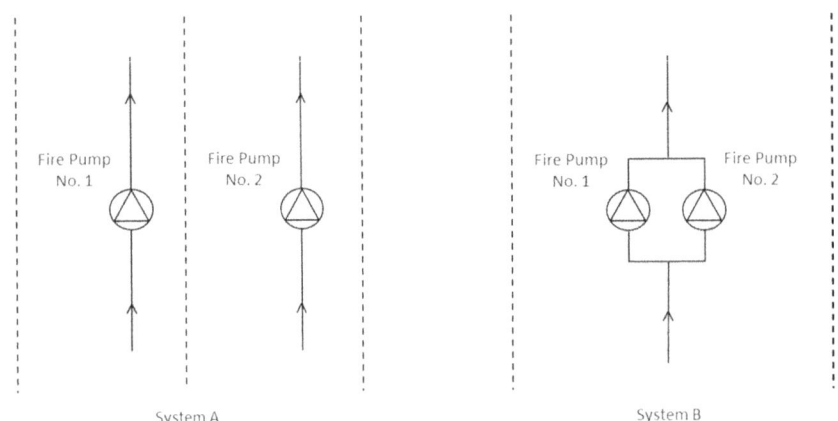

Figure 3 Two Types of Fire Water Supply Systems

5.2 Reliability Analysis for Internal Failure

Regarding internal failures, the two systems are identical, because the location of equipment does not affect the reliability of the system if they are operated in the same environment. Therefore, specifications and HEF of System A and System B are exactly the same when analyzing internal failures.

Failure data of the fire pump and other parameters for System A and System B are given in Table 5. The failure rate of the fire pump is from the OREDA handbook[15] and the beta-factor is calculated from IEC61508 [16] with expert judgment. The test interval of the fire pump is from the guideline of IMO [17] and the repair time is assumed to be negligible. The demand rate is from historical data of IMO [18] and the demand duration rate is based on the IMO requirement of fire pumps [19]. Renewal time is based on expert judgment.

Table 5 Specifications for Internal Failure

Notation	System A	System B	Description
λ_{DU}	7.10 x 10^{-7} per hour	7.10 x 10^{-7} per hour	DU-failure rate
B	0.02	0.02	Beta factor
μ_{DU}	1.39 x 10^{-3} per hour	1.39 x 10^{-3} per hour	DU repair rate
λ_{de}	1.05 x 10^{-6} per hour	1.05 x 10^{-6} per hour	Demand rate
μ_{de}	2 per hour	2 per hour	Demand duration rate
m	1.39 x 10^{-2} per hour	1.39 x 10^{-2} per hour	Renewal rate

According to Jin, et al. [14], the $HEF(t)$ is equal to the visit frequency to state 0, from any other state:

$$HEF(t) = \sum_{i=1}^{5} P_i(t) \cdot a_{i0} \quad (3)$$

where $P_i(t)$ is the probability that the system is in state i at time t, and a_{i0} is the transition rate from state i to state 0. $HEF(t)$ therefore means how often hazardous events occur for a given system design and demand rate [14]. HEF for internal failure may be obtained by solving (3) with Markov models in Figure 1. The calculation result is given in Table 6.

Table 6 HEF for Internal Failure of System A and B

System	HEF
System A (Physically Separated)	9.86×10^{-8} per year
System B (Physically Not-Separated)	9.86×10^{-8} per year

5.3 Reliability Analysis for External Failure

Regarding external failures, the two systems have one significant difference; whether the components are located in separated compartments or not. It is reflected on the value of β_{EF}. The external failure rate and the external survival rate can be calculated from (1) and (2). The demand rate, demand duration, and renewal time is the same as in Section 5.2. The conditional probability, EF, is derived from historical data for the ratio of fire origin location [20] and the area of each fire pump room [21]. All the other specifications and parameters are same for System A and System B as given in Table 7.

Table 7 Specifications for External Failure

Notation	System A	System B	Description
λ_{SE}	1.05×10^{-6} per hour	1.05×10^{-6} per hour	Survival rate by external event
λ_{EF}	2.04×10^{-9} per hour	2.04×10^{-9} per hour	External failure rate
β_{EF}	1	0	Beta factor for external failure
λ_{de}	1.05×10^{-6} per hour	1.05×10^{-6} per hour	Demand rate
μ_{de}	2 per hour	2 per hour	Demand duration rate
m	1.39×10^{-2} per hour	1.39×10^{-2} per hour	Renewal rate

HEF for external failure of System A and System B may be obtained by the same way with Section 5.2, and the calculation result is given in Table 8.

Table 8 HEF for External Failure of System A and B

System	HEF
System A (Physically Separated)	1.35×10^{-9} per year
System B (Physically Not-Separated)	1.79×10^{-5} per year

6. DISCUSSION AND CONCLUDING REMARKS

This paper introduces Markov models with HEF for both internal failures and external failures of marine systems, and two case studies have been carried out: one for internal failures and the other for external failures. Each of these case studies has calculated HEF for a physically separated system and a physically not-separated system. The study shows some interesting results.

First, for the physically separated system, the HEF of internal failures is about 100 times as high as that of external failures. On the contrary, for physically not-separated system, the HEF of internal failures is about 100 times as low as that of external failures. However, we cannot conclude that we do not need to consider external failures for physically separated systems nor that internal failures are meaningless to include for physically not-separated systems, because this case study has analyzed only a small part of the entire system. The contribution of each type of failures may vary when the entire system's components are included in the model. Instead, the case study provides a good ground for the necessity to pay much more attention to external failures with a feasible reliability analysis method.

Second, the HEF values are very small in these case studies. The reason for this is as follows: (i) the HEF value is neither a failure frequency nor a demand frequency, but a combination of them. It is a frequency of occurrence for both equipment failure and demand. The HEF value therefore may be very small compared to the failure rates or demand rates. (ii) This paper only covers fire pumps in the fire water supply system, which is a very small part of the entire system. Reasonable HEF values are expected if the entire system is included in the analysis. Future work should include all components of the fire water supply system into the analysis model investigating the sensitivity of contribution of two types of failures.

Last, this paper analyzes internal failure and external failure through its own respective models. One model is for internal failure and the other is for external failure. Merging them into a single model and investigating the results should be further addressed.

7. REFERENCES

[1] IMO, Guidelines for vessels with dynamic positioning systems, MSC/Cric.645, (1994).
[2] DNV, Special equipment and systems additional class - Dynamic positioning systems, Part 6 Chapter 7, (2011).
[3] DNV, Special equipment and systems additional class - Redundant propulsion, Part 6 Chapter 2, (2012).
[4] IMO, Casualty threshold, safe return to port and safe areas, SOLAS Chapter II-2. Regulation 21, (2010).
[5] IMO, System capabilities after a flooding casualty on passenger ships, SOLAS Chapter II-1. Regulation 8-1, (2010).
[6] IMO, Design criteria for systems to remain operational after a fire casualty, SOLAS Chapter II-2. Regulation 22, (2010).
[7] M. Rausand and A. Høyland, *System reliability theory; models, statistical methods, and applications*, 2nd ed. Wiley,(2004), Hoboken, NJ
[8] M. Rausand, *Risk Assessment; Theory, Methods, and Applications*. Wiley,(2011), Hoboken, NJ
[9] J. V. Bukowski, "Incorporating process demand into models for assessment of safety system performance," in *Reliability and Maintainability Symposium, RAMS'06. Annual*, pp. 577-581, (2006).
[10] W. M. Goble and H. Cheddie, *Safety Instrumented Systems verification: practical probabilistic calculations*. Isa,(2005)
[11] F. Innal, Contribution to modelling safety instrumented systems and to assessing their performance critical analysis of IEC 61508 standard, Ph.D. thesis, University of Bordeaux, Bordeaux, France, (2008).
[12] Y. Misumi and Y. Sato, "*Estimation of average hazardous-event-frequency for allocation of safety-integrity levels,*" Reliability Engineering and System Safety, vol. 66, pp. 135-144, (1999).
[13] H. Jin, M. A. Lundteigen, and M. Rausand, "*Reliability performance of safety instrumented systems: A common approach for both low- and high-demand mode of operation,*" Reliability Engineering & System Safety, vol. 96, pp. 365-373, (2011).
[14] H. Jin, M. Rausand, and M. A. Lundteigen, "*New Reliability Measure for Safety Instrumented Systems,*" International Journal of Reliability, Quality and Safety Engineering, vol. 20, p. 1350005, (2013).
[15] OREDA, *Offshore Reliability Data Handbook*. Det Norske Veritas Industri Norge AS,(2009), No 1322 Høvik, Norway
[16] IEC61508, Functional safety of electrical/electronic/programmable electronic safety-related systems, parts 1–7. International Electrotechnical Commission, 2010, Geneva
[17] IMO, Guidelines for the maintenance and inspection of fire-protection systems and appliances, MSC Circ.850, (1998).
[18] IMO, FSA – Cruise ships Details of the Formal Safety Assessment, MSC 85/INF.2, (2008).
[19] IMO, Revised guidelines for the approval of equivalent water-based fire-extinguishing systems for machinery spaces and cargo pump-rooms, MSC Circ.1165, (2005).
[20] O. V. Nilsen, FSA for Cruise Ships - Subproject 4.1, SAFEDOR, D-4.1.1, (2005).
[21] H. Kim, Methodology for Quantitative Analysis of Ship's Outfitting Systems Applying Safe Return to Port Rule, The Graduate School, Naval Architecture and Ocean Engineering, Seoul National University, Korea, (2010).

Heterogeneous Redundancy Analysis based on Component Dynamic Fault Trees

Jose Ignacio Aizpurua[a*], Eñaut Muxika[a], Ferdinando Chiacchio[b], and Gabriele Manno[c]

[a] University of Mondragon, Mondragon, Spain
[b] University of Catania, Catania, Italy
[c] Strategic Research & Innovation DNV GL, Høvik, Norway

Abstract: The aggregation of hardware components to add recovery capabilities to a system function may result in high costs. Instead of adding redundancies with homogeneous nature aimed at providing recovery capabilities to a predefined system function, there is room in some scenarios to take advantage of over-dimensioning design decisions and overlapping structural functions using heterogeneous redundancies: components that, besides performing their primary intended design function, restore compatible functions of other components.

In this work, we present a methodology to evaluate systematically the effect of failures of alternative redundancy and reconfiguration strategies, fault detection, and communication implementations on system dependability. To this end, a modeling approach called Generic Dependability Evaluation Model and its probabilistic analysis paradigm using Component Dynamic Fault Trees are presented. Application to a railway example is presented showing tradeoffs between dependability and cost when deciding to implement possible redundancy and reconfiguration strategies. Finally, details of the experiment prototype implemented using real railway communication elements are described so as to validate the design concepts treated throughout the paper.

Keywords: Heterogeneous redundancies, Dependability, Design methodology, Monte Carlo simulations, Cost reduction.

1. INTRODUCTION

Traditional design strategies to improve fault tolerance of a system are based on the replication of hardware components in redundancy configurations. Generally the nature of the backup components is homogeneous, i.e., they provide recovery capabilities to a predefined system function; accordingly they are known as homogeneous redundancies.

However, in some scenarios, it is possible to make use of heterogeneous redundancies, consisting of components that, besides performing their primary intended design function, are able to restore compatible functionality of other components. This is the case of Massively Networked Scenarios (MNS), systems characterized by several replicas of system functions throughout the physical structure, e.g., a train has replicated functions throughout its cars; a building has replicated functions throughout its floors and rooms. Nowadays, in such architectures, the comparison between homogeneous and heterogeneous redundancies is gaining the interest of researchers and industrial stakeholders. In fact, design strategies based on heterogeneous redundancies have shown potential to improve system dependability cost-effectively [1, 2, 3].

In previous work [4], an adaptive dependable design methodology, the D3H2 (aDaptive Dependable Design for systems with Homogeneous and Heterogeneous redundancies), for the dependability

---

Acronyms & Abbreviations: (C)DFT: (Component) Dynamic Fault Tree; FD_R_SF: Fault detection of the R_SF; FD_SF: Fault Detection of the SF; GDEM: Generic Dependability Evaluation Model; (E)GFM: (Extended) Generic Functional Model; MF: Main Function; MVB: Multi-function Vehicle Bus; PL: Physical Location; PU: Processing Unit; R_SF: Reconfiguration of the SF; SF: Subfunction.
[*] Contact E-mail address: jiaizpurua@mondragon.edu

assessment of MNS was presented. Applying modeling and analysis techniques, the methodology identifies heterogeneous redundancies systematically; integrates redundancies in the *extended HW/SW architecture* supporting necessary functions and implementations including reconfiguration, fault detection, and communication; and performs dependability and cost analyses of the *extended HW/SW architecture*. In this paper, we present an extension of the methodology overcoming previous static logic limitations by means of Generic Dependability Evaluation Model (GDEM) and Component Dynamic Fault Trees (CDFTs), and quantify the effect of the failure of redundancy and reconfiguration strategies on the *extended HW/SW architecture*. The GDEM defines failure relationships between system functions and implementations and allows identifying systematically the combinations of faults that lead the *extended HW/SW architecture* to fail; while the CDFT enables the probabilistic assessment of the GDEM. Besides, importance measurements are performed to obtain robustness indicators of design strategies. To test the feasibility of the approach, its key design concepts are implemented in a practical case of the railway industry: reconfiguration capabilities have been added to hardware train network components to reuse already existing elements.

Related Work: The evaluation of the influence of design decisions on dependability and cost is an ongoing research challenge. While many works have concentrated on addressing the influence of homogeneous redundancies on system dependability and cost, approaches focusing on the evaluation of the influence of heterogeneous redundancies on system dependability are scarce.

The concept of heterogeneous redundancies have been addressed in the literature with different names but with the same underlying design goal: reuse of system components to provide a compatible functionality. Shelton and Koopman worked on the concept of functional alternatives [1]; Wysocki and Debouk presented a methodology for assessing architectures using shared redundancies [2]; and Adler et al. presented a methodological support for characterizing an adaptation model while meeting availability-cost requirements [3]. However, to fully exploit the potential of heterogeneous redundancies without incurring in additional penalties on the system architecture, there exist assumptions and activities that should be addressed: (1) identification of heterogeneous redundancies should be done systematically rather than relying only on the ability of the designer; (2) the use of homogeneous/heterogeneous redundancies in MNS requires fitting the system with health management implementations, i.e., fault detection (FD) and reconfiguration (R). Consequently, alternative health management strategies and their influence on dependability should be addressed so as to avoid unexpected consequences. For further details and discussion in this area see [5].

Contribution and overview of the paper: as a result, with respect to relevant approaches, we propose a methodology that: (1) identifies systematically heterogeneous redundancies and integrates them in the system architecture; (2) constructs alternative system architectures comprehending different redundancy and reconfiguration strategies; and (3) performs the dependability assessment of system architectures systematically and exhaustively with the goal of extracting design indicators and identifying weaknesses and strengths of the analyzed system architectures. The remainder of this paper is organized as follows: Section 2 overviews the D3H2 methodology and introduces the application example; Section 3 describes in detail the dependability analysis within the D3H2 methodology, and its application to a railway example; Section 4 validates D3H2's design concepts in a experiment prototype; and finally, Section 5 presents conclusions and our future research goals.

2. D3H2 METHODOLOGY: OVERVIEW & APPLICATION EXAMPLE

In this section we present an overview of the main activities of the D3H2 methodology [4] and apply them to an application example.

2.1. Overview

The D3H2 methodology characterizes a system of interest as a set of HW, SW, and communication components taking into account their interfaces and provided functionality. The main goal of D3H2 methodology is to create a system architecture that meets dependability and cost requirements, and evaluate the influence of redundancy (homogeneous, heterogeneous) and reconfiguration (centralized, distributed) strategies on system dependability and cost systematically (see Figure 1). The methodology integrates the following main modeling and analysis activities:

(1) Construction of the *functional model* based on the next *tokens:* (1) *Main Function* (MF) (e.g., Temperature Control); (2) the *Physical Location* (PL) in which the MF is performed; (3) necessary set of input (I), control (C) and output (O) *Subfunctions* (SFs) to perform the MF (e.g., Temperature Measurement); and (4) *implementations* of the SF characterized by the HW, SW, and communication resources they use and ordered with respect to their implementation priority.

(2) *Compatibility analysis*: systematic identification of heterogeneous redundancies based on the physical location of the SF and the compatibility of the SF they perform; and extraction of *reconfiguration strategies* and priorities.

(3) Construction of the *extended HW/SW architecture* by adding health management implementations to the *preliminary HW/SW architecture*, i.e., fault detection (FD_SF), reconfiguration (R_SF), and reconfiguration's fault detection (FD_R_SF) implementations.

(4) And finally, *dependability and cost analysis* of the *extended HW/SW architecture* to validate dependability and cost requirements. The ins and outs of these activities are described in [4].

Figure 1: D3H2 Methodology [4]   **Figure 2: Train Car and Network Structure**

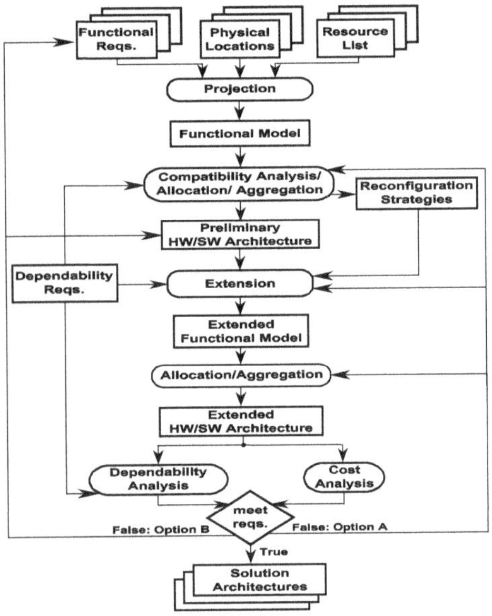

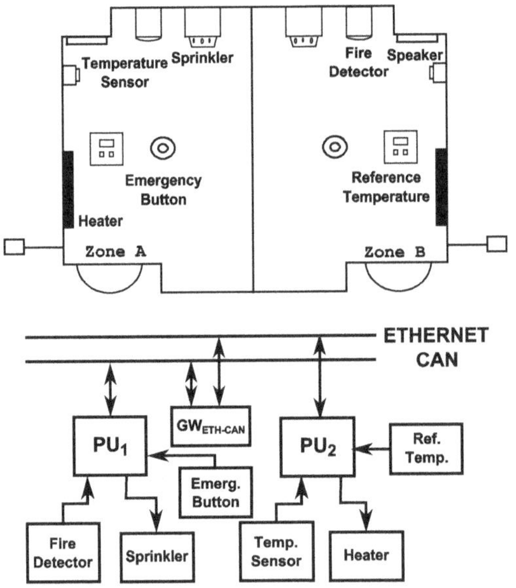

2.2. Application Example

In this subsection, we apply and describe the D3H2's main activities in an application example of the railway industry. Namely, we concentrate on analyzing Fire Protection Main Function and possible heterogeneous redundancies arising from the analysis of a train car.

System: in the studied train car fire detectors are distributed strategically to detect a fire. Usually, there is a fire detector (smoke sensor) for each train car's compartment (see $Zone_A$ - $Zone_B$ in Figure 2) and an emergency button for passengers. Potential fire conditions are validated by a fire control algorithm and if confirmed, sprinklers neutralize the hazardous situation.

Functional Model: among different Main Functions performed in a train car (cf. Table 1), the functional model of the Fire Protection and part of the Main Functions located at compatible physical location are described (Fire Protection and Temperature Control). There exists other Main Functions located at the same physical location, but for the sake of clarity, we have not taken them into account.

Compatibility Analysis & Reconfiguration Strategies: the compatibility analysis is based on 2 compatibility cases: (1) implementation of the same SF in compatible PLs; and (2) different SF's I/O implementations located at compatible PL. The first case is performed automatically comparing SFs and PL, while the second case requires evaluating if they may fulfill additional compatible SFs. From the second case, the compatibility analysis automatically suggests a list of implementations located at a Fire Protection's compatible PL (Train.Car$_1$.Zone$_A$) and the designer intuitively finds a temperature sensor able to indicate the presence of fire using temperature value thresholds (see grey cells at the functional model in Table 1) (see [4] for more information of the compatibility analysis).

Extended Functional Model (EFM): based on design decisions, the EFM is constructed by adding necessary fault detection and reconfiguration mechanisms to the SFs with redundant implementations to evaluate their influence on dependability and cost. For instance, the *extended HW/SW architecture* shown in Table 2 is based on a model with single FD_SF, duplicated R_SF implementations, and each R_SF implementation with its FD_R_SF implementation.

Table 1: Functional Model

MF	PL	SF	Resources
Fire Protection	Train. Car1. Zone$_A$	User Emerg. Signal (UES)	Emergency Button, SW$_{UES}$, PU1
		Fire Detection	Fire Detector, SW$_{FireDet}$, PU1
		Fire Control	UES, Fire Detection, PU1, SW$_{FireControl}$
		Fire Extinction	Fire Control, PU1, SW$_{FireExtinction}$, Sprinkler
Temp. Control	Train. Car1. Zone$_A$	Temperature Measurement	Temperature Sensor, SW$_{Temp}$, PU2
		...	...

Table 2: Extended HW/SW Architecture

MF	PL	SF	Resources
Fire Prot.	Train. Car1. Zone$_A$	UES	Emergency Button, SW$_{UES}$, PU1
		Fire Detection	Fire Detector, SW$_{FireDet}$, PU1
			Temperature Sensor, SW$_{FireDet}$, PU2, Comm. CAN, Comm. ETH, GW$_{ETH-CAN}$
		FD_FireDet	PU1, SW$_{FD\_FireDet}$, Comm. CAN, Comm. ETH, GW$_{ETH-CAN}$
		R_FireDet	PU1, SW$_{R\_FireDet}$, Comm. CAN
		R_FireDet	PU2, SW$_{R\_FireDet}$, Comm. CAN, Comm. ETH., GW$_{ETH-CAN}$
		FD_R_FireDet	PU2, SW$_{FD\_R\_FireDet}$, Comm. CAN, Comm. ETH, GW$_{ETH-CAN}$
		FD_R_FireDet	PU1, SW$_{FD\_R\_FireDet}$, Comm. CAN, Comm. ETH, GW$_{ETH-CAN}$
		Fire Control	UES, Fire Detection, PU1, SW$_{FireControl}$, Comm. CAN, Comm. ETH, GW$_{ETH-CAN}$
		Fire Extinction	Fire Control, SW$_{FireExtinction}$, PU1, Sprinkler

- **Heterogeneous Redundancies (Semi-Automatic)**
- **Health Monitoring Implementations (Systematic)**
  o FD_SF, R_SF and FD_R_SF

As for the dependability analysis, in [4] we focused on automating all the D3H2 activities using the Component Fault Tree paradigm [6], keeping the traceability between the *dependability model* and *extended HW/SW architecture* as manageable as possible. However, so as to adhere to the *extended HW/SW architecture*'s dynamic failure logic, it is necessary an approach which captures systems dynamic failure logic and does not hamper the readability of the *dependability model*. Hence, to address these goals, Section 3 defines compositional Generic Dependability Evaluation Model (GDEM) and Component Dynamic Fault Tree (CDFT) paradigms.

3. D3H2 METHODOLOGY: DEPENDABILITY ANALYSIS

The key introductory concepts for dependability analysis are defined in § 3.1. Then, the GDEM (cf. § 3.2) and its analysis approach based on CDFTs (cf. § 3.3) are defined.

3.1. Concepts and Notation

The objective of the GDEM is the generic, systematic and complete failure modeling of *extended HW/SW architectures*. The failure model of the *extended HW/SW architectures* comprehends the possible failure modes of its implementations: FD implementations (FD_SF, FD_R_SF) fail in omission (O) when it does not detect a failure when it occurs and false positive (FP) when it detects a failure when it does not exist; the reconfiguration (R) implementation fails in omission when it fails to reconfigure an implementation when it is required; and failure of SF implementations cover omission and wrong value failure modes.

The failures of all system subfunction implementations (SF, FD_SF, R_SF, FD_R_SF) are defined at the implementation level (i.e., [MF].[PL].[SF].[Impl] Failure) according to the failures of the implementation's resources. Combining implementation level failures, SF level failures are defined systematically ([MF].[PL].[SF] Failure). For the sake of clarity, in subsequent characterizations we omit the generic common part ([MF].[PL]). Table 3 defines the notations of the failure events and working events according to their SF and failure modes (omission and false positive).

Table 3: Notation of Failure and Working Events

Notation	Failure Logic	Notation	Failure/Working Logic
F_X	X failure	F_R	[R_SF] failure
F_{SF}	[SF] failure	$F_{R_i O}$	[R_SF].[Impl$_i$] omission
F_{SF_i}	[SF].[Impl$_i$] failure	$F_{FD\_R_i FP}$	[FD_{[R_SF].[Impl$_i$]}] false positive
F_{FD_i}	[FD_SF].[Impl$_i$] failure	$F_{R_i O/FP}$	[R_SF].[Impl$_i$] omission or FP = **OR**($F_{R_i O}$, $F_{FD\_R_i FP}$)
F_{FD}	[FD_SF] failure	$F_{FD\_R_i}$	[FD_{[R_SF].[Impl$_i$]}] failure
$F_{FD\ FP}$	[FD_SF] false positive	W_X	X working
$F_{SF_i FP}$	[SF].[Impl$_i$] failure or FP = **OR**(F_{SF_i}, $F_{FD\ FP}$)	W_{SF_i}	[SF].[Impl$_i$] working = **NOT**(F_{SF_i})
$F_{FD_i O}$	[FD_SF].[Impl$_i$] omission		

The stochastic failure characterization of each resource is characterized randomly sampling the failure times according to their cumulative probability distribution functions (CDFs) along the system lifetime. The methodology supports any CDFs, but for the sake of simplicity, without losing the generality of the approach, in subsequent probabilistic characterizations exponential failure distributions are assumed. Hence, the failure characterization of system resources is defined according to their failure rates (λ_{Res}). The failure characterization of a SF's i-th implementation ([SF].[Impl$_i$]) with N resources is specified as follows:

$$F_{SF_i} = \mathbf{OR}(F_{Res_1}, F_{Res_2}, ..., F_{Res_N}) \quad (1)$$

The same characterization holds for FD_SF, R_SF, and FD_R_SF implementations.

3.2. Component Dynamic Fault Trees (CDFT)

To quantify the failure probability of the GDEM, we have analyzed existing dynamic and compositional fault tree paradigms looking for the following characteristics: (1) component-based failure characterization: embed the failure logic of a set of related events/components and (re)use it where needed instead of characterizing the system failure behavior in a single flat model; (2) dynamic gates: capture the system failure logic accounting for the time-ordered events; (3) NOT gates: address the influence of functional (NOT failed) events; (4) support for any probability density function; (5) possibility of modeling repeated Basic Events (BEs); and (6) repeated components.

The integration of static fault trees and compositional characterization is not new: Component Fault Trees [6] addressed this concept prominently. However, to the best of our knowledge, there is no approach which addresses explicitly the integration of DFTs and component oriented characterization. There exists combinations of combinatorial and state-based approaches which do address the compositional characterization (e.g., [7]), but state-based approaches are not considered in this work in order to avoid possible state explosion and manageability issues.

Addressing all the aforementioned characteristics, we define the *concept* of Component Dynamic Fault Trees (CDFTs): while a BE characterizes self-contained simple failure logic, a component encloses any-complexity failure logic (with possibly multiple I/O dependencies) specified using BEs, gates, and even further components. The CDFT paradigm makes it possible to embed in a component the dynamic failure logic of a (sub)system and (re)use it where needed addressing repeated components and repeated BEs. Figure 3 characterizes a hypothetical CDFT model with repeated components (C_2) and CDFT gates. Each component (C_1, C_2) may have as inputs further gates, or (repeated) BEs and/or other components. The behavior of the CDFT gates are characterized according to its inputs events (A, B), which are extendible to an arbitrary number of input events.

Figure 3: Component Dynamic Fault Tree Overview

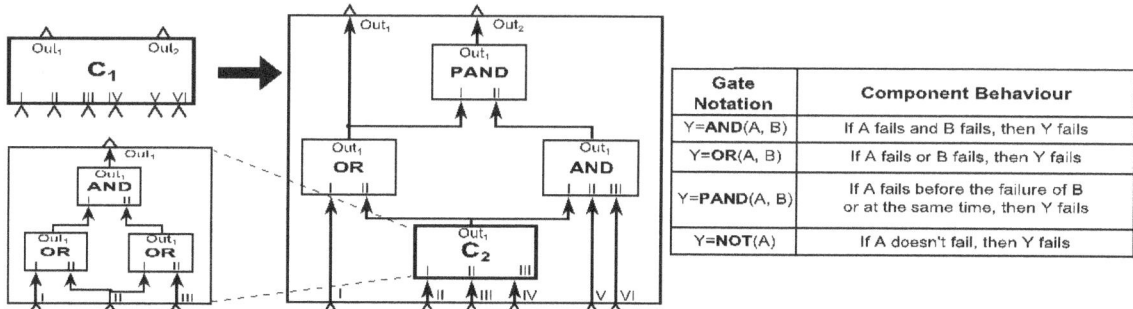

To *implement* the CDFT paradigm and analyze the (un)reliability of a system, Monte Carlo simulations are performed on the system's CDFT structure calculating the time to failure of BEs according to their cumulative probability distribution function. Connected gates/components use this information to determine their outcome (functional or failed state). When a failure at the output of a gate/component occurs, the failure time information is passed to the next gate/component so that the system's dynamic failure logic is tracked from BEs to system-level Top Event (TE). The logic of CDFT gates comprehend combinatorial (AND, OR, NOT) and time-ordered (PAND) failure logic. To analyze CDFTs, the MatCarloRe tool [8] has been extended with NOT gates.

3.3. Generic Dependability Evaluation Model (GDEM)

The GDEM defines the generic but implementable dependability evaluation algorithm, defining the dynamic failure behavior of systems which use fault detection and reconfiguration implementations covering all possible failure situations for the specified HW/SW architectures. It allows evaluating design decisions consequence on system dependability. Resulting equations characterize the failure of such systems compositionally so that the failure logic is kept clear for complex systems.

To this end, the GDEM characterizes combinations of SF's implementation failures that prevent the *extended HW/SW architecture* from performing its intended MF. The failure of any SF necessary for a MF provokes its immediate failure. Hence, from this point onwards we will analyze the failure of a SF. The subfunction will fail when all its implementations have failed ($F_{All\ Impl.}$), an implementation fails and reconfiguration does not happen (failure unresolved, $F_{Unresolved}$), or its input dependencies ($F_{Dependencies}$) have failed (cf. Equation 2):

$$F_{SF} = OR(F_{All\ Impl.}, F_{Unresolved}, F_{Dependencies}) \qquad (2)$$

Assuming that there exist M implementations of the subfunction, the $F_{All\ Impl.}$ event happens when each implementation fails or is detected as failed (false positive):

$$F_{All\ Impl.} = AND(F_{SF_1\ FP}, ..., F_{SF_M\ FP}) \qquad (3)$$

The failure unresolved ($F_{Unresolved}$) occurs when the working implementation fails and either the fault is not detected or the reconfiguration itself fails. For each implementation there are different failure unresolved events ($F_{Unr.\ Impl_i}$) because each implementation may have different failure probabilities, however, note that the last implementation's failure probability can not be solved:

$$F_{Unresolved} = OR(F_{Unr.\ Impl_1}, ..., F_{Unr.\ Impl_{M-1}}) \qquad (4)$$

To define $F_{Unr.\ Impl_i}$, let us introduce two new events. The first event occurs when the i-th implementation of the subfunction fails and the reconfiguration has failed but after successfully reconfiguring previous i-1 implementations (reconfiguration sequence failure, $F_{R\ Seq._i}$). Assuming $F_{SF_{1..i-1}\ FP} = AND(F_{SF_1\ FP}, ..., F_{SF_{i-1}\ FP})$ indicates the failure or false positive from 1 to i-1 implementations:

$$F_{R\ Seq._i} = PAND(F_{SF_{1..i-1}\ FP}, F_R, F_{SF_i\ FP}) \qquad (5)$$

Probabilistic Safety Assessment and Management PSAM 12, June 2014, Honolulu, Hawaii

The second event occurs when the *i*-th implementation of the SF fails and the fault detection of the SF has failed but after detecting correctly previous *i*-1 implementations failures (fault detection sequence failure, $F_{FD\ Seq._i}$). Note that fault detection's false positive and omission failures are mutually exclusive:

$$F_{FD\ Seq._i} = \mathbf{PAND}(F_{SF_{1..i-1}}, F_{FD}, F_{SF_i}) \tag{6}$$

Due to the characterization of time-ordered failures, Equations 5 and 6 can not be further simplified. Accordingly, *i*-th implementation's failure unresolved event ($F_{Unr.\ Impl_i}$) occurs when either the fault detection sequence ($F_{FD\ Seq._i}$) or the reconfiguration sequence ($F_{R\ Seq._i}$) fails:

$$F_{Unr.\ Impl_i} = \mathbf{OR}(F_{R\ Seq._i}, F_{FD\ Seq._i}) \tag{7}$$

Dependencies address Input (I) and Control (C) subfunctions influence on control and Output (O) SFs respectively. Control SF failure impacts the output SF failure directly (C→O); and the influence of input SF on control SF depends if the system's control configuration is operating in Closed Loop (C_CL) or Open Loop (C_OL):

$$F_{Dependencies} = \mathbf{OR}(F_{Dep.\ C\_CL}, F_{Dep.\ C\_OL}) \tag{8}$$

Assuming that $W_{C\_X} = \mathbf{OR}(W_{C\_X_1}, ..., W_{C\_X_Q})$ means that all Q implementations of C_X subfunction are working, equations in 9 describe the different input subfunctions that affect each control configuration (I_CL→C_CL, I_OL→C_OL). Usually, the $F_{Dep.\ C\_OL}$ event will not happen because the open loop control generally does not have input dependencies:

$$F_{Dep.\ C\_CL} = \mathbf{AND}(W_{C\_CL}, F_{I\_CL}) \qquad F_{Dep.\ C\_OL} = \mathbf{AND}(W_{C\_OL}, F_{I\_OL}) \tag{9}$$

The reconfiguration failure is a special subfunction and therefore F_R is developed like Equation 2, except that there are no additional dependencies:

$$F_R = \mathbf{OR}(F_{All\ R\ Impl.}, F_{R\ Unresolved}) \tag{10}$$

$F_{All\ R\ Impl.}$ indicates the failure of all reconfiguration implementations, and $F_{R\ Unresolved}$ designates the reconfiguration's failure unresolved condition. Assuming P reconfiguration implementations:

$$F_{All\ R\ Impl.} = \mathbf{AND}(F_{R_1\ O/FP}, ..., F_{R_P\ O/FP}) \tag{11}$$

$F_{R\ Unresolved}$ event happens when P-1 implementations of the FD_R subfunctions fail:

$$F_{R\ Unresolved} = \mathbf{AND}(F_{FD\_R_1}, ..., F_{FD\_R_{P-1}}) \tag{12}$$

Equation 12 boils down to our design choice: all reconfiguration's fault detection implementations (FD_R_SF) are active and homogeneous redundancies (heartbeat implementation). Accordingly, the false positive occurs when all FD_R_SF implementations raise the false positive condition simultaneously. Although the system may operate correctly when a false positive occurs, it has to assume that the information provided by the fault detection implementation is correct, since there is no mechanism to detect the incorrect operation of fault detection.

The fault detection failure F_{FD} depends on the operation of the destination subfunction (SF_{DEST}), because the FD implementation is located at the same PU. Hence, F_{SF_DEST} influences directly F_{FD}. When the FD implementation fails, the change of SF_{DEST}'s implementation determines its reconfiguration. We assume that the change of destination SF's implementation activates the corresponding FD implementation and the previous one is deactivated. Equation 13 describes the FD_SF failure case when FD_SF has K implementations:

$$F_{FD} = \mathbf{OR}(F_{FD\_Dest_1}, ..., F_{FD\_Dest_K}) \tag{13}$$

As for the *i*-th fault detection implementation's failure ($F_{FD\_Dest_i}$), it happens when the first 1 to *i*-1 implementations of the destination SF fail and reconfigure correctly ($F_{SF\_DEST_{1..i-1}}$), and then the *i*-th implementation of the fault detection or destination SF fails:

$$F_{FD\_Dest_i} = \textbf{PAND}(F_{SF\_DEST_{1..i-1}}, \textbf{OR}(F_{SF\_DEST_i}, F_{FD_i O})) \tag{14}$$

3.4. Experiments

Based on the example described in Section 2, simulations are performed to evaluate the influence of redundancy and reconfiguration strategies on system's (un)reliability and cost. The cost of health monitoring SW components (SW_HM: SW_FD, SW_R, SW_FD_R) is quantified considering their development cost. Regarding their λ values, hypothetical reasonable values are assumed, considering them lower than less critical components' failure rates. The assessment of the reliability and cost of SW components is outside the scope of this work, see [9] for an application on SW reliability methods. Regarding sensor's cost, human cost related with mounting and testing tasks is considered assuming 10 minutes/sensor. PU element characterizes all PU elements, and communication includes CAN and Ethernet communication protocols and their gateway. We are aware that the cost of SW components is greater than adding sensors, but in this example it has been assumed that SW development costs will be paid off in 4 years.

Table 4: Failure Rates & Cost of HW/SW/Communication Components

Component	Fail. Rate (year^{-1})	Cost (€)
SW_HM and SW_FD_FP	1 E-2	80 each
Fire Detector [10] / Temp. Sensor [11]	3.77 E-2 / 1.49 E-2	20 + 60€/hour / -
PU [9]	3.87 E-2	30
Communication and Gateway	5 E-3	200

Figure 4 describes Fire Protection configurations' relative unreliability and cost with respect to the configuration without redundancies described in Table 1. Alternative *extended HW/SW architectures* are analyzed adding a heterogeneous redundancy (*HeR*) (see Table 2) or homogeneous redundancy (*HoR*) to the Fire Detection SF. In case of homogeneous redundancies, the fire detector has been replicated with 2 alternative configurations: connect both fire detectors to PU1 (*HoR A*) or connect each fire detector to a different PU (*HoR B*). As Figure 4 shows, heterogeneous redundancies are more economical than homogeneous redundancies and their unreliability is slightly higher than homogeneous redundancies due to the added mechanisms (SW) to make implementations compatible.

Figure 4: Relative Unreliability and Cost of Fire Protection Configurations (10^6 iterations)

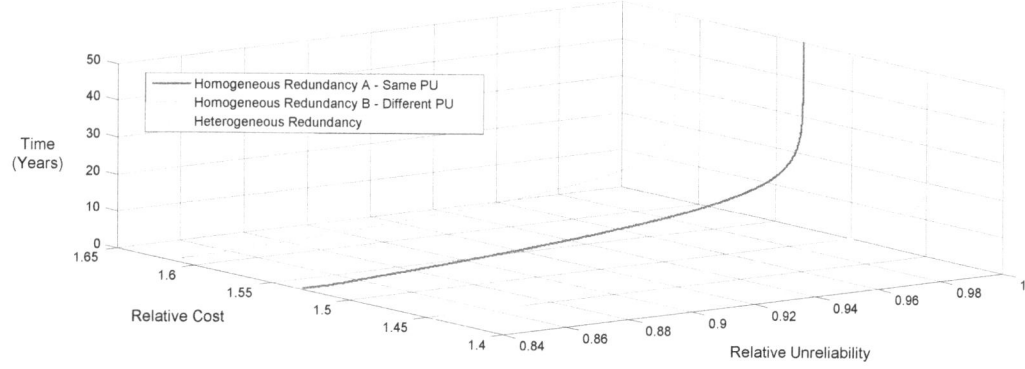

To analyze further the differences between *HoR* and *HeR*, the Failure Criticality Index (FCI) evaluation has been implemented [12] calculating for the component *i*: the ratio between the number system failures caused by component *i* to the total number of system failures. Table 5 shows the impact of the failure of redundancy and reconfiguration (centralized, distributed) implementations on the Main Function failure. The shown values are the influence on the Fire Protection MF of Fire Detection SF's redundancy ($F_{FireDet_2}$) and its reconfiguration strategies (logic sum of $F_{R\,Seq_1}$ and $F_{R\,Seq_2}$).

Table 5: Failure Criticality Index Values (10^6 iterations)

Causes\Configs.	HoR A – Cen.	HoR B – Cen.	HeR – Cen.	HoR A – Dist.	HoR B – Dist.	HeR – Dist.
Redundancy type	0,339027	0,174606	0,179927	0,276643	0,154960	0,170669
Reconfiguration	0,177554	0,171728	0,174496	0,114232	0,107315	0,106994

FCI values provide indicators about bottleneck influences on system (un)reliability: heterogeneous and homogeneous B redundancies perform better than homogeneous A redundancy configuration due to the bottleneck influence on causing the top event, i.e., PU1 performs as a common cause failure and its failure incurs the simultaneous failure of other SF implementations. The same logic applies to the reconfiguration distribution strategies: distributed reconfiguration implementations perform better than centralized implementations due to the bottleneck influence on system (un)reliability.

4. VALIDATION OF THE D3H2 METHODOLOGY: PROOF OF CONCEPT

To proof the feasibility of the D3H2 methodology in real applications, a key application concept in our methodology has been validated: we have added reconfiguration capabilities to existing hardware train network components to recover the system from failures at runtime using heterogeneous redundancies.

Trains have a standard form of data communication specified in the Train Communication Network (TCN) standard [13]. TCN is a real-time data network comprised of an architecture inter-connecting train vehicles and equipments within a vehicle. The TCN standard specifies Wire Train Bus (WTB) for the inter-connection of vehicles and Multi-function Vehicle Bus (MVB) for intra-vehicle device communication. In this work we focus on the communication within a vehicle using MVB.

MVB operates in master-slave configuration connecting devices in a vehicle. Class 2 or higher devices are considered: intelligent devices participating in the message communication with administration capabilities or connected I/O elements. The master guarantees deterministic medium access managing periodic and sporadic access to the bus. The communication in MVB follows the publisher/subscriber paradigm: a publisher broadcasts variables and this information is distributed to the subscribers. To this end, a traffic store is implemented; each device holds the variables it produces/consumes in a shared memory that is a partial copy of the whole network's distributed database.

As for the implementation of the reconfiguration process, we identify two phases: construction of the *reconfiguration table*, i.e., statically/dynamically determined reconfigurations; and activation of *reconfiguration strategies*, i.e., design-time/runtime reconfigurations. Dynamic resolution of the *reconfiguration table* allows gaining flexibility, but requires exploring the architecture dynamically. For safety and predictability purposes, statically determined reconfiguration design decisions are adopted. Regarding the (de)activation of *reconfiguration strategies*, while design-time reconfigurations reduce design complexity, runtime reconfigurations reduce processing cost and bandwidth overhead activating redundant communication threads exclusively when their need arises.

In a train there are safety-critical functions which must meet hard real-time constraints (e.g., door control) and some of these functions are transmitted through MVB. Besides, other communication protocols coexist in the train; for instance, Ethernet communication protocol transports non-critical infotainment data. Ethernet provides more flexibility to perform architectural modifications at runtime at the expenses of losing predictability with respect to MVB. There exist other communication networks in a train (e.g., CAN), but this proof of concept has been focused on MVB and Ethernet.

Therefore, the following design decisions have been adopted: MVB has been used for design-time reconfigurations and runtime reconfigurations have been implemented in Ethernet. On one hand, design-time reconfigurations in MVB are performed by assigning reconfiguration routes at design-time and activating them from the outset. The bandwidth consumption of these redundant communications is constant and their processing is activated solely when their need rise up. On the other hand, (dynamic) runtime modifications in Ethernet are effectuated using UDP communication threads in client-server-like configurations. Communication threads are created and deleted as their need arises, so that the bandwidth and processing costs increase exclusively in case of reconfiguration.

The following reconfiguration scenarios *(SC)* have been tested: *(SC1)* sensor-level heterogeneous redundancy reconfigurations; *(SC2)* communication-level heterogeneous redundancy reconfigurations; and *(SC3)* processing unit-level homogeneous redundancy reconfigurations.

Two additional reconfiguration attributes define the reconfiguration space of these scenarios: reconfiguration *granularity* comprehends node or task level reconfigurations, indicating implementation's reactivation by changing the whole PU and its allocated tasks, or a single task respectively. Reconfiguration *object* addresses SW, HW and communication (Comm.) level reconfigurations: SW reconfigurations modify the SW implementation changing its parameters or structure; HW reconfigurations involve changing the complete HW device; and communication reconfiguration modifies nominal communication routes with alternative ones. Figure 5 describes the reconfiguration space of the tested scenarios: *SC1* deals with sensor failures using communication level reconfigurations, *SC2* switches the communication protocol to handle communication failures, and *SC3* replaces the PU and communication routes to cope with PU failures. For instance, all the scenarios perform task-level and communication-level reconfigurations, but only *SC3* addresses node-level and communication-level reconfigurations.

Figure 5: Reconfiguration Space

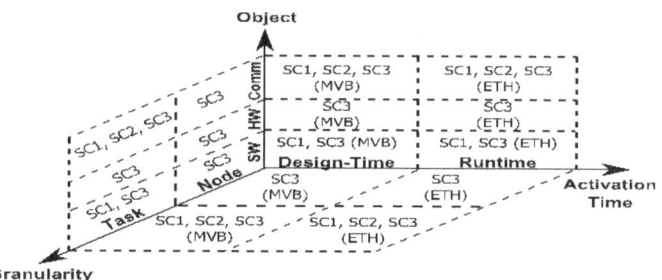

SC1: without losing the applicability of the scenario, *SC1* focuses on the example presented in [14]. A train car vehicle may have different compartments (cf. Figure 6 $Zone_A$ - $Zone_B$) and independent Temperature Control Main Function implementations at each compartment. Assume that 2 PUs are connected to perform the Temperature Control in each vehicle's compartment: one PU (PU_1 or PU_3) measures the temperature (*SF_1: TempMeasure*) using a sensor (S_1 or S_2) and gets the reference temperature (*SF_2: RefTemp*) using a reference button (R_1 or R_2), and the other PU (PU_2 or PU_4) acts as a controller (*SF_3: TempControl*) and actuator (*SF_4: Heating*) heating the compartment using heaters (H_1 or H_2). *TempMeasure*'s nominal communication route (*Rt*) in each compartment is as follows: Rt_1: $S_1 \rightarrow PU_1 \rightarrow ETH \rightarrow PU_2 \rightarrow H_1$; Rt_2: $S_2 \rightarrow PU_3 \rightarrow ETH \rightarrow PU_4 \rightarrow H_2$. Given that one sensor of any compartment fails, we reuse the already existing one in the same car, but in different compartment.

Figure 6: Reconfiguration Scenarios

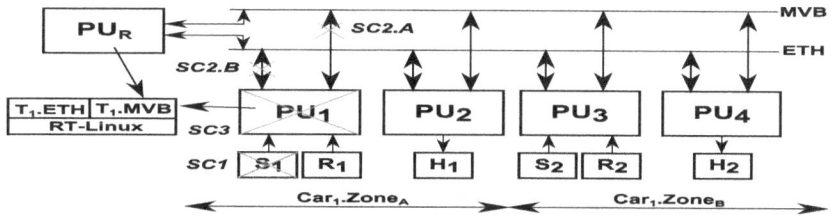

Focusing on the reconfiguration of *TempMeasure* at $Car_1.Zone_A$, its value-based fault detection is located in the destination PU_2. When incorrect values are received at PU_2, the reconfiguration implementation (located with the *TempMeasure*'s fault detection), acknowledges to the faulty component so that it stops sending data. The reconfiguration implementation checks the IP address and the UDP port of the next priority implementation of *TempMeasure* in its *reconfiguration table*, establishing the communication with S_2. This process changes the communication route from Rt_1 to Rt_{12}: $S_2 \rightarrow PU_3 \rightarrow ETH \rightarrow PU_2 \rightarrow H_1$. The design of the devices identified as heterogeneous redundancies enables them to redirect their information to different information sinks dynamically when a reconfiguration signal is received. During the reconfiguration, source and sink PUs synchronize and S_2 continues sending data towards PU_2 until S_1 is restored. Implemented reconfiguration mechanisms are applicable to input SF implementations operating with heterogeneous redundancies (e.g., Fire Protection example). MVB reconfigurations apply the same process, with the difference that Rt_{12} is activated from the outset.

SC2: since a train incorporates different communication protocols, there is room to benefit from heterogeneous redundant communications. Despite bidirectional communications have been implemented between PU_1 and PU_2, for simplicity the following unidirectional routes are considered: Rt_1: $T_1.MVB \rightarrow PU_1 \rightarrow MVB \rightarrow PU_2$; Rt_2: $T_1.ETH \rightarrow PU_1 \rightarrow ETH \rightarrow PU_2$ where *T1.MVB* and *T1.ETH* identify MVB and Ethernet tasks respectively (cf. Figure 6). When a communication link is down, the general communication-level reconfiguration process is as follows: (1) the application located in the sink PU detects the communication failure (time-based fault detection), (2) subsequently, it reconfigures itself creating a server to continue receiving data using the operating communication protocol, and (3) it informs the source PU about the communication failure; (4) finally, the source PU is also reconfigured switching from the faulty to the operative communication. Hence, when MVB is disconnected (*SC2.A*, cf. Figure 6), UDP communication threads are created dynamically to continue sending MVB data via Ethernet changing communication routes from Rt_1 to Rt_{12}, where Rt_{12}: $T_1.MVB \rightarrow PU_1 \rightarrow ETH \rightarrow PU_2$; and vice versa, when Ethernet is disconnected (*SC2.B*, cf. Figure 6) the communication route is changed from Rt_2 to Rt_{22}, where Rt_{22}: $T_1.ETH \rightarrow PU_1 \rightarrow MVB \rightarrow PU_2$.

SC3: point to point unidirectional communication from PU_1 to PU_2 is considered with the next communication routes: Rt_1: $T_1.MVB \rightarrow PU_1 \rightarrow MVB \rightarrow PU_2$; Rt_2: $T_1.ETH \rightarrow PU_1 \rightarrow ETH \rightarrow PU_2$. The tasks that PU_1 is performing are rearranged in another compatible PU to deal with PU_1 failures. A higher level reconfiguration implementation (PU_R) has been added to redirect all the data that the failed PU was sending from its communication interfaces. PU_R monitors the performance of both PUs (PU_1, PU_2) and when it detects that any of them is down (time-based fault detection); it is reconfigured sending the data that was sending before through MVB and Ethernet. Consequently, Rt_1 is replaced by Rt_{12}: $T_1.ETH \rightarrow PU_R \rightarrow ETH \rightarrow PU_2$; and Rt_2 switches to Rt_{22}: $T_1.MVB \rightarrow PU_R \rightarrow MVB \rightarrow PU_2$.

All in all, the integration of the three scenarios in a single architecture results in a fault-tolerant architecture which copes with sensor, communication and PUs failures reusing already existing elements. Note that these scenarios have been tested isolated from the other functions comprising a real train, and hence, we do not have to deal with possible memory and bandwidth issues.

5. CONCLUSIONS & FUTURE GOALS

In massively networked scenarios there is room to optimize system architectures to improve system's dependability and reduce the overall cost. The proposed methodology provides the designer with indicators to support tradeoff design decisions between dependability and cost when deciding to optimize the use of system resources and allocation of system tasks on Processing Units (PUs). Generic Dependability Evaluation Model and Component Dynamic Fault Trees have been presented as a means to perform tradeoff analyses between dependability and cost, evaluating the influence of alternative redundancy and reconfiguration strategies.

Heterogeneous redundancies reuse system resources to provide compatible functionalities to a system function. Conversely, homogeneous redundancies add additional resources to replicate system functions. In the developed application example heterogeneous redundancies reduce the overall cost and improve system dependability with respect to the architecture without redundancies, while performing almost as well as homogeneous redundancies. Regarding the allocation of tasks on PUs, the impact of centralized and distributed reconfigurations on system (un)reliability has been analyzed showing that the criticality of centralized reconfigurations is higher than distributed reconfigurations

For our future goals, we plan to integrate repair concepts and uncertainty analyses in the methodology to evaluate the influence of unknown (SW) failure rate values on system (un)availability. Moreover, we plan to address the automatic optimization of system architectures based on requirements specified as dependability and cost values to extract the best combination of homogeneous and/or heterogeneous redundancies. Lastly, the evaluation of the degradation of the functionality with heterogeneous redundancies may be addressed analyzing other factors than system failure probability.

References

[1]   C.P. Shelton & P. Koopman. *"Improving System Dependability with Functional Alternatives"*, In Proceedings of Dependable Systems and Networks 2004, pp. 295-304, (2004).

[2]   J. Wysocki & R. Debouk. *"Methodology for Assessing Safety-Critical Systems"*, International Journal of Modelling & Simulations, vol. 27, no. 2, pp. 99-106, (2007).

[3]   R. Adler, D. Schneider, & M. Trapp. *"Engineering Dynamic Adaptation for Achieving Cost-Efficient Resilience in Software-Intensive Embedded Systems"*, In Proceedings of Engineering of Complex Computer Systems, pp. 21-30, (2010).

[4]   J. I. Aizpurua & E. Muxika. *"Functionality and Dependability Assurance in Massively Networked Scenarios"*, In Proceedings of ESREL 2013, pp. 1763-1771, (2013).

[5]   J. I. Aizpurua & E. Muxika. *"Model Based Design of Dependable Systems: Limitations and Evolution of Analysis and Verification Approaches"*, International Journal on Advances in Security, vol. 6, pp. 12-31, (2013).

[6]   B. Kaiser, P. Liggesmeyer, & O. Mäckel. *"A New Component Concept for Fault Trees"*, In Proceedings of Safety Critical Systems & Software 2003, pp. 37- 46, (2003).

[7]   H. Boudali, P. Crouzen, & M. Stoelinga. *"A Compositional Semantics for Dynamic Fault Trees in Terms of Interactive Markov Chains"*, In Proceedings of Automated Technology for Verification & Analysis, vol. 4762, pp. 441-456, (2007).

[8]   G. Manno, F. Chiacchio, L. Compagno, D. D'Urso, & N. Trapani. *"MatCarloRe: An Integrated Monte Carlo Simulink Tool for the Reliability Assessment of Dynamic Fault Tree"*, Expert Systems with Applications, vol. 39, no. 12, pp. 10334-10342, (2012).

[9]   G. Vinod, T. Santosh, R. Saraf, & A. Gosh. *"Integrating Safety Critical Software System in Probabilistic Safety Assessment"*, Nuclear Eng. & Design, vol. 238, no. 9, pp. 2392-2399, (2008).

[10]   SINTEF Industrial Management. *"Offshore Reliability Data Handbook"*, (2009).

[11]   IAEA-TECDOC478. *"Component Reliability Data for Use in Probabilistic Safety Assessment"*, Technical Report, (1988).

[12]   W. Wendai, J. Loman, & P. Vassiliou. *"Reliability Importance of Components in a Complex System"*, In Proceedings of IEEE Reliability and Maintainability Symposium, pp. 6-11, (2004).

[13]   International Electrotechnical Committee. *"Train Communication Network, IEC 61375"*, (2007).

[14]   J.I. Aizpurua & E. Muxika. *"Dependable Design: Trade-off Between the Homogeneity of Functions and Resources"*, In Proceedings of DEPEND 2012, volume 1, pp. 13-17, (2012).

Probabilistic Analysis of Asteroid Impact Risk Mitigation Programs

Jason C. Reinhardt, Matthew Daniels, and M. Elisabeth Paté-Cornell
Stanford University, Stanford, United States of America

Abstract: Encounters with near-Earth asteroids (NEAs) are rare, but can have significant consequences for humanity. Probabilistic analysis of asteroid impact risks is important to fully understand the danger that they pose. This work builds on the prior development of a method and model to simulate the distribution of asteroid impact magnitudes on the Earth's surface over a 100-year period. This approach enables analysis of the full distribution of impact events, including those that are large and infrequent. Results of this approach have shown some of the greatest risks to life and property over the next century are posed by objects in the 300-to-1000-meter diameter range, which impact the Earth more frequently than those greater than 1 kilometer in diameter, and can still produce impact events with global effects. This paper extends previous work to assess NEA risk mitigation efforts. We compare three types of possible space missions to alter the orbits of hazardous asteroids: kinetic impactors, standoff nuclear explosions, and gravity tractors. Each type of mission is assessed in terms of its reduction of impact risks. The analytic framework and results of this work can serve as input to a wide set of decisions including technology investments in potential countermeasures.

Keywords: Asteroids, NEAs, Asteroid Detection, Asteroid Risk Mitigation, Probabilistic Risk Assessment

1. INTRODUCTION

A globally cataclysmic asteroid impact is believed to be a low probability event. It is sufficiently consequential, however, that mitigation options have been proposed and are being studied. These range from "civil defense" measures to space missions that alter the trajectory of asteroids on a collision course with Earth. In this paper, we compare the effectiveness of three asteroid deflection options based on their risk reduction potential. Those options are kinetic impactors, standoff nuclear explosions, and gravity tractors.[*]

In previous work, a method was proposed to assess the risk of NEA impacts given no risk mitigation measures, and that method has been used to perform some preliminary risk assessments [3]. This paper builds upon those results to assess the effectiveness of NEA impact countermeasures. Our results, along with further development, can help to inform decision makers in government, research centers, and industry about risk mitigation priorities.

2. ASSESSING THE RISK OF ASTEROID IMPACTS

A significant amount of work has been done in recent decades to examine the risk of NEA impacts on the Earth [4-7]. These studies, however, assumed average values for many relevant NEA properties such as density, velocity, and angle of impact, each of which influences the magnitude of the impact effects. Effects calculations are typically nonlinear, so using mean values as input does not necessarily produce the mean output. Therefore, many of these approaches do not account for dense, fast, high-angle strikes, and the correspondingly more intense effects.

---

[*] The use of nuclear devices in space could violate the terms of the 1967 Outer Space Treaty. It could, however "be possible to negotiate exceptions to the treaties so as to make a planetary defense system legal under international law" [1]. It has been argued that "asteroids, comets, and meteors that would be targeted are non-living, completely natural objects with no aspects of human input or control in their genesis and direction," and therefore, that the use of nuclear devices to protect mankind would be peaceful, non-aggressive actions "for the benefit and in the interests of all countries" [2].

Many risk metrics for asteroid impacts have been developed. Most focus on either the risks associated with specific objects or estimates of average annual death rates [8,9]. While object-specific data are valuable, the probability of impact for specific NEAs does not provide a sufficient measure of total risk, even when likely consequences are considered. This is because these risk estimates only consider NEAs that have been observed, and do not account for those that have not been observed yet but are believed to exist. Mean annual death rate estimates are based on the frequencies of large past events and their potential effects on the current global population. Like all low-probability high-consequence event risk measures, these average estimates are problematic. The loss distributions themselves are far more informative.

Finally, recent studies have suggested that NEAs in 10-meter to 1.5-kilometer range pose a greater risk to human life than previously thought [10]. These objects are large enough to cause damage and fatalities, and impact more frequently than very large asteroids. In light of the incomplete detection of asteroids in this range, key questions persist: what are the risks posed by these NEAs? How effective are different risk mitigation measures?

2.1 Overview of the Project Fox Method

This study is built on a simulation tool known as Project Fox [3]. The objective of Project Fox was to design and construct an analysis method and computation tool to assess the aggregate risk of NEA impacts over the next 100 years. The Project Fox approach uses probability distributions for key NEA encounter properties: diameter (we assume that they are spherical), relative velocity with the Earth at time of impact, angle of impact, density of NEA material, NEA type (stony or metallic), location of impact, and ground density at point of impact. Distributions for each of these parameters were assessed from relevant literature in the field, or through interviews with experts.

The Project Fox method differentiates two NEA impact effect regimes. First, it considers *primary effects*, which result directly from the impact of the NEA itself and include blast waves, thermal radiation, cratering, fireballs, and seismic waves. Primary effects are estimated using the computations of Collins, Melosh, and Marcus [11]. In contrast to primary effects, a *cataclysm* is the regime in which global effects are feasible (in this case, sub-micron dust loading of the upper atmosphere). This is assessed through calculations of the resulting mass ejecta and total energy release. The key thresholds for testing whether or not a cataclysm may occur are based on studies of catastrophic climate events [12]. The population that can be affected by a particular NEA impact is estimated using geography-specific population data from the LandScan database [13].

The Project Fox simulation tool estimates the complementary cumulative distribution of primary effect fatalities over 100 years without impact risk mitigation actions. The simulation also estimates the probability of a potentially cataclysmic event (as defined above) over 100 years. The original Project Fox results are shown in Figures 4 and 5 (solid and dotted black curves) as a comparison to the results of this study. The first figure shows a plot indicating that the probability that in the next 100 years at least 1 person is killed from a NEA encounter is about 3×10^{-4} and that more than 1 million people are killed is about 2×10^{-6}. The initial estimation of the probability of an impact that could produce a cataclysm is $q = 8 \times 10^{-4}$, a value that is greater than, but consistent with findings from Chapman and Morrison [5]. Finally, as shown in Figure 5, Project Fox found that the majority of cataclysm risk over the next 100 years comes from NEAs in the 300-to-1000-meter diameter range.

3. NEA DEFLECTION: MODELS OF ORBIT MODIFICATION

We study three methods for deflecting an asteroid: kinetic impactors, standoff nuclear explosions, and gravity tractors. These are summarized in Table 1. All three approaches are well known and were discussed as plausible countermeasures by the National Academies [14]. Each approach is based on causing a change of velocity (ΔV) of the asteroid, either slowing it down or speeding it up on its trajectory. This causes a change in position (Δs) at a future time. We use the convention that an asteroid must be deflected by at least one Earth-diameter by the time of its encounter with Earth.

Several time parameters are used as part of the analysis for this paper. These are illustrated in Figure 1. For an asteroid that might impact the Earth, the total warning time (t_W) is the time from discovery of the asteroid to the time of possible impact. For any asteroid deflection option, there is a preparation time for a mission (t_P), which may include development, build, assembly, and test activities; a transit time (t_T) for the spacecraft to reach the asteroid after launch from Earth; and an effect lead time (t_E), the time elapsed between the initial effect on the asteroid and its original impact time. For a gravity tractor, there is also a "dwell time" (t_G), during which the spacecraft maintains proximity to the asteroid (we assume 100 meters for this analysis).

	Insensitive to NEA material properties?	Potentially feasible for 50m–1.5km objects in less than 10 years?	Relevant technology demonstrations?	Diagram *Notation defined in sections 3.1–3.3*
Gravity Tractor	Yes	No	NASA's NEAR-Shoemaker mission rendezvoused, orbited, and landed on the near Earth asteroid 433 Eros in 2000–2001.	
Kinetic Impactors	No	Yes	NASA's Deep Impact Mission impacted comet Temple 1 in 2005. (The spacecraft went on to a flyby of Comet Hartly 2 in 2010).	
Stand-Off Nuclear Detonation	No	Yes	The Fishbowl nuclear test series by the U.S. in 1962 demonstrated effects of nuclear detonations in space.	

Table 1: Summary of NEA Deflection Alternatives

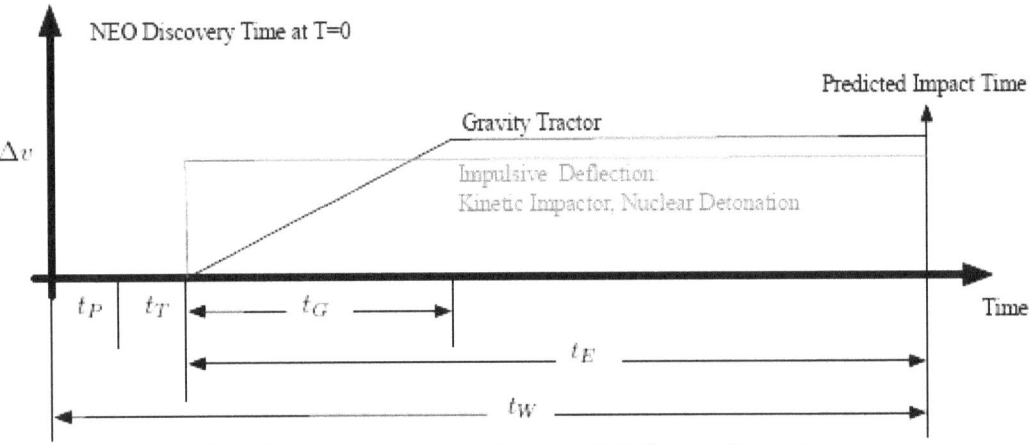

Figure 1: Time Parameters Associated with Asteroid Deflection Alternatives

For all three deflection methods, we assume a simple relationship between ΔV and Δs for velocity changes of an asteroid parallel to its direction of motion, based on the work of Ahrens and Harris [15]:

$$\Delta s \approx 3 \cdot t_E \cdot \Delta V \qquad \text{Eqn. 1}$$

where t_E is the lead time in seconds, ΔV is the change in velocity in meters per second, and Δs is the change in position in meters.[†]

3.1. Gravity Tractors

A gravity tractor is based on simple physics: a relatively large spacecraft maintains position close to an asteroid, and gravitational attraction between the two objects gradually alters the asteroid's velocity. The acceleration applied per second by a gravity tractor spacecraft on an asteroid is:

$$a = \frac{Gm}{r^2} \qquad \text{Eqn. 2}$$

where $G = 6.67 \times 10^{-11}$ m³ kg⁻¹ s⁻² is the gravitational constant, m is the mass of the spacecraft in kilograms, and r is the distance in meters between the center-of-mass of the spacecraft and the asteroid.

The gravity tractor approach requires a timescale of decades for most objects. One of its significant benefits is that it generally does not depend on the properties of the asteroid material.

The technical challenges associated with gravity tractors center on the operation of a propulsion system for maintaining proximity to the asteroid over a few decades. This may be within the technical capabilities of proposed nuclear-electric propulsion systems (e.g., see [18]). Generally, the effectiveness of this approach increases with the mass of the spacecraft and the amount of time it can stay near the asteroid. This analysis assumes that the spacecraft continues to act on the NEA body until the end of its lifetime, t_G, or the end of the effect lead time, t_E, whichever is shorter.

The change in position for the NEA resulting from a gravity tractor (Δs_G) can be approximated with the formula (see Appendix 1 for a derivation, which is consistent with [14]):

$$\Delta s_G = \left(2 t_E t_G - t_G^2\right) \frac{6 G m}{(\phi + 2 d_G)^2} \quad \text{where } t_G \leq t_E \qquad \text{Eqn. 3}$$

where t_G is the dwell time of the spacecraft near the NEA body, expressed in seconds. The parameter G is the gravitational constant in cubic meters per kilogram per second squared, m is the mass of the spacecraft in kilograms, ϕ is the radius of the NEA body in meters, ρ is the density of the asteroid, and d is the distance in meters maintained between the spacecraft and the NEA surface.

3.2. Kinetic Impactors

A kinetic impactor changes the momentum of an asteroid by impacting its surface. This method is effective on a much shorter timescale than a gravity tractor, but the change in velocity depends on properties of the asteroid material. Generally, these properties are difficult to determine until a spacecraft arrives at the object, making the exact effectiveness of a kinetic impactor uncertain until that time.

An asteroid's change in velocity after an impact is a simple physics problem, and we use the convention of the 2010 National Academies Committee Report [14]:

[†] Chelsey and Spahr [16] use a different approach, solving for the ΔV required to deflect an object by one Earth-radius, using a geometric mean of equations from Carusi et al. [17]. Chelsey and Spahr note that their approach corresponds within a factor of two to the Ahrens and Harris approach. Therefore, for simplicity, we adopt the Ahrens and Harris equation:

$$\Delta V = \beta_K \frac{mU}{M} \qquad \text{Eqn. 4}$$

where m is the mass of the spacecraft in kilograms, U is the relative velocity of the spacecraft and the NEA in meters per second, M is the mass of the NEA (also in kilograms), and β_K is a parameter that represents the amplification effects of impact-ejecta on the asteroid's momentum (β_K has a lower bound of 1 and we assume an upper bound of 5 to be consistent with the literature [14]). We assume that the relative velocity of the NEA and the spacecraft is $U = 10$ kilometers per second in all analyses presented in this paper.

With Equations 1 and 4, we can approximate the position change of the NEA resulting from a kinetic impactor (Δs_K) by the formula:

$$\Delta s_K = 3 t_E \beta_K \frac{6mV}{\pi \phi^3 \rho} \qquad \text{Eqn. 5}$$

Where t_E is the effect lead time in seconds, ϕ is the NEA diameter in meters and ρ is the density of the NEA in kilograms per cubic meter.

3.3 Stand-Off Nuclear Detonations

The third alternative considered in this paper is the use of nuclear explosions near the surface of an asteroid to change its trajectory (we do *not* consider methods to break up an asteroid by using nuclear explosives). Neutron radiation from the detonation heats the surface of the NEA body, ejecting material and changing the asteroid's momentum.

Scientists at the Lawrence Livermore National Laboratory (LLNL) have conducted preliminary studies of nuclear detonations for asteroid deflection by using detailed numerical simulations (e.g., see [19]). LLNL researchers have provided an approximation formula to estimate an asteroid's velocity change resulting from a standoff nuclear explosion [20]. Using the relationship between ΔV and Δs from Equation 1, we approximate the change in position for an NEA resulting from a standoff nuclear explosion (Δs_N) in meters, by using the following formula:[‡]

$$\Delta s_N = 3 t_E \beta_N \frac{8A}{\phi^3} \sqrt{\left(\frac{Y \phi d_N^2}{\phi + 2d_N}\right)\left(1 - \frac{\sqrt{\left(1+\frac{2d_N}{\phi}\right)^2 - 1}}{1+\frac{2d_N}{\phi}}\right)\left(\frac{\phi}{d_N}\left(1+\ln\left(\frac{Y}{Bd_N^2}\right)\right) - \left(1+\frac{\phi}{d_N}\right)\ln\left(1+\frac{\phi}{d_N}\right)\right)} \qquad \text{Eqn. 6}$$

where t_E is the effect lead time is seconds, A and B are constant terms (that fit the results of the LLNL numeric simulations) with $A = 57.5 \frac{m^3}{s \cdot \sqrt{kT}}$ and $B = 3.16 \times 10^{-4} \frac{kT}{m^2}$. Yield from the nuclear device, Y, is expressed in kilotons of TNT equivalent (kT) and the NEA body diameter, ϕ, is expressed in meters. Finally, as in Equation 5, the parameter β_N represents an amplification factor resulting from the force of ejecta, similar to the effect described in the case of the kinetic impactor. Typical values of β_N are assumed to be in the range of 1 to 3. This study assumes that the standoff distance, d_N, in meters is chosen for each mission to maximize Δs_N and that a 1-megaton-yield device is used for all deflection missions.

[‡] Equation 6 is a reasonable estimation under the specific circumstances of this analysis: range and stand-off distances, yield of the device, and the asteroid diameters that we consider. We also assume solid asteroids. However, it should not be used without careful consideration of these assumptions and of the approximations made in its derivation, and cannot be generalized to other cases or uses.

4. SIMULATION AND ANALYSIS

We make several assumptions, beyond those already discussed, for the purpose of simplifying the modeling and analysis. These simplifications allow us to develop a high-level method to estimate the risk mitigation performance of the options examined:

- **Perfect Observation** – It is assumed that all asteroids that will impact the Earth in a 100-year period have been discovered at the start of that timeframe.
- **Instant Mitigation** – For all three deflection options, it is assumed that the preparation time for a mission (t_P) and transit time (t_T) for the spacecraft to reach the asteroid after a launch from Earth are both zero.
- **Perfect Launch, Transit, and Rendezvous Reliability** – It is assumed that the mission will launch, transit to the NEA, and initiate its actions successfully.
- **NEA Cohesion** – It is assumed that the size of an asteroid relative to either a kinetic impactor or standoff nuclear detonation is such that the asteroid will not break up after deflection.
- **Spacecraft Mass** – For kinetic impactors and gravity tractors, we are assuming a spacecraft mass of 10,000 kg. This mass is consistent with the recent National Academies' assumptions to approximate current launch capabilities [14].
- **Spacecraft Lifetime** – For gravity tractors, we model the lifetime of the spacecraft (t_G) as an uncertain factor that is characterized by an exponential distribution with a mean of 50 years.

Many of these assumptions can be relaxed in future studies. Under the current assumptions, the results of this analysis should be interpreted as an estimate of the upper bound of risk mitigation performance (given the parameters used). The purpose here is to demonstrate the method developed and provide a coarse understanding of the effectiveness of deflection methods.

4.1 Modeling the Effectiveness of Mitigation Measures

Each deflection option is examined in terms of the estimated change in position, Δs, at the time when the asteroid would have struck the Earth. The position shift is approximated using one of the formulations shown in Equations 3, 5, and 6. Each of these formulas can be written as:

$$\Delta s = 3 \cdot g(t_E, \beta) \cdot h(\phi, \rho, \ldots) \text{ for stand-off nuclear explosions or kinetic impactors, and}$$

$$\Delta s = 3 \cdot g(t_E, t_G) \cdot h(\phi, \rho, \ldots) \text{ for gravity tractors.}$$

The function g takes two random variables as its arguments. For standoff nuclear explosions or kinetic impactors, they are the effect lead time, t_E, and the β amplification factor. For gravity tractors, they are the effect lead time and the spacecraft lifetime, t_G. In all cases, the g function variables are assumed to be independent. The joint distribution is calculated and discretized for selected values, resulting in a joint probability mass function (PMF) over the range of outputs of the g function.

Once a PMF estimate of the g function is determined, assessing the probability that a NEA is deflected is a matter of calculating the probability that Δs is greater than a threshold that corresponds to the minimum position shift that would avoid a collision with the Earth. This probability is noted as p. We use the convention that an asteroid with an Earth-impact trajectory must be deflected by approximately one Earth-diameter, about $D = 12,800 \text{km}$, at the time of its initially anticipated encounter with Earth. It is assumed that no impact occurs if $\Delta s \geq D$. This is equivalent to calculating the value of the complementary cumulative distribution function (CCDF) of g at the argument $\frac{D}{3 \cdot h(\cdot)}$, which is:

$$p = \Pr(NEA\ Deflected) = \Pr\left(g(\cdot) \geq \frac{D}{3 \cdot h(\cdot)}\right) \qquad \text{Eqn. 7}$$

4.2 Preliminary Comparison of Mitigation Measures

The three mitigation measures examined in this paper have varying levels of effectiveness for different combinations of NEA diameter and effect lead time. Figure 2 shows the probability of deflection, p, given the asteroid diameter and the effect lead time for each method. These results also assume that the mission is successful in its rendezvous with the asteroid. At the longest lead times considered here, a kinetic impactor (left in Figure 2) is most effective against objects smaller than 500 meters in diameter and less effective at shorter lead times. By contrast, gravity tractor options (right in Figure 2) are somewhat effective for larger NEAs, but only at long lead times. Strikingly, however, standoff nuclear explosions (center in Figure 2) are the most effective option offering high probabilities of deflection for large objects at shorter effect lead times.

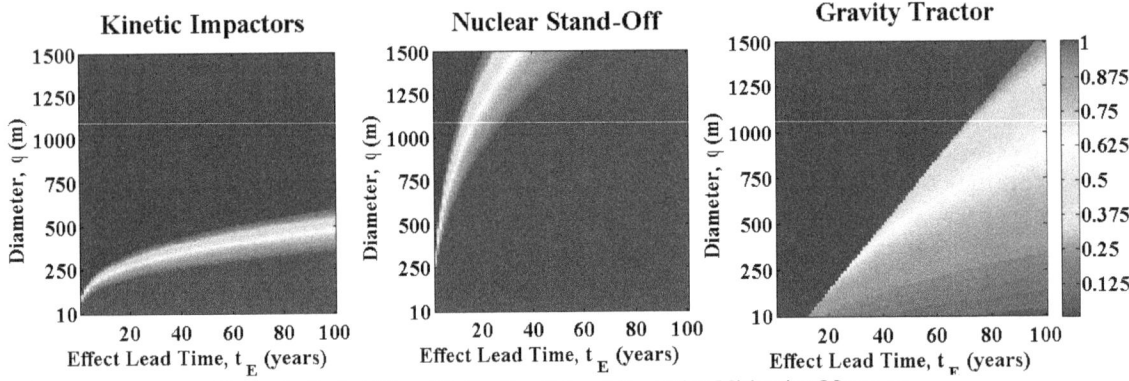

Figure 2: Probability of Deflection Using Alternative Mitigation Measures

Legend: The scale on the right shows the probabilities of successful deflection, and each part of the figure shows its variations for each deflection method, given the effect lead time and the asteroid diameter. NEO density is assumed to be $\rho = 3,000$ kilograms per cubic meters.

The kinetic impactor and nuclear stand-off explosion effectiveness depend on amplification factors, β_K and β_N. The gravity tractor's efficacy depends on the spacecraft lifetime, t_G. The β factors are assumed to be uniformly distributed over the ranges discussed previously (1 to 5 and 1 to 3 respectively), and t_G is assumed to be exponentially distributed with a mean of 50 years. Figure 3 illustrates the expected mitigation performance for each method as a function of the asteroid diameter. The expectation of the probability of a successful deflection, $E_{t_E}\{p(\phi, t_E)|\phi\}$, is taken over the effect lead time, t_E. Standoff nuclear detonations outperform kinetic impactors and gravity tractors at every considered diameter. Gravity tractors are superior to kinetic impactors for diameters larger than approximately 400 meters. For NEAs with diameters less than 400 meters, kinetic impactors are fairly effective.

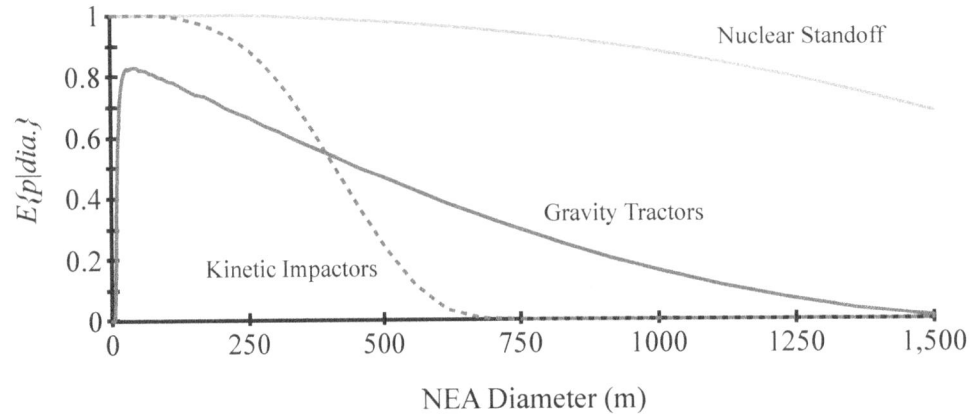

Figure 3: Mean Probability of Deflection Given Diameter

Probabilistic Safety Assessment and Management PSAM 12, June 2014, Honolulu, Hawaii

4.3 Simulation

In order to assess the different risk mitigation options, the Project Fox simulation was modified to include possible deflection of different types of NEAs. For each impact in the simulation, the set of asteroid characteristics (diameter, density, velocity, impact angle, and ground density) are drawn according to the joint distribution calculated from the probabilistic inputs of each parameter. Once an object is selected, the g and h functions are calculated for a particular mitigation option, and the probability p that the NEA is deflected is calculated using Equation 7. On the order of ten-million 100-year periods are used in the simulation for each mitigation option to get sufficient samples to assess the resulting cumulative complementary distribution function of the primary effect fatalities and cataclysm probability.

5. PERFORMANCE OF MITIGATION MEASURES

We use intermediate results discussed in Section 4 to evaluate several possible policies for asteroid responses described in Figures 2 and 3. These policies use the three deflection options examined and are (from most to least plausible):

Policy A: Use the most effective method against any NEA greater than 10 meters.
Policy B: Use the most effective non-nuclear method against any NEA greater than 10 meters.
Policy C: Use the most effective non-nuclear method for asteroids greater than 10m and less than 500 meters, and use whatever method provides the highest probability of effectiveness for asteroids greater than 500 meters.
Policy D: Use only kinetic impactors on any NEA larger than 10 meters.
Policy E: Use only standoff nuclear explosions on any NEA larger than 10 meters.
Policy F: Use only gravity tractors on any NEA larger than 10 meters.

Figure 4 illustrates the simulated risk reduction of these mitigation policies. The solid black curve corresponds to the case of no mitigation option. The dotted black curve illustrates a quantile-parameterized distribution (QPD) of the number of potential, primary-effect casualties that fit the empirical simulation results. This allows extrapolation of these results to estimate the probability of larger outcomes. Policy F (solid blue curve) uses only gravity tractors to mitigate the hazard of all asteroids, and does provide some risk reduction. However, because encounters with smaller NEAs happen more frequently, the mean effect lead time is short. It does not provide enough time for gravity tractors to change the course of the asteroid enough for effective deflection. Policy D (dotted blue curve) provides significantly more risk reduction because it enables much higher probabilities of deflection for NEA objects up to 500 meters. Policy B (solid red curve) allows for any non-nuclear option to be used for any asteroid size and provides slight more risk reduction than kinetic impactors alone. This is because Policy B selects the most effective of two options that perform differently for different size NEOs. Finally, Policy C (dotted red curve) allows for nuclear explosive devices to be used for any objects larger than 500 meters, and provides greater risk reduction than any of the non-nuclear options. This is because standoff nuclear explosions are very effective against NEAs in the diameter ranges considered, as shown in Figure 2. For potentially large fatality outcomes, the CCDFs of each of the non-nuclear policies tend to converge towards a limit of effectiveness because none of the non-nuclear mitigation options are very effective against large diameter NEAs.

The curves associated with policies A and E, which allow for the use of standoff nuclear explosions, are conspicuously absent from the results shown is Figure 4 because no casualty occurred across all simulation runs. This does not imply that nuclear explosive devices are a perfectly effective, especially if we do not have enough lead time. For example, for large NEAs that are going to impact the Earth imminently, even nuclear explosives (the 1 megaton explosive considered in this work) cannot prevent the impact. Therefore, according to these preliminary results, standoff nuclear explosions appear to be relatively more effective at reducing the likelihood of primary-effect deaths than the other mitigation options considered here. This is a promising result, as it implies that the world already possesses the basic technology to greatly reduce the risk of NEA impacts on the Earth.

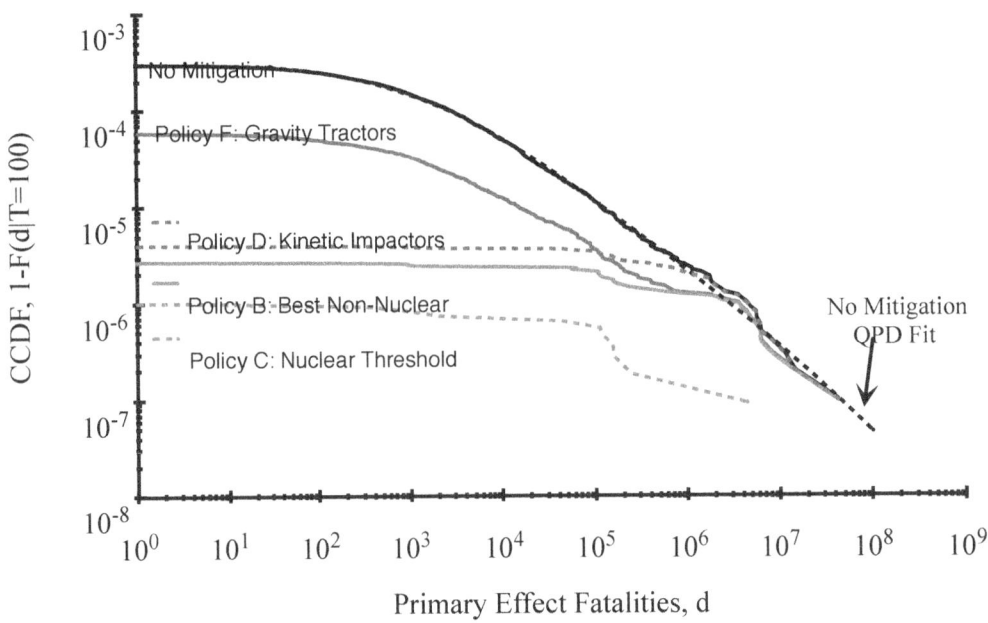

Figure 4: Reduction of Primary Effect Fatality Risk by Proposed Mitigation Policies

Consistent with the scope of Project Fox, this study defines a cataclysm as the regime in which global effects can occur. Specifically, this condition exists when the energy released in an impact is larger than 200MT and the mass ejecta exceeds 1.28×10^{11} metric tons. Figure 5 illustrates the cataclysm risk reduction effectiveness of the proposed policies.

Based on the initial Project Fox results (solid black curve), the probability of a cataclysm over a one hundred year period is significant for NEA diameters greater than 300 meters. The total probability q of a global-effect scenario over the next 100 years was found to be about 8×10^{-4} without risk mitigation measures.

Policy D (dotted blue curve) uses only kinetic impactors and reduces this risk by 7% to $q = 7.4 \times 10^{-4}$. This is because kinetic impactors are decreasingly effective for NEAs above 500 meters.

Policies B and F (solid red curve) provide a 29% risk reduction, resulting in $q = 5.7 \times 10^{-4}$. Policy B employs only non-nuclear options, while Policy F uses only gravity tractors. These policies have similar results given that gravity tractors are the most effective for NEAs larger than 400 meters, and the probability of cataclysms resulting from smaller NEAs is small.

Policy C (dotted red line) uses non-nuclear options for asteroids below 500 meters in diameter and all options—including nuclear devices—for asteroids above 500 meters. This policy is very effective, with a risk reduction of 82%, resulting in $q = 1.5 \times 10^{-4}$. Under Policy C, however, there is still a small residual risk of cataclysms for NEAs between 300 and 500 meters. If the threshold used for the use of nuclear explosive were lowered to 300 meters, this policy would likely have an effectiveness closer to those of the policies where nuclear use is permitted for all diameters.

Policies A and E (solid green curve) address the residual risk of Policy C, providing a total risk reduction of 84%, resulting in $q = 1.2 \times 10^{-4}$. Here again, standoff nuclear detonations are very effective, and become the dominant risk mitigation measure.

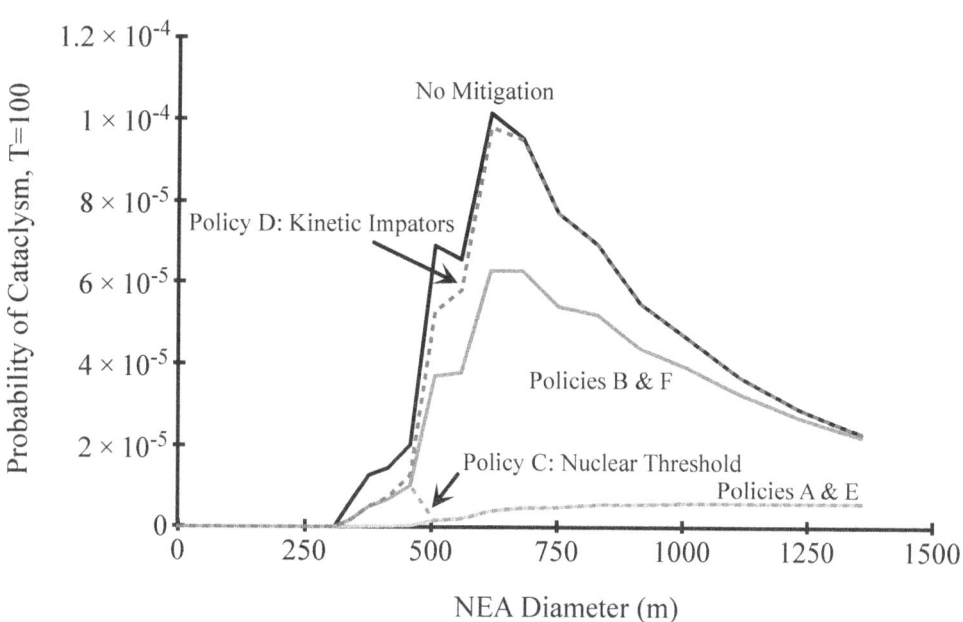

Figure 5: Reduction of Cataclysm Risk by Proposed Mitigation Policies

6. CONCLUSION

This study extends simulations of asteroid impact risks to include the effects of three risk mitigation options based on asteroid deflection. In order to keep the models simple and calculations feasible, reasonable approximation formulas are used to assess the performance of gravity tractors, kinetic impactors, and standoff nuclear detonations. Results indicate that kinetic impactors outperform gravity tractors in averting primary-effects fatalities, but gravity tractors outperform kinetic impactors for reducing global cataclysm risk. This is reasonable, as gravity tractors are more effective than kinetic impactors for large diameter NEAs. While all NEA deflection policies provide some risk reduction, deflection policies that include the use of nuclear explosive devices outperform all deflection policies that do not, because they can deflect larger asteroids with a shorter lead time.

Under the assumptions made in this paper, the results represent an upper bound of risk reduction for all the mitigation options examined. The delay for preparing technologies, the transit time from launch to the asteroid, the probability of failure during NEA rendezvous, the chances of spacecraft launch failures, equipment failures, and a host of other concerns would reduce the benefits of all mitigation options. Perhaps the most optimistic assumption in each simulation is that all NEAs that could impact the Earth over the next 100 years are already known. It is probably not the case at this time, especially for asteroids less than 1000 meters in diameter.

The results of this analysis suggest that we most likely have the technology to successfully mitigate most of the risk from asteroid impacts, given sufficient time between NEA discovery and a potential Earth impact. This suggests that an important next step is to improve and expand NEA discovery and observation missions, especially those that can provide data on objects in the 300-to-1000-meter range, and those that are difficult to observe, for example because they are aligned with the sun. Ideally, observation missions would be effective down to the low 10s of meters in diameter.

It is likely that the technology exists to deflect many moderate-sized asteroids given sufficient lead-time. The problem is that we may not be able to see them coming.

Acknowledgements

We would like to thank Paul Miller, Joe Wasem, and the Lawrence Livermore National Laboratory research staff for their generous support of time, insights, and assistance. We extend a special thanks to Brian Muirhead at the Jet Propulsion Laboratory for his time and counsel. Finally, we also would like to thank our colleagues Xi Chen, Wenhao Liu, and Petar Manchev, who helped develop the original Project Fox risk assessment method and simulation.

References

[1] M.B. Gerrard and A.W. Barber. "*Asteroids and comets: US and international law and the lowest-probability, highest-consequence risk*", NYU Environmental Law Journal, Vol. 6, No. 4 (1997).
[2] J.C. Kunich. "*Planetary defense: the legality of global survival*", Air Force Law Review, Vol. 41, pp. 119-162 (1997).
[3] J.C. Reinhardt, X. Chen, W. Liu, P. Manchev, and M.E. Paté-Cornell. "*Project Fox: Assessing asteroid risks*", Risk Analysis, Submitted for publication.
[4] G. Neukum and B. A. Ivanov. "*Crater size distributions and impact probabilities on Earth from lunar, terrestrial-planet, and asteroid cratering data*", Hazards due to Comets and Asteroids 359 (1994).
[5] C.R. Chapman and D. Morrison. "*Impacts on the Earth by asteroids and comets: Assessing the hazard*", Nature, Vol. 367, pp. 33-40, (6 January, 1994).
[6] P. Brown, R.E. Spalding, D.O. ReVelle, E. Tagliaferri, and S.P. Worden. "*The flux of small near-Earth objects colliding with the Earth*", Nature, Vol. 420, pp. 294-296, (21 November, 2002).
[7] G.H. Stokes et al. "*Study to determine the feasibility of extending the search for near-Earth objects to smaller limiting siameters: Report to the Near-Earth Object Science Definition Team*", National Aeronautics and Space Administration, (22 August 2003).
[8] *Small-Body Database*, Jet Propulsion Laboratory, http:// ssd.jpl.nasa.gov/sbdb.cgi, accessed February 2014.
[9] C.R. Chapman. "*The hazard of near-Earth asteroid impacts on earth*", Earth and Planetary Science Letters, Vol. 222, pp. 1-15, (2004).
[10] P.G. Brown, et al. "*A 500-kiloton airburst over Chelyabinsk and an enhanced hazard from small impactors*", Nature, Vol. 503, pp. 238-241, (14 November 2013).
[11] G.S. Collins, H.J. Melosh, and R.A. Marcus. "*Earth Impact Effects Program: A web-based computer program for calculating the regional environmental consequences of a meteoroid impact on Earth*", Meteoritics and Planetary Science, Vol. 40, Num. 6, pp. 817-840, (2005).
[12] O.B. Toon, R.P. Turco, and C. Covey. "*Environmental perturbations caused by the impacts of asteroids and comets*", Reviews of Geophysics, Vol. 35, No. 1, pp. 41-78, (February 1997).
[13] Oak Ridge National Laboratory. "*LandScan Global 2011: Global Population Database,*" (2012).
[14] Committee to Review Near-Earth Object Surveys and Hazard Mitigation Strategies. "*Defending Planet Earth: Near-Earth Object Surveys and Hazard Mitigation Strategies: Final Report*", National Academies Press, 2010.
[15] T.J. Ahrens and A.W. Harris. "*Deflection and fragmentation of near-Earth asteroids*", Nature, Vol. 360, No. 6403, pp. 429-433, 1992.
[16] S.R. Chesley and T.B. Spahr. In *Mitigation of hazardous comets and asteroids* (eds. M.J.S. Belton et al.), Chapter 2, pp. 22-37, Cambridge University Press, Cambridge, (2004).
[17] A. Carusi, G. B. Valsecchi, G. D'Abramo, and A. Boattini. "*Deflecting NEAs in route of collision with the Earth*", Icarus, Vol. 159, No. 2. pp. 417-422, (2002).
[18] E.T. Lu and S.G. Love. "*Gravitational tractor for towing asteroids*", Nature, Vol. 438, pp. 177-178, (2005).
[19] M.B. Syal, D. S.P. Dearborn, and Peter H. Schultz. "*Limits on the use of nuclear explosives for asteroid deflection,*" Acta Astronautica, Volume 90, Issue 1, pp. 103-111 (2013).
[20] J. Wasem and P. Miller. personal correspondence with the authors, February 27, 2014. Not for quotation or attribution at this time.

Appendix 1: Derivation of Gravity Tractor Formula

The approximation for position shift for any deflection method is given by:

$$\Delta s = 3\, \Delta s'$$
$$\Delta s' = \Delta v \cdot t$$
$$\Delta v = a \cdot \Delta t$$

The acceleration induced by a massive spacecraft on a NEA body is given by:

$$\Delta a = \frac{Gm}{r^2}$$

The acceleration depends only on the mass of the spacecraft, and is constant for a given encounter. There are two cases that must be examined: 1) the lifetime of the spacecraft is less than the effect lead time, or $t_G \leq t_E$, and 2) the lifetime of the spacecraft lifetime is longer than the effect lead time, or $t_G > t_E$. In the first case, we have the following:

$$\Delta s' = \frac{Gm}{r^2} t_G \left(\frac{t_G}{2} + (t_E - t_G) \right)$$

$$\Delta s' = \frac{Gm}{r^2} \left(t_E t_G - \frac{t_G^2}{2} \right)$$

$$\Delta s' = \frac{Gm}{2r^2} \left(2 t_E t_G - t_G^2 \right)$$

$$\Delta s' = \frac{Gm}{2 \left(\frac{\phi}{2} + d_G \right)^2} \left(2 t_E t_G - t_G^2 \right)$$

$$\Delta s' = \frac{2Gm}{(\phi + 2 d_G)^2} \left(2 t_E t_G - t_G^2 \right)$$

Substituting back into position shift approximation, we have:

$$\Delta s = 3\, \Delta s'$$
$$\Delta s = \frac{6 G m}{(\phi + 2 d_G)^2} \left(2 t_E t_G - t_G^2 \right)$$

In the case where $t_G > t_E$, the effect lead time, t_E, is essentially equal to t_G, and the result above holds.

Physics-based Entry, Descent and Landing Risk Model

Ken Gee*[a], Loc Huynh[b], and Ted Manning[a]
[a] NASA Ames Research Center, Moffett Field, USA
[b] Science and Technology Corporation, Moffett Field, USA

Abstract: A physics-based risk model was developed to assess the risk associated with thermal protection system failures during the entry, descent and landing phase of a manned spacecraft mission. In the model, entry trajectories were computed using a three-degree-of-freedom trajectory tool, the aerothermodynamic heating environment was computed using an engineering-level computational tool and the thermal response of the TPS material was modeled using a one-dimensional thermal response tool. The model was capable of modeling the effect of micrometeoroid and orbital debris impact damage on the TPS thermal response. A Monte Carlo analysis was used to determine the effects of uncertainties in the vehicle state at Entry Interface, aerothermodynamic heating and material properties on the performance of the TPS design. The failure criterion was set as a temperature limit at the bondline between the TPS and the underlying structure. Both direct computation and response surface approaches were used to compute the risk. The model was applied to a generic manned space capsule design. The effect of material property uncertainty and MMOD damage on risk of failure were analyzed. A comparison of the direct computation and response surface approach was undertaken.

Keywords: EDL, physics-based risk model, TPS, Monte Carlo, response surface.

1. INTRODUCTION

The Engineering Risk Assessment Team at NASA Ames is currently developing a High Fidelity Mission Risk (HFMR) model to assess the overall mission risk of manned space systems [1]. The HFMR model evaluates risk for the ascent, on-orbit and entry, descent and landing (EDL) phases of a space system mission. In a nominal mission, shown schematically in Figure 1, a launch vehicle is used to place a manned space vehicle into Earth orbit. The manned vehicle remains in orbit for some mission duration and then returns to Earth in the EDL phase. In the EDL phase, the vehicle must slow from its orbital velocity to its landing velocity at the end of the entry trajectory. Atmospheric drag is used to slow the vehicle down during the entry trajectory. However, friction between the vehicle and the a*tmosphere creates high aerodynamic heating that the can destroy the vehicle if not for its thermal protection system (TPS). The TPS of the vehicle is typically composed of a stack-up of materials that can withstand the expected high heating rates and temperatures during the entry. At the end of the entry trajectory, the vehicle can deploy parachutes or use retro-rockets to further reduce its velocity. Landings can occur in water or on land.

Figure 1. Schematic of manned space mission analyzed by HFMR model.

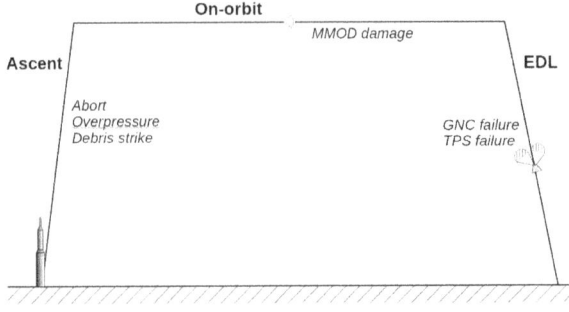

*Ken.Gee-1@nasa.gov

In each phase there are failure scenarios that can lead to a loss of mission and possibly a loss of crew. Failure scenarios in each phase may include loss of control of the launch vehicle or catastrophic failures of the engines during ascent, damage due to micrometeoroid/orbital debris impacts while in orbit and TPS failures in the EDL phase. One possible failure of the TPS occurs when the temperature at some point in the system exceeds a given limit. For example, if the maximum allowable temperature of the adhesive used to bond the TPS material to the underlying structure is exceeded, the TPS material may separate, exposing the structure to the aerothermodynamic heating environment. This can lead to structural failure and a loss of crew. TPS failures can also result from the damage caused by MMOD impacts while on-orbit, higher-than-expected heating due to trajectory and/or atmospheric uncertainties, or larger-than-expected variations in the material properties of the TPS itself.

Physics-based models are used to evaluate the crew risks associated with failure scenarios and the resulting failure environments. To date, models have been developed to assess the risk of blast overpressure [2,3] and debris strike [4,5] on the crew module resulting from a catastrophic failure of the launch vehicle. The failure scenarios are modeled using a level of fidelity necessary to resolve the physics involved. For example, the debris strike model defined the debris field in terms of the number of pieces and the mass, reference area, drag coefficient and imparted velocity of each piece. The model then computed the trajectory of each debris piece using a three degree-of-freedom trajectory tool. The probability of at least one piece of debris striking the crew vehicle was computed by comparing the relative position of the debris field and crew vehicle during an ascent abort. The conservative approach assumed that any debris strike resulted in a loss of crew. A vulnerability criterion was developed that related the penetration velocity as a function of the debris mass. Applying the criterion, only debris pieces that had a relative velocity greater than the penetration velocity were counted in the strike probability. The probability of loss-of-crew due to these failure scenarios are computed using the physics-based models in a Monte Carlo simulation. The Monte Carlo approach has been used to analyze the effect of dispersions or uncertainties on the entry trajectories for Apollo [5] through to Stardust [6]. Monte Carlo methods have been used to determine the effect of TPS material property uncertainties on the thermal response [7].

A physics-based model has been developed to assess some of the risks associated with the EDL phase. The model included modules to compute the entry trajectory, the aerothermodynamic environment during entry and the thermal response of the TPS to the heating environment. The model also included the capability to model the effects of MMOD damage on the TPS response. The Monte Carlo approach was used to account for dispersions and uncertainties in the vehicle state at the start of the entry phase, the aerothermodynamic environment and the TPS material properties. Failure criteria based on bondline temperature were used to determine the success or failure of each Monte Carlo sample. The following sections describe the EDL risk model in more detail and an application of the model to assess the EDL risks for a generic capsule-based manned vehicle design.

The purpose of the TPS was to protect the underlying structure and crew from the heating environments encountered during the reentry. Without the TPS in place, the heating environment would cause structural failure and loss of crew. Potential failure scenarios for the TPS include manufacturing defects resulting in materials that did not have the expected thermal properties, entry trajectories resulting in heating environments more severe than expected and damage to the TPS from mishandling or MMOD impacts. The risk model was developed to assess the effects of the dispersions in material properties and vehicle state and the damage due to MMOD impacts on the ability of the TPS to protect the crew. The probability of failure is dependent upon the criteria used to measure success or failure of the system.

2. EDL RISK MODEL

The physics-based portions of the EDL risk model computed the aerodynamic and aerothermodynamic database for a given vehicle design, the entry trajectories and the TPS thermal response to the heating environments. The physics-based tools are integrated with an algorithm to perform the Monte Carlo analysis and manage the resulting data sets. The analysis data flow of the physics-based portion of the model is shown in Figure 2.

Figure 2. Flowchart of EDL risk model.

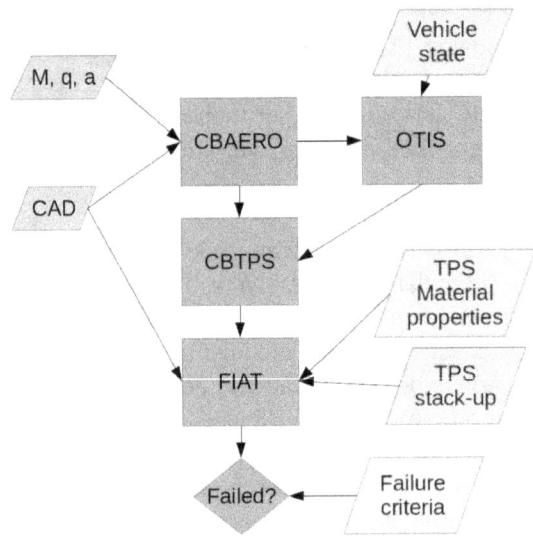

CBAERO [8], an engineering level aerothermodynamics tool, was used to predict the aerodynamic and aerothermodynamic environment about the spacecraft during entry. Input data include a surface grid of the vehicle being analyzed and a range of values for Mach number, dynamic pressure and angle of attack to define the parameter space encompassing the entry trajectories. A component of CBAERO called CBTPS was used to determine the aerothermodynamic heating time history at specific points on the vehicle, given a particular trajectory.

The trajectories were computed using OTIS, a three-degree-of-freedom trajectory optimization tool [9,10], starting from entry interface (EI) and ending at the designated landing site. Lift and drag data for the vehicle were obtained from the CBAERO database. Optimized solutions were obtained using distance to a specified landing site as the objective function. Constraints were used to ensure that dynamic pressure, heat flux and total acceleration were within design limits. The optimization procedure may at times fail to find an optimized or feasible solution given the initial vehicle state and constraints. Currently, only successfully optimized trajectories are used in the risk analysis.

The heating time history computed from the trajectory and aerothermodynamic database was used as input to compute the thermal response of the TPS stack-up. The thermal response of the TPS material stack-up was computed using FIAT, a one-dimensional transient analysis tool [11]. FIAT also required definitions of the TPS stack-ups and a database file containing the TPS material properties. FIAT computed the thermal response of the stack-up to the heating environment, generating output in the form of time histories of temperature and heat flux at specified locations in the stack-up.

The risk of failure of the TPS was computed by comparing the data obtained from the FIAT analysis against a set of failure criteria, such as temperature limits at a specific depths or layers. Probabilities were computed using a Monte Carlo simulation. Dispersions in the vehicle state at the start of the entry trajectory, material properties of the TPS, the heating environment and TPS stack-up thickness were used to define the parameter space of the Monte Carlo simulation.

A representation of a TPS stack-up is shown in Figure 3. The stack-up can be composed of multiple layers of materials, such as high-emissivity coatings, toughened materials to withstand impact damage, substrate layers and adhesive layers that attached the TPS stack-up to the underlying structure. The stack-up was defined within a FIAT input file and included the thickness of each material layer.

Figure 3. Schematic of TPS stack-up in heat shield.

The damage due to an MMOD strike was modeled by modifying the TPS stack-up to reflect the loss of material, as shown schematically in Figure 4. The damage was modeled as an hemisphere with radius equal to the cavity depth. To model the effect of damage on the bondline temperature using FIAT, an area-weighted average of the conduction heat flux, accounting for the regions of undamaged and damaged tile, was used. Failure risk was computed for two impact scenarios. The conservative approach assumed that all MMOD strikes occurred at the point of maximum heating. A more realistic approach randomly selected the point of impact in each Monte Carlo sample.

Figure 4. Schematic of TPS stack-up damage due to MMOD strike.

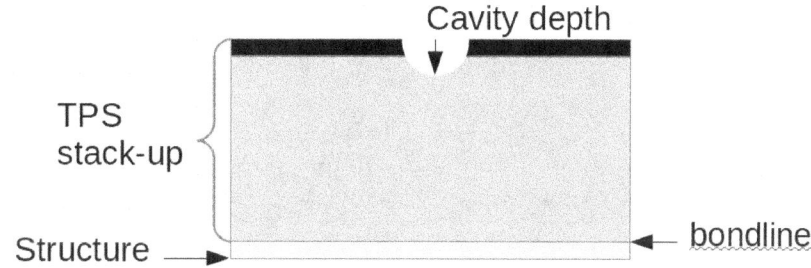

Figure 5. Side view of generic manned capsule.

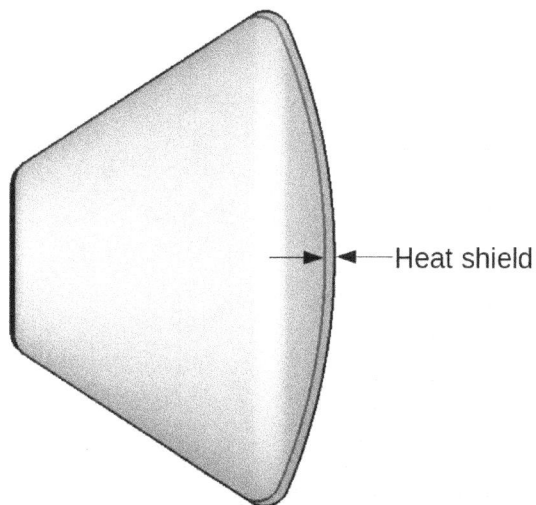

3. EXAMPLE APPLICATION OF THE RISK MODEL

The EDL risk model was applied to the analysis of a manned capsule designed for missions to low Earth orbit (LEO). The generic capsule design, shown in Figure 5, utilized a heat shield consisting of a reusable TPS material similar to that used on the Space Shuttle [12]. The TPS stack-up consisted of a high-emissivity coating, a toughened material layer, a substrate layer and an adhesive layer to attach the TPS stack-up to the structure. In this example, the failure criterion was set as the maximum temperature that the adhesive used to attach the TPS material to the structure could withstand before debonding occurred. If this temperature was exceeded, it was assumed that the TPS debonded from the structure, causing failure of the TPS. A conservative assumption was made that any failure of the TPS would lead to a loss of crew, either through subsequent failure of the vehicle structure due to the direct exposure to the heating environment or through loss of aerodynamic control due to changes in the outer mold line of the vehicle.

Figure 6. Nominal entry trajectory of generic capsule returning from LEO.

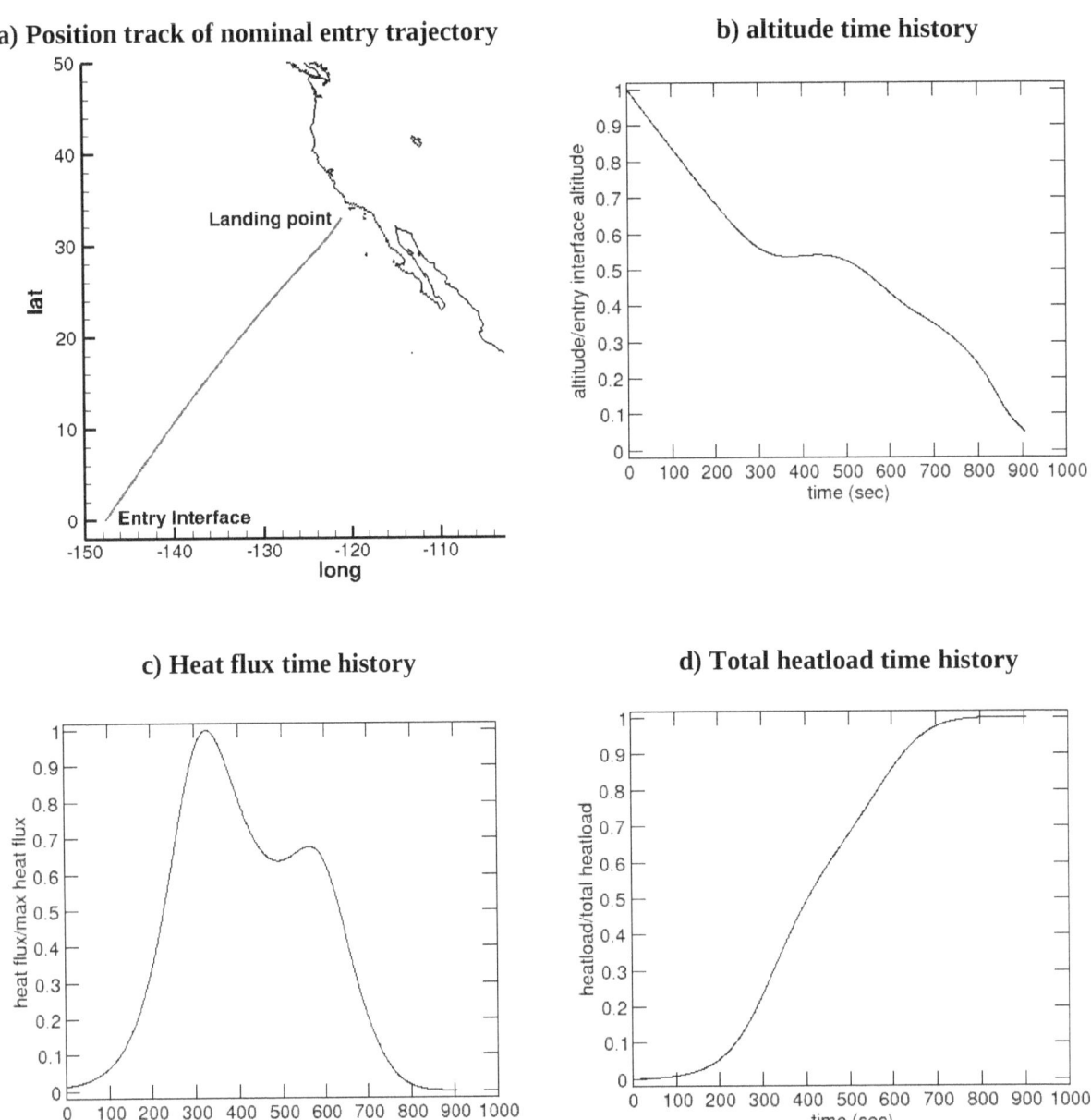

CBAERO was used to compute lift and drag coefficients of the capsule assuming a center of gravity offset from the centerline of the vehicle. A nominal entry trajectory was computed using OTIS assuming a return from the low Earth orbit (LEO) and a landing in the San Clemente Water Landing Area off the coast of California [13], as shown in Figure 6a. The nominal altitude, heat flux and total heat load time histories are plotted in Figure 6b through 6d, respectively. Typical of this type of entry, peak heating occurred early in the trajectory and decreased to near zero at the end of the trajectory as the vehicle velocity decreased. Since heat load was the time integral of the heat flux, the maximum occurred at the end of the trajectory.

Monte Carlo simulations were computed using the uncertainty bounds for the vehicle state at entry interface listed in Table 1 and the TPS material properties listed in Table 2. Uniform distributions were assumed for all parameters. The risk model was used to assess the failure probability as a function of the design thickness of the heatshield. Given a nominal heatshield design, the risk model was used to assess the failure probability due to MMOD impact damage. In addition, the feasibility of using a response surface approach to reduce the computational expense of the Monte Carlo simulations was assessed.

Table 1. Uncertainty bounds in vehicle state at entry interface.

Parameter	Range
velocity	±0.1%
azimuth	±0.05 deg
FPA	±0.25 deg
longitude	±0.2 deg
latitude	±0.2 deg
mass	±3%

Table 2. Uncertainty bounds in TPS material properties.

Parameter	Range
Heating augmentation factor	1.0 – 1.25
Density	±4%
Specific heat	±5%
Thermal conductivity	±5%
Cavity depth	0 – 1.5 in

3.1. Risk Assessment of the Heatshield Design Thickness

The need to account for uncertainty and dispersions in the design can be illustrated in the following example. Using the nominal entry trajectory and nominal material property values, a minimum thickness for the heatshield, for which the temperature at the bondline no longer violated the maximum allowable value, can be obtained, as shown in Figure 7. Applying the risk model to include the effects of entry state and material property uncertainties on the bondline temperature indicated that additional thickness was required to reduce the failure probability to acceptable levels, as shown in Figure 8.

The failure probabilities in Figure 8 were obtained using Monte Carlo simulations consisting of 2000 samples. As the failure probability decreased to zero, additional samples were required to fully resolve the probability of failure. Results from Monte Carlo simulations consisting of 10000 samples are listed in Table 3 and indicate that a thickness of 2.6 in. was required to reduce the bondline temperature failure probability to a 1 in 10000 level.

Figure 7. Bondline temperature as function of heatshield thickness for nominal conditions.

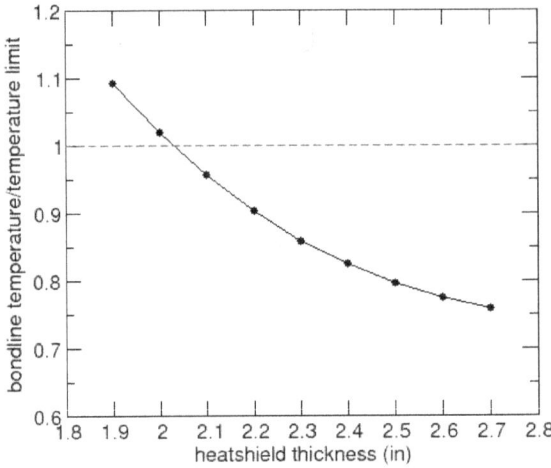

Figure 8. Bondline temperature failure probability as a function of heatshield thickness

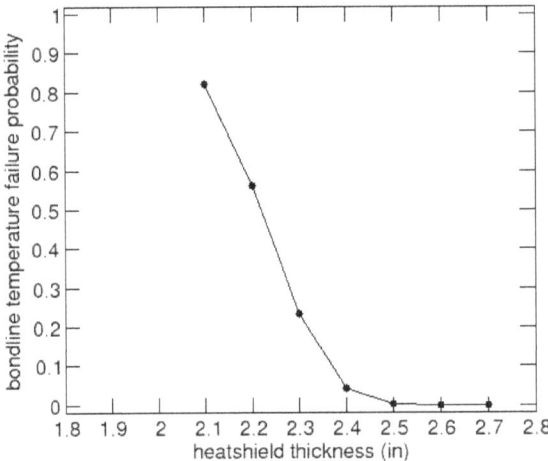

Table 3. Bondline temperature failure probabilities as function of heatshield thickness and number of MC simulation samples

Thickness (in)	Failure probability using 2000 samples	failure probability using 10000 samples
2.5	0.001	0.0044
2.6	0	0.0001
2.7	0	0

3.2. Risk Assessment of MMOD Impact Damage

Assuming that a failure probability on the order of 1e-4 was an acceptable risk, the nominal heatshield design was set to a thickness of 2.6 in. The risk model was then used to assess the failure probability of the heatshield design due to MMOD impact damage. The model assumed that an MMOD impact caused a hemispherical divot in the TPS stack-up with the penetration depth equal to the radius of the divot, as shown in Figure 4. For each Monte Carlo sample, the vehicle state at entry interface and the

TPS material properties were randomly generated based on the bounds listed in Tables 1 and 2. Bondline temperatures were computed using this set of values for penetration depths ranging from 0.1 to 1.5 in. for the two impact scenarios discussed in Section 2. There were 10000 samples in each Monte Carlo simulation. Failure probabilities for MMOD impacts that always occurred at the point of highest heating are listed in Table 4. Failure probabilities for MMOD impacts at randomly selected points on the heatshield are listed in Table 5. Assuming that all MMOD impacts occurred at the point of maximum heating yielded higher failure probabilities, confirming the conservatism of this assumption. In this example, the additional local heating in the damage cavity and the growth of the cavity during reentry were not modeled.

Table 4. Probability of exceeding bondline temperature limit for MMOD impacts at point of maximum heating.

cavity depth (in)	0	0.1	0.2	0.3	0.4	0.5	0.75	1.0	1.25	1.5
failure probability	0.0001	0.0001	0.0001	0.0002	0.0004	0.0005	0.0094	0.2421	0.9338	1.0000

Table 5. Probability of exceeding bondline temperature limit for MMOD impacts at randomly selected points on heatshield.

cavity depth (in)	0	0.1	0.2	0.3	0.4	0.5	0.75	1.0	1.25	1.5
failure probability	0.0000	0.0000	0.0000	0.0000	0.0000	0.0000	0.0000	0.0001	0.0057	0.4929

The choice of a bondline temperature limit as the failure criterion was appropriate for assessing the reliability of the system. A reusable TPS system would need to retain its integrity over a number of missions. This would not be possible if the TPS material did not remain attached to the vehicle. However, using the bondline temperature limit to assess the probability of TPS failures leading to a loss of crew may be too conservative. Typically, the maximum bondline temperature for each sample was reached at or near the end of the trajectory, mirroring the heatload time history shown in Figure 6c.. Even if the adhesive were to fail, causing the TPS material to debond and exposing the structure to the environment, the heat flux near the end of the trajectory, as shown in Figure 6b, was low enough that structural failure due to heating would be a remote possibility. Loss of aerodynamic control due to changes to the outer mold line may not be an issue, either, especially if the capsule was under parachutes when the bondline temperature limit was reached. The development of appropriate failure criteria for loss of crew assessments is an area of active research.

3.3. Response Surface Approach

Each of the Monte Carlo simulations discussed in the previous section used the direct computation approach and required about 1800 CPU hours to complete on the NASA Pleiades supercomputer [14] using its Ivy Bridge nodes. A less computationally intensive approach was to generate a response surface defining the relationship between the bondline temperature and the input parameters in the parameter space. Once defined, the response surface was used to generate the values for the Monte Carlo sampling. The parameter space was defined by the variables listed in Tables 1 and 2. Kriging [15] was selected as the multivariate interpolation scheme used to create the response surface. Kriging required a training data set to determine the interpolation coefficients used in the scheme. A Latin hypercube design was used to populate the parameter space with the training data points. A total of 2000 data points were used in the training set, requiring about 300 CPU hours to compute. The response surface generated from the training data was used to predict the bondline temperature in Monte Carlo simulations consisting of 10000 samples. MATLAB was used to generate the response surface and perform the Monte Carlo simulation. This took about 10 minutes on a single CPU. Table 6 lists the failure probabilities as a function of penetration depth predicted using the response surface and the direct computation approach for the nominal heatshield thickness, assuming all impacts were at the point of maximum heating. For small penetration depths (< 0.5 in.), the response surface was not able to resolve the low failure probabilities. The response surface was better able to resolve the higher failure probabilities for the larger penetration depths. Since the response surface method was faster, used less computational resources and provided reasonable estimates of the failure probabilities, the method would be useful in quickly providing risk information about early vehicle designs. Once the design matured, the direct computation method would be used to generate more refined risk data.

Table 6. Probability of bondline temperature failure due to MMOD strike damage predicted using response surface.

Cavity depth (in)	0	0.1	0.2	0.3	0.4	0.5	0.75	1	1.25	1.5
Response surface	0.0000	0.0000	0.0000	0.0000	0.0000	0.0000	0.0082	0.4232	0.9613	0.9987
Direct computation	0.0001	0.0001	0.0001	0.0002	0.0004	0.0005	0.0094	0.2421	0.9338	1.0000

4. CONCLUSION

A physics-based model was developed to predict the probability of a loss of crew due to failure scenarios that can occur during the entry, descent and landing phase of a manned space mission. The model integrated an engineering-level aerothermodynamics tool to compute the lift, drag and heating coefficients for a given vehicle, a three-degree-of-freedom trajectory tool to compute the entry trajectory and a one-dimensional transient response code to predict the thermal response of the vehicle thermal protection system. The model accounted for damage to the TPS due to MMOD impacts. Monte Carlo simulations were used to predict the effects of dispersions and uncertainties in the vehicle state at the start of the entry trajectory, the heating environment during the trajectory and the material properties of the TPS on the ability of the vehicle to operate within its design limits. Both direct computation and response surface methods were used in the Monte Carlo simulations.

The model was used to analyze a generic manned capsule designed for missions to low Earth orbit. A temperature limit at the bondline between the TPS and the underlying structure was used as the failure criterion. Exceeding the temperature limit would cause the TPS material to debond from the structure. The model was used to determine the nominal heatshield thickness based on the failure probabilities due to the dispersions in trajectory, aerothermodynamic heating and material properties. For the nominal heatshield design, the model was used to predict the failure probabilities due to MMOD damage. Conservative values were obtained by assuming the MMOD impact always occurred at the point of maximum heating. More realistic values were obtained by assuming the impacts occurred at random locations on the heatshield.

A response surface approach was also developed to predict the failure probabilities in a less computationally intensive manner. The response surface was built using Kriging as the interpolation scheme and a Latin hypercube design to populate the parameter space. The response surface provided results that compared well with the data obtained using the direct computation method and certainly can be used in the preliminary design phase to quickly provide risk information.

The selection of a bondline temperature limit was appropriate for assessing the reliability of the TPS design. Using the criterion to predict the probability of loss of crew due to TPS failures may be too conservative for the sample problem, since bondline temperature failures typically occurred late in the trajectories, when heating may not be significant enough to cause structural damage. The development of appropriate failure criteria for the assessment of loss of crew probabilities during the EDL phase is an ongoing research topic.

References

[1] Go, S., Mathias, D., Lawrence, S. and Gee, K., "An Integrated Reliability and Physics-based Risk Modeling Approach for Assessing Human Space Launch Systems," 12th International Conference on Probabilistic Safety Assessment and Management (PSAM12), Honolulu, HI, June, 2014.

[1] Lawrence, S. L., Mathias, D., Gee, K. and Olsen, M., "Simulation Assisted Risk Assessment: Blast Overpressure Modeling," PSAM-0197, 8th International Conference on Probabilistic Safety Assessment and Management (PSAM8), New Orleans, LA, May, 2006.

[2] Lawrence, S. L. and Mathias, D., "Blast Overpressure Modeling Enhancements for Application to Risk-Informed Design of Human Space Flight Launch Abort Systems," RAMS 06B-3, 2008 Reliability and Maintainability Symposium, Las Vegas, NV, January, 2008.

[3] Gee, K. and Mathias, D., "Assessment of Launch Vehicle Debris Risk During Ascent Aborts," RAMSRM-312, The 54th Annual Reliability and Maintainability Symposium, Las Vegas, NV, January, 2008.

[4] Gee, K. and Lawrence, S. L., "Launch Vehicle Debris Models and Crew Vehicle Ascent Abort Risk," Reliability and Maintainability Symposium (RAMS), Orlando, FL, January, 2013.

[5] Marx, M. H., "A Computer Program to Predict the Impact Dispersions of Debris Resulting From the Forced Entry of Apollo Vehicles," NASA-TM-X-69808, October, 1968.

[6] Desai, P. N., Lyons, D. T., Tooley, J. and Kangas, J., "Entry, Descent, and Landing Operations Analysis for the Stardust Re-Entry Capsule," AIAA Paper 2006-6410, 2006.

[7] Sepka, S. A. and Wright, M., "A Monte Carlo Approach to FIAT Uncertainties – Improvements and Applications for MSL," AIAA Paper 2009-4234, 41st AIAA Thermophysics Conference, San Antonio, TX, June, 2009.

[8] Kinney, D. J. and Garcia, J. A., "Predicted Convective and Radiative Aerothermodynamic Environments for Various Reentry Vehicles Using CBAERO," AIAA Paper 2006-659, 44th AIAA Aerospace Sciences Meeting and Exhibit, January, 2006.

[9] Hargraves, C. R. and Paris, S. W., "Direct Trajectory Optimization Using Nonlinear Programming and Collocation," AIAA Journal of Guidance, Control and Dynamics, Vol. 10, No. 4, 1987, pp. 338-342.

[10] Riehl, J., Paris, S. and Sjauw, W., "Comparison of Implicit Integration Methods for Solving Aerospace Trajectory Optimization Problems," AIAA Paper 2006-6033.

[11] Chen, Y. K. and Milos, F. S., "Ablation and Thermal Response Program for Spacecraft Heatshield Analysis," Journal of Spacecraft and Rockets, Vol. 36, No. 3, May-June, 1999, pp. 475-483.

[12] Rodriguez, A. C. and Snapp, C. G., "Orbiter Thermal Protection System Lessons Learned," AIAA Paper 2011-7308, AIAA SPACE 2011 Conference & Exposition, Long Beach, CA, September, 2011.

[13] Tigges, M. A., Bihari, B. D., Stephens, J., Vos, G., Bilimoria, K. D., Mueller, E. R., Law, H. G., Johnson, W., Bailey, R. E., and Jackson, B., "Orion Capsule Handling Qualities for Atmospheric Entry," AIAA Paper 2011-6264, AIAA Guidance, Navigation and Control Conference, Portland, OR, August, 2011.

[14] http://www.nas.nasa.gov/hecc/resources/pleiades.html#url

[15] Lophaven, S. N., Nielsen, H. B., and Sondergaard, J., "DACE: A MATLAB Kriging Toolbox," Technical Report IMM-TR-2002-12, Informatics and Mathematical Modelling, Technical University of Denmark, 2002

Physics-Based Fragment Acceleration Modeling for Pressurized Tank Burst Risk Assessments

Ted A. Manning[a,*], Scott L. Lawrence[a]
[a] NASA Ames Research Center, Moffett Field, CA, USA

Abstract: As part of comprehensive efforts to develop physics-based risk assessment techniques for space systems at NASA, coupled computational fluid and rigid body dynamic simulations were carried out to investigate the flow mechanisms that accelerate tank fragments in bursting pressurized vessels. Simulations of several configurations were compared to analyses based on the industry-standard Baker explosion model, and were used to formulate an improved version of the model. The standard model, which neglects an external fluid, was found to agree best with simulation results only in configurations where the internal-to-external pressure ratio is very high and fragment curvature is small. The improved model introduces terms that accommodate an external fluid and better account for variations based on circumferential fragment count. Physics-based analysis was critical in increasing the model's range of applicability. The improved tank burst model can be used to produce more accurate risk assessments of space vehicle failure modes that involve high-speed debris, such as exploding propellant tanks and bursting rocket engines.

Keywords: Physics-based Risk Assessment, Computational Fluid Dynamics, Fluid Structure Interaction, Tank Burst.

NOMENCLATURE

$(..)_{00}$	Initial conditions inside tank	γ	Ratio of specific heats (C_p/C_v)
$(..)_0$	Time-evolving bulk conditions enclosed by fragment field	a	Speed of sound
		τ	Nondimensional time
$(..)_\infty$	Far-field, ambient conditions	g	Nondimensional fragment position
$(..)_{int}$	Fragment wall interior conditions	p	Nondimensional pressure
$(..)_{ext}$	Fragment wall exterior conditions	ρ	Density
N	Number of fragments	T	Temperature
ψ	Fragment half-angle	V_0	Volume enclosed by fragment field
M_t	Tank mass	A_{gap}	Fragment gap area (total)
A_{frag}	Fragment frontal area	C_d	Discharge Coefficient
R_{frag}	Inner radius of tank fragment	F	Baum burst scaling parameter
x	Radial position of tank fragment	Δt	Acoustic passing time over fragment half-width.
t	Time		
Θ	Characteristic time	n	Expansion term phase-in factor
X	Characteristic length		

1. INTRODUCTION

Highly pressurized tanks, which are pervasive in a wide range of industrial and aerospace applications, pose a significant risk to surrounding personnel, structures, and equipment in the event of a structural failure. While the release of the stored content is itself often inherently problematic, high-speed impacts from fragments of the failed tank wall and support structure may be the source of the greatest damage. Crewed space launch vehicles contain a number of pressurized vessels, including cryogenic propellant tanks, helium ullage pressurization tanks, and high-pressure combustion chambers. The burst of any one of these has the potential to damage adjacent systems and compromise the crew module, either directly or through a chain of failures and bursts. Understanding the details of the pressurized tank burst process is therefore important in assessing the risk to crew. As part of comprehensive efforts to develop physics-based risk assessment techniques for space systems at NASA, we carried out numerical simulations to investigate mechanisms for accelerating tank fragments in a two-dimensional representation of pressure vessels found in space launch vehicles. We

---
* Ted.A.Manning@nasa.gov

compared these results to an industry-standard explosion model and proposed some modifications that bring the model more in line with the simulation results.

The use of physics-based modeling in estimating debris strike probability is well established in space launch vehicle risk assessments. Reference [1] describes an approach in which a "catalog" of propellant tank debris assigned with mass, position, and velocity forms the initial condition for physics-based analysis to find the probability of debris penetration and loss of the crewed abort vehicle. Although such debris strike predictions use three-degree-of-freedom trajectory and nonlinear structural dynamics analyses, the initial debris catalog is based on engineering judgment and empirical data from a specific explosion scenario.

To obtain fragment velocity estimations that more directly reflect the configurations in question, engineering models have been developed to estimate debris velocities from simplified representations of bursting tanks. These models are typically based on one of three physical approaches: energy partition, momentum conservation, or force driven. In energy partition methods, the expansion work of the enclosed fluid is allocated to the blast wave, tank wall fragments, and other energy manifestations (e.g., thermal energy). For example, Baum [2] obtains the upper-limit fragment velocity based on an energy scaling of the tank's initial operating conditions. In momentum conservation methods, which are typical for liquid storage tanks, the speed of fragments from a full burst is sized from the fluid escape velocity one would obtain in a large but localized breach [3]. In force-driven methods, the evolution of pressure-accelerated wall fragments is directly modeled.

Baker, et al. [4] devised a force-driven cylindrical tank burst model that draws from a number of earlier works for spherical and cylindrical tanks [5,6,7]. The model assumes that infinitely long cylindrical wall fragments are driven by the time-varying pressure of the internal fluid, which decreases according to expansion of the enclosed volume and flow through the gaps between the fragments (Figure 1). The model neglects the effect of an external fluid, which may be valid if the internal pressure is very high, but raises concerns regarding its usefulness for sea-level bursts of tanks of more modest pressure. Given its force-driven formulation, the Baker model is readily examined using computational fluid dynamics (CFD) analysis.

Figure 1: Simplified geometry of cylindrical tank burst model ($N = 12$ fragments).

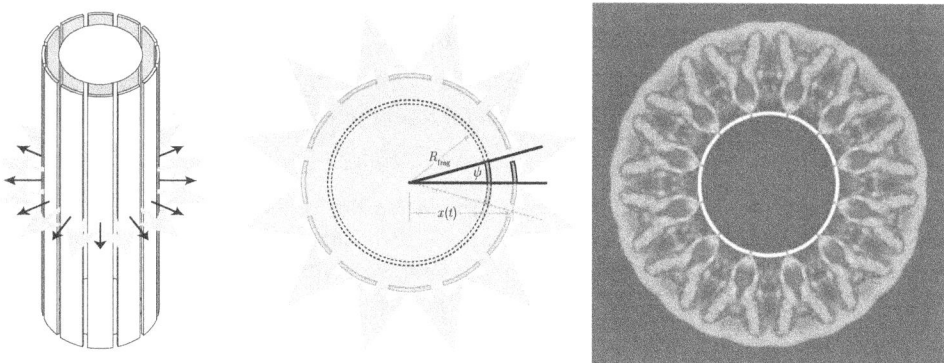

In the present work, we conducted time-accurate CFD simulations of bursting cylindrical tanks to evaluate the Baker model and guide any improvements. We utilized the CFD code OVERFLOW [8], and invoked its rigid-body dynamics module to capture fluid interaction with accelerating tank fragments. As in the Baker model, we defined a two-dimensional domain based on the cross-sectional slice of an infinitely long cylindrical tank. While CFD is commonly used to assess blast overpressure risk [9], there are fewer examples in debris acceleration. Wang et al. [10] conducted cylindrical tank burst simulations, but the fracture in this work was considerably different than the present effort, involving only a single break at the mid-plane, perpendicular to the tank axis.

Based on our initial, cylindrical tank simulations, we modified the Baker model to account for external ambient pressure, inner and outer wall expansion and compression, and varying fragment size. Through a detailed comparison of the modeled and simulated peak fragment velocities from bursts of several launch vehicle pressure vessel configurations, we show that, despite the complexity of the tank burst flow, Baker-type models predict maximum fragment velocity reasonably well when internal-to-external pressure ratios are high. We then show that our modified model provides improved fragment velocity predictions over a wider range of conditions, particularly in low-pressure-ratio cases where the original Baker model over-predicts the peak fragment velocity by several times. The modified model will form the basis for constructing new debris catalogs for physics-based risk assessments of crewed launch vehicle failure modes that involve debris strike. Due to subtle geometric inconsistencies found in the Baker model formulation [4], derivations and results shown in this paper that are attributed to the original Baker model reflect the corrected rather than strictly original formulation.

In Section 2, beginning with the simplified, two-dimensional representation of a cylindrical tank, we first present the formulation of the Baker model, and follow this with our proposed modifications. In Section 3 we describe the CFD tools that we used to simulate the flow of a bursting tank using the same simplified geometry as the Baker model. In Section 4, we compare the peak velocities from the classic Baker model, modified Baker model, and simulation results to an energy-partition-based velocity limit prediction for five pressure vessel configurations. Finally, we evaluate the detailed fragment velocity and internal pressure trends for two of the five tank configurations.

2. BAKER MODEL AND PROPOSED MODIFICATIONS

In this section, the Baker model formulation is described in terms of geometric assumptions and processes affecting pressure evolution, and important steps in the derivation of the governing equations are highlighted. The modifications to the Baker model, which involve adjusting key terms in the original formulation, are then discussed.

2.1. Baker Model

The premise of the Baker model is that a cylindrical tank can be approximated using an infinitely long cylinder with fragments that also extend the length of the cylinder. The fragment dynamics can then be assessed in a two-dimensional framework. There are N fragments of equal subtended angle 2ψ, such that the fragment half-angle is $\psi = \pi/N$ (Figure 1). When the energy of fracture is small compared to the energy of the pressurized fluid [4], fracture dynamics may be neglected, allowing one to consider only the fragment motion due to the pressure field distribution arising from the explosively escaping fluid.

The Baker model considers a fragment that is driven by the pressure enclosed within the ring of fragments traveling away from the center of the tank. The internal pressure is assumed to be high enough to neglect the effects of the fluid outside the tank, such that the external pressure is considered zero (a vacuum). The evolution of the decreasing interior pressure is then governed by two effects: 1) fluid loss through the gaps between the fragments and 2) the enlarging volume enclosed by the fragments. The motion of the fragment is then simply governed by a force balance ($F = ma$). The resulting equations can be solved to obtain fragment positions and velocities as a function of time. The asymptotic or peak velocity that arises from this solution is then typically taken as the terminal velocity of the fragment. The force balance for each fragment is represented in the expression

$$\frac{M_t}{N} x''(t) = (p_{\text{int}} - p_{\text{ext}}) A_{\text{frag}} \qquad (1)$$

where M_t/N is the fragment mass, $x(t)$ is the distance of the fragment from the original tank center, A_{frag} is the projected frontal area of the fragment in the x direction, and p_{int} and p_{ext} are the average pressure on the interior and exterior faces of the tank wall fragment, respectively. The nondimensional variables for time (t), fragment position (x), and bulk interior pressure (p_0), are normalized as

$$\tau = t/\Theta, \quad g(\tau) = x/\mathbf{X}, \quad p(\tau) = p_0(\tau)/p_{00} \qquad (2)$$

to characteristic temporal and spatial reference quantities

$$\Theta = \sqrt{\frac{2}{\gamma - 1}\left(\frac{M_t a_{00}}{A_{\text{frag}} p_{00}}\right)}, \quad \mathbf{X} = \frac{2}{\gamma - 1}\left(\frac{M_t a_{00}^2}{A_{\text{frag}} p_{00}}\right) \qquad (3)$$

and initial interior pressure (p_{00}). These nondimensional parameters reduce Equation 1 to

$$g''(\tau) = Np(\tau)\left(\frac{p_{\text{int}}}{p_0} - \frac{p_{\text{ext}}}{p_0}\right). \qquad (4)$$

The subscript "00" denotes initial conditions inside the tank, a_{00} is the interior speed of sound, and γ is the ratio of specific heats in the tank interior. The choice of interior and exterior pressure is the modified model's first point of departure from the original Baker model and will be further discussed in the next section.

In Baker, the interior fragment pressure is assumed to be related to the velocity of the fluid at the fragment wall, $x'(t)$, through a one-dimensional isentropic perfect gas relation, whereas the exterior fragment wall pressure, p_{ext}, is taken as 0. In nondimensional form, the interior and exterior wall pressures are then

$$\frac{p_{\text{int}}}{p_0} = \left[1 - \frac{g'^2}{p^{\frac{\gamma-1}{\gamma}}}\right]^{\frac{\gamma}{\gamma-1}}, \quad \frac{p_{\text{ext}}}{p_0} = 0 \qquad (5)$$

giving Baker's nondimensional force-balance equation

$$g''(\tau) = Np(\tau)\left[1 - \frac{g'^2(\tau)}{p^{\frac{\gamma-1}{\gamma}}}\right]^{\frac{\gamma}{\gamma-1}}. \qquad (6)$$

To augment the force-balance equation, Baker derives a pressure equation from the differential form of the ideal gas equation, which is expressed in terms of the time-varying bulk thermodynamic properties of pressure p_0, volume V_0, mass m_0, gas constant R_{gas}, and temperature T_0 inside the tank:

$$\frac{1}{p_0(t)}\frac{dp_0(t)}{dt} = \frac{1}{m_0(t)}\frac{dm_0(t)}{dt} + \frac{1}{T_0(t)}\frac{dT_0(t)}{dt} - \frac{1}{V_0(t)}\frac{dV_0(t)}{dt}. \qquad (7)$$

This equation is of the form A = B + C − D. The density form of the ideal gas law and the adiabatic assumption of constant p/ρ^γ allow us to express the term C as a function of pressure and combine it with A. The rate at which the bulk pressure falls inside the tank depends on mass loss through the gaps (term B) and volume expansion of the tank as defined by the envelope of fragments (term D). If we assume that the mass flow through the gaps satisfies the conditions for choked (sonic) flow (modulated by a discharge coefficient C_d), then the mass flow can be related to the interior conditions and the gap area A_{gap}. Using the nondimensionalizations of Eqs. 2 and 3, we then obtain Baker's pressure equation:

$$\frac{p'(\tau)}{p(\tau)} = \underbrace{-\frac{\alpha\left[g(\tau) - g(0)\right]p^{\frac{\gamma-1}{2\gamma}}(\tau)}{g^2(\tau)}}_{\text{Mass loss effects}} \underbrace{-2\gamma\frac{g'(\tau)}{g(\tau)}}_{\text{Expansion effects}} \qquad (8a)$$

where

$$\alpha = 2C_d\gamma \left(\frac{2}{\gamma+1}\right)^{\frac{\gamma+1}{2(\gamma-1)}} \left(\frac{2}{\gamma-1}\right)^{-\frac{1}{2}}. \qquad (8b)$$

Eqs. 6 and 8 are the governing equations of the classic Baker model. Eq. 8b reduces the published Baker definition of α by a factor of two by correcting the definition of fragment gap area; we use this corrected definition in both the Baker and modified Baker model results.

2.2. Modifications to the Baker Model

Several modifications to the Baker model are proposed based on comparisons with computational results. The modifications are summarized as follows:

Force balance equation:
1. Use of one-dimensional receding piston relation to define interior wall pressure
2. Addition of ambient static external pressure
3. Use of one-dimensional advancing piston relation to define exterior wall pressure

Pressure equation:
1. Initial suppression of expansion effect term, followed by phase-in.

With respect to the force-balance equation (Eq. 4), the modification to the method for evaluating the interior wall pressure p_{int} (Figure 2) was guided by the CFD simulation results, which indicated that the internal pressure on the fragments would be better represented using the relation for a receding piston rather than the isentropic formula of Eq. 5. Additionally, the modified method incorporates the external wall pressure p_{ext} (which is omitted in Baker's model) in terms of an ambient pressure p_∞ and a compression wave imparted by the moving fragment.

Figure 2: Pressure evolution due to fragment motion in modified model (one-dimensional view).

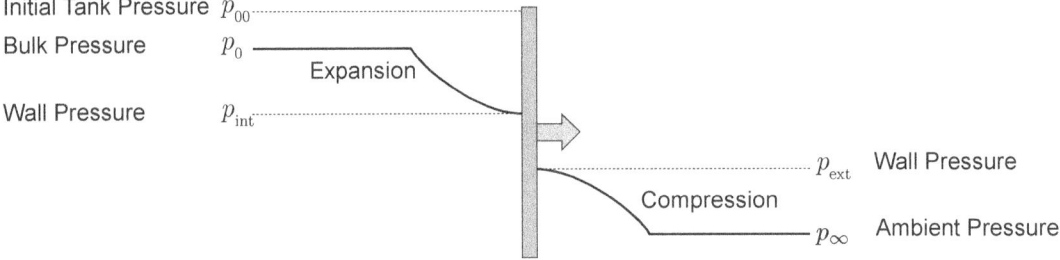

The moving piston relations (see, for example [11]) can be expressed in nondimensional form as:

$$\frac{p_{\text{int}}}{p_0} = \left[1 - \sqrt{\frac{\gamma-1}{2}}\frac{g'(\tau)}{p^{\frac{\gamma-1}{2\gamma}}}\right]^{\frac{2\gamma}{\gamma-1}}, \quad \frac{p_{\text{ext}}}{p_0} = \frac{1}{p(\tau)}\frac{p_\infty}{p_{00}} G\left[1 + \sqrt{1 - \frac{H}{G^2}}\right] \qquad (9a)$$

$$G = 1 + \frac{\gamma(\gamma+1)}{2(\gamma-1)}\frac{a_{00}^2}{a_\infty^2}(g')^2, \quad H = 1 - \gamma\frac{a_{00}^2}{a_\infty^2}(g')^2. \qquad (9b)$$

Examination of the CFD simulation results also provided some insight in modeling Baker's pressure equation (Eq. 8a). The dynamic nature of the expansion process suggested that the term representing expansion effects (term D in Eq. 7) might not be appropriate, at least in the initial phases of the expansion. For this reason, we introduced a phase-in of the expansion effects term, initiated after some delay following initialization. In the present implementation, the delay and phase-in time intervals are based on a simple inverse relationship with the number of fragments. That is, the dynamic effects are assumed to take longer to settle when fewer gaps are available to foster communication between the external and internal conditions.

The governing equations of the modified Baker model are obtained by substituting Eqs. 9 into Eq. 4 to obtain the force-balance equation and by multiplying the expansion term of Eq. 8 by a phase-in function to obtain the pressure equation. Both the Baker and modified Baker governing equations are coupled sets of ordinary differential equations, which we solve using the ODE45 function in MATLAB [12].

3. COMPUTATIONAL FLUID DYNAMICS SIMULATIONS

We used the high-fidelity CFD code, OVERFLOW [8], to evaluate the suitability of the Baker model. Where differences were found, the high-fidelity results were used to guide the modifications to the Baker model described in the previous section. OVERFLOW was used to assess the maximum velocity of a tank fragment for the same simplified tank geometry utilized in the Baker and modified Baker analyses. While fluid mechanical simplifications were made regarding the use of two-dimensionality and omission of viscosity, a number of high-fidelity flow properties were retained, including compressibility and multiple gas species.

OVERFLOW is an aerospace industry-standard CFD code that is developed and maintained at NASA. It is typically used to compute flow fields about complex aerospace vehicles, including space launch vehicles, and can include energetic flow features such as rocket plumes. To obtain flow fields, OVERFLOW solves unsteady, compressible Navier-Stokes equations on overlapping, structured grids, known as overset meshes, which can accommodate complex and time-varying geometries. It also permits moving bodies to be modeled according to the laws of rigid body dynamics, which are needed in the present research to simulate fragment acceleration. OVERFLOW is used in its two-dimensional mode, and viscosity and turbulence modeling are disabled. For the scale of tanks considered, the dominant forces are governed by inviscid (i.e., pressure) rather than viscous effects.

The computational domain is simplified in the manner Baker has taken, where a single fragment is considered within a two-dimensional representation of the cylindrical tank cross-section. Due to the bilateral symmetry of the fragment, the domain is further reduced to a pie-wedge-shaped region with angle ψ, i.e., as bounded by the black lines in Figure 1. The fragment is an annular piece of the tank wall and is assumed to be fractured at the upper boundary. To simplify the simulation, this pre-fractured portion is represented as an initial gap between the fragment and the domain boundary, typically at 2% of the fragment arc. The lower boundary lies on the bilateral symmetry plane. Figure **3** shows a schematic of the computational domain and the overset mesh around a fragment gap.

Figure 3: **Computational domain and overset grid [13] detail at fragment gap (***N* = 6**).**

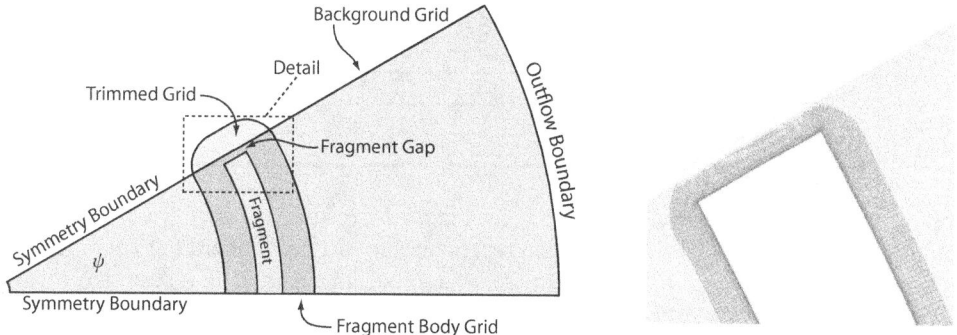

The domain is initialized with a quiescent flow field set to prescribed thermodynamic equilibrium conditions inside and outside the tank. The interface between the high-pressure inner region and lower-pressure outer region is taken up entirely by the tank wall, except at the gap between the tank fragment and the upper domain boundary, where the transition occurs abruptly at the centerline of the fragment. Because the external vacuum conditions assumed in the Baker model are not possible in OVERFLOW, a minimal low-pressure external fluid is specified. A series of sensitivity studies were performed to ensure that results are independent of grid resolution and time-step.

An important parameter in both the Baker and modified Baker models is the discharge coefficient C_d, which appears in the mass flow terms of the pressure equation (Eq. 8). Ranging from 0 (no flow) to 1 (maximum flow), the discharge coefficient is an empirical parameter that characterizes the efficiency, compared to an ideal channel, with which an orifice such as the gap permits fluid to pass. Higher values of C_d will therefore allow more mass flow, cause more rapid reduction in tank pressure, and consequently lower peak fragment velocity.

Estimates for the discharge coefficient have been extracted from the CFD simulations in an effort to provide some insight into reasonable values for this model parameter. This extraction process is challenging due to the fragment motion and the rapid increase in the orifice size. A range of values was obtained in each tank configuration outlined in Section 4, though typically a plateau was observed prior to reverse flow. This plateau varied from case to case, but most often fell between 0.7 and 0.85. For the modeling comparisons provided in subsequent sections, a value of 0.8 was used in both the original and modified Baker models. Some sensitivity runs that were performed indicated that the predicted peak velocities are not extremely sensitive to C_d for values between 0.5 and 0.9.

4. COMPARISON OF SIMULATION TO MODEL RESULTS

A total of five tank configurations (Table 1) were examined, covering a range of operating conditions applicable to crewed launch vehicles: a Space Shuttle external tank (STS ET), a Space Shuttle main engine (SSME) main combustion chamber (MCC), a solid rocket booster (SRB), an RL10 MCC, and a tank first introduced in [4]. The first four are found on either the Space Shuttle or NASA's planned Space Launch System (SLS). The fifth configuration is a high-pressure tank that is similar to helium storage start tanks used for launch vehicle engine spin start and ullage management.

Table 1: Tank configurations modeled.

	STS ET	SSME MCC	SRB	RL10 MCC	Baker 6
Tank contents	Air	H_2, O_2, H_2O	APCP gas	H_2, O_2, H_2O	Air
Tank material	Aluminum	Inconel	Steel	Inconel	Aluminum
Radius (cm)	420	22.3	184	11	25.4
Thickness (cm)	0.21	1.24	1.2	1.17	0.68
p_{00} {Δp} (psi)	{22}	3000	800	500	10,000
T_{00} (K)	293	3400	3430	3400	272
p_∞ (atm)	0.001~1.0	1	1	1	1
Pressure Ratio	1500~2.5	205	54	9	680
Internal spec. heat ratio γ	1.4	1.37	1.155	1.37	1.4
Fragment count: N	3~24	3~24	4~24	3,12	3~24
Phase-in $\Delta t\, a_0/\psi\, R_{frag}$	1	0.1	1	1	1
Phase-in factor n	8	8	8	8	8
Discharge coeff. C_d	0.8	0.8	0.8	0.8	0.8

For brevity, the subsections below present the analysis results in detail for only the first two configurations (highlighted in Table 1), which represent very different operating conditions in terms of pressure, temperature, species, and tank dimensions. First, however, it is instructive to examine the landscape of tank burst peak velocities obtained for all configurations in the context of the energy partition framework. As mentioned in the introduction, Baum defined the upper limit of velocity fragments [2] based on a nondimensional energy scaling parameter F that depends on a combination of tank case mass M_t, initial pressure p_{00}, initial volume V_{00}, and internal speed of sound a_{00}.

$$F = \frac{p_{00} V_{00}}{\frac{1}{2} M_t a_{00}^2} \qquad (9)$$

When fitted to experimental results, Baum's upper limit for cylindrical tank peak velocity is

$$\frac{V_{\text{peak}}}{a_{00}} = 0.88 F^{0.55}. \qquad (10)$$

We find that the limit line is consistent with the upper bound of the cases considered in this study (Figure 4), both experimentally and modeled. This adds evidence that the simplified representation of the cylindrical tank assumed in this paper is reasonably valid.

Figure 4: Tank burst velocity peaks compared to velocity limit of Baum [2]. Black symbols are experimental results, colored symbols are simulation and model results.

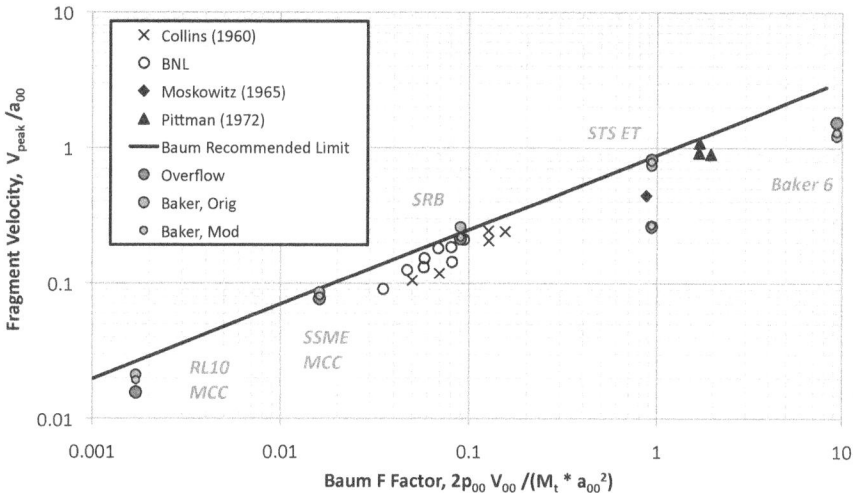

The burst of any of the tanks considered would generate tank wall fragments that either directly threaten the crew module, even during abort, or begin a cascade of bursts. In each of the two cases presented in detail below, we begin by describing the findings of the CFD simulations. We then compare the velocity and pressure histories of the modified Baker analysis to those of the original Baker formulation and the CFD analyses. The sensitivity of the velocity peak to the fragment count N and external pressure is also presented for the STS ET case.

The first configuration is modeled after the oxygen tank of the Space Shuttle external tank, which is similar to the first stage of the planned Space Launch System. The aluminum walls of the ET are very thin compared to the radius and the internal pressure is only a small multiple of standard atmospheric pressure. The second case is the main combustion chamber of the Space Shuttle main engine. It is characterized by very high pressure and temperature. In both cases, we model the internal fluid as an ideal gas, with specific gas constants and ratios of specific heat selected based on weighted averages of the constituents.

The CFD simulations show that, in all tank burst cases, the flow field is characterized by jets, preceded by cylindrical shock waves, issuing out of the narrow gaps between the fragments. The fragments accelerate radially from the tank center, a compression wave rises from the leading face of the fragments, the gaps and jets widen, and the internal pressure falls. The shocks then coalesce into a single front and part of the flow recirculates back toward the centerline of the fragments. The shape of the jets is a strong function of the initial pressure ratio; for pressure ratios above the critical value for choked flow, the flow is dominated by a supersonic jet that expands rapidly with respect to radial distance, whereas the jet remains subsonic and narrow for low pressure ratios (Figure 5). Meanwhile, expansion waves originating from the widening gaps and the trailing edge of the accelerating fragments travel within the tank interior, causing the interior pressure to fall. The internal pressure eventually falls below the external pressure, and the flow relative to the fragments reverses and causes fragment deceleration.

Figure 5: STS ET simulation density contours for $N = 12$, p_{ext} = 0.01 atm (choked, left) and 1 atm (unchoked, right).

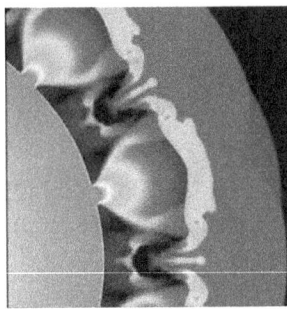
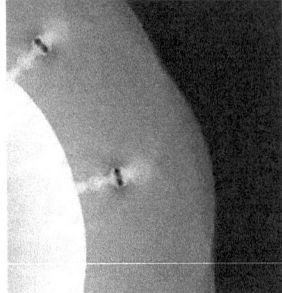

4.1. Space Shuttle External Tank

To retain a tractable albeit approximate basis for comparing simulation and model results in the low-pressure-ratio and low-tank-wall-mass regimes of the STS ET, we forgo the physics of phase change and assume the tank is filled with ambient-temperature air instead of cryogenic liquid propellant and ullage gases. The ET pressure is only a small multiple of standard atmospheric pressure (1.5 atmospheres above the external pressure), and the tank wall is a very small fraction of the initial tank radius. As a result, the external pressure plays an important role in the fragment dynamics. Despite being relatively thin, the tank walls were assumed to be rigid and deformation of the tank wall was not modeled. The STS ET oxygen tank is rated to 22-psi gauge pressure, such that the gauge pressure ($\Delta p = p_{00} - p_\infty$) is held constant with decreasing external pressure during normal ascent, giving decreasing internal pressures with altitude. The internal pressure in the simulations and the modified Baker model is set according to $p_{00} = \Delta p + p_\infty$. However, because the original Baker model omits the external fluid, we set the Baker model pressure to $p_{00} = \Delta p$ for all p_∞. The values of the parameters in the modified Baker model (phase-in Δt, n) were selected to produce the best match with simulation results.

Simulation and model results for STS ET fragment peak velocities as functions of fragment count N and external pressure p_{ext} (in terms of equivalent altitude) are presented in Figure 6 and Figure 7, respectively. The simulation results indicate that the variation of peak fragment velocity at p_{ext} = 1 atm (sea level) is only a weak function of fragment count, reaching a maximum of just below 100 m/s for 10 or more fragments and rising only slightly at smaller counts (Figure 6a). The modified Baker result captures this trend well, whereas the classic Baker model vastly over-predicts the peak and produces the opposite trend in small N. The time history of the $N = 12$ case at sea level (Figure 6b) shows an early velocity peak followed by a decay in the OVERFLOW results. The modified Baker result reproduces the velocity rise and peak, but diverges slightly on the decay. The internal pressure trace (Figure 6c) shows a rapid fall punctuated by periodic waves that, based on the simulations, can be attributed to the multiple passings of the expansion wave from the initial opening of the fragment gap. The modified Baker model captures the internal pressure well up to the second expansion wave passing, at which point the solutions diverge slightly. The simulation shows that the internal pressure falls below the external pressure (below $p_0/p_{00} \approx 0.4$), whereas the modified Baker model decays more slowly. The classic Baker results exhibit large error at p_{ext} = 1 atm due to its inability to account for the external pressure.

As effective altitude increases (lowering external pressure), the simulation results predict that the fragment velocity increases toward an asymptotic maximum of around 270 m/s. The modified Baker model captures this trend well. As shown in Figure 7a, the modified Baker and simulation results agree in peak velocity over a range of effective altitudes. The original Baker result, which is independent of altitude, agrees best with the simulation result at higher altitudes. For example, at the effective altitude of 48 km (p_{ext} = 0.001 atm, Figure 7b and c), both Baker models nearly reproduce the simulations results in velocity and internal pressure, with a slight edge to the modified model.

Figure 6: STS ET results at p_{ext} = 1 atm: (a) Peak velocity sensitivity to N, (b) velocity at $N = 12$, and (c) internal pressure history at $N = 12$.

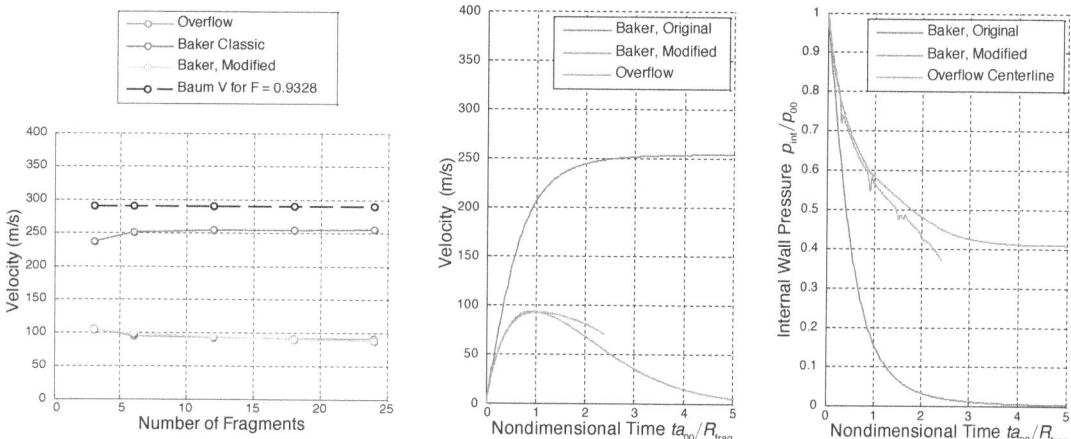

Figure 7: STS ET results at $N = 12$: (a) Peak velocity vs. effective altitude, (b) velocity at p_{ext} = 0.001 atm, and (c) internal pressure history at p_{ext} = 0.001 atm.

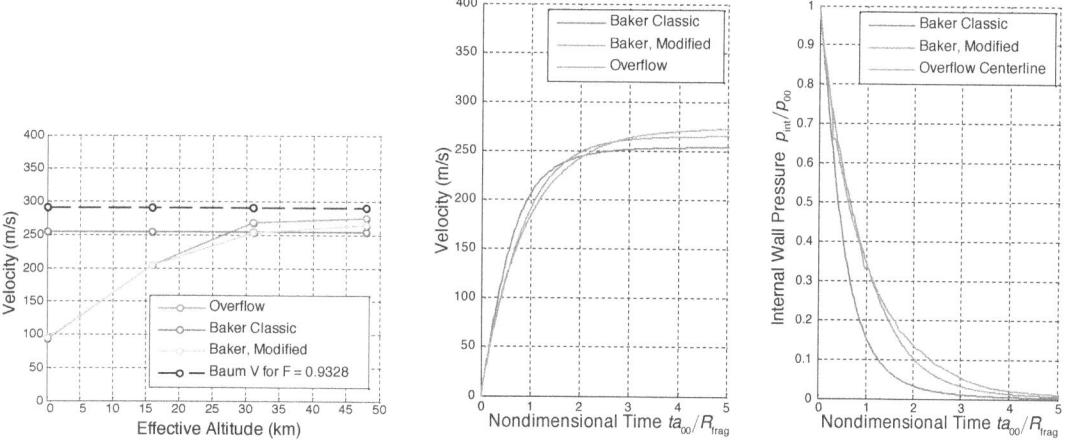

4.2. Space Shuttle Main Engine Main Combustion Chamber

The Space Shuttle main engine, which is also now called RS-25 on the core stage of the SLS, operates its main combustion chamber at very high pressures and temperature due its staged combustion cycle. A burst of this chamber could send fragments into critical thrust vector control systems, high-pressure helium tanks, propellant feed-lines, or the core stage tanks directly. Unlike the STS ET, the internal fluid is modeled as a separate species from the external air, since the low molecular weight of the hydrogen fuel will affect the gas properties. While the actual, roughly cylindrical combustion chamber walls contain coolant channels, the walls in the model were simplified to solid annuli that achieve the same axial and azimuthal mass distribution. The expansion effect phase-in parameters for the modified Baker model were retuned to the values indicated in Table 1 to achieve better agreement with simulation results.

Visualizations of the OVERFLOW simulation results (Figure 8) again show supersonic jets issuing from the fragment gaps. There are propagating shocks inside the jet and numerous vortical structures. Despite this complexity, the simulations suggest again that the peak velocity is only a weak function of the fragment count, though with a slight decrease at low N instead of the increases seen in the other configurations. With retuned expansion term phase-in parameters (Table 1), the modified Baker model peak velocities agree better with the simulation results than the classic Baker, at least in the high-N limit. Both models predict a stronger dependence on fragment count at low N. At $N = 12$, the velocity over-prediction by the classic Baker model, despite the very good agreement in internal pressure, is attributed to the lack of external pressure.

Figure 8: SSME MCC simulation Mach number contours for $N = 12$, with reverse flow shown in fourth image.

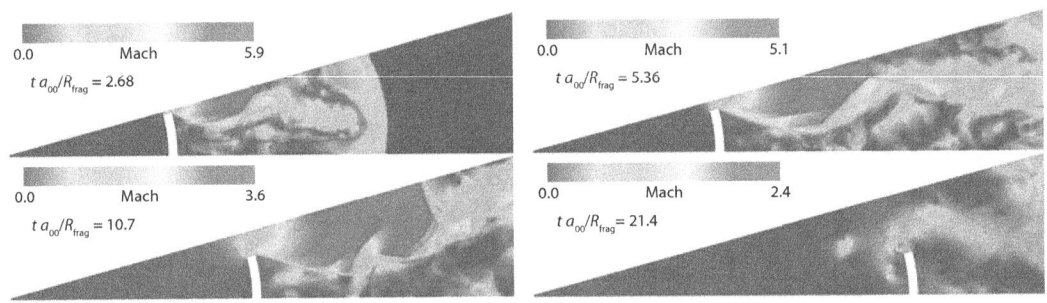

Figure 9: Velocity, internal pressure, and peak velocity results for SSME MCC.

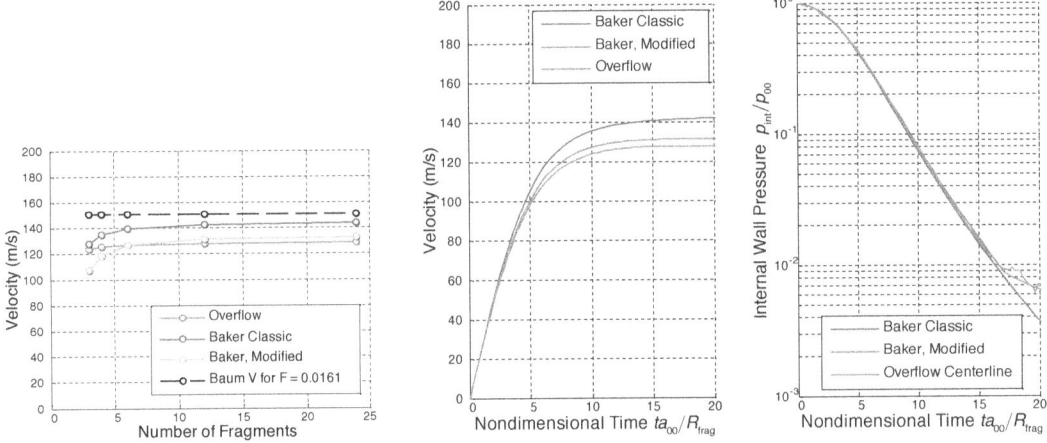

5. CONCLUSIONS

In an effort to introduce high-fidelity, physics-based analysis to catastrophic failure propagation modeling in crewed launch vehicle risk assessments, the Baker engineering model for estimating the maximum velocities of fragments from bursting pressurized cylindrical tanks was updated using the results of coupled CFD and rigid body dynamic simulations. Time-accurate, two-dimensional, inviscid flow simulations were carried out on five pressurized tank configurations over a range of fragment sizes and external pressures. The tank configurations varied in terms of internal pressure regime, temperature, gas properties, and tank wall mass and thickness. The modifications to the engineering model included the addition of external atmospheric pressure, local internal and external effects based on one-dimensional piston theory, and measures to account for fragment curvature. Simulation results were also used to calibrate modeling parameters such as the discharge coefficient.

As a plausibility check for the simplified tank representation used in the models and simulations covered in this paper, the velocity peaks of all five simulated and modeled tank configurations were scaled and found to comply with a velocity bound from an energy-partition-based prediction. Among the two tank configurations presented in detail, the modified Baker model was found to agree better with the simulation-based predictions for maximum fragment velocity than the original model in nearly all of the conditions examined, although this advantage diminished at high pressure ratios. The modified model was found to be extremely reliable in the low internal-to-external pressure ratio case of the propellant tank across all ranges of fragment counts at sea level, where the original model did not produce a useful result. The modified model was still found to be a reliable predictor in the high fragment count limit of the high-pressure, high-temperature case of the Space Shuttle main engine combustion chamber, but fared less well in the low fragment count limit, where the original Baker model also over-predicted a drop in peak velocity.

Future work on the modified model will be directed at improving the low fragment count (large fragment curvature) predictions in high-temperature and high-pressure cases. The effect of evaporating cryogenic propellants will also be considered. The modified engineering model will be used to help improve the definition of fragment velocities in debris catalogs used for debris strike risk assessments of crew modules and other vulnerable crew launch vehicle systems.

References

[1] K. Gee. and S. L. Lawrence, "*Launch Vehicle Debris Models and Crew Vehicle Ascent Abort Risk,*" Reliability and Maintainability Symposium (RAMS), Orlando, FL, January 2013.

[2] M. R. Baum, "*Disruptive Failure of Pressure Vessels: Preliminary Design Guidelines for Fragment Velocity and the Extent of the Hazard Zone,*" ASME Pressure Vessels and Piping Conference, San Diego CA, June 1987.

[3] W. M. Hinckley, D. L. Lehto, N. L. Coleburn, A. J. Gorechlad, J. M. Ward, and J. Petes, "*Space Shuttle Range Safety Command Destruct System Analysis and Verification. Phase II – Ordnance Options for a Space Shuttle Range Safety Command Destruct System,*" NSWC Technical Report 80-417, March 1981.

[4] W. E. Baker, J. J. Kulesz, R. E. Ricker, R. L. Bessey, P. S. Westine, V. B. Parr, and G. A. Oldham, "*Workbook for predicting pressure wave and fragment effects of exploding propellant tanks and gas storage vessels,*" Contractor Report CR-134906, NASA (1977).

[5] W. E. Baker, V. B. Parr, R. L. Bessey, and P. A. Cox, "*Assembly and Analysis of Fragmentation Data for Liquid Propellant Vessels,*" Contractor Report CR-134538, NASA (1974).

[6] R. L. Bessey, "*Fragment Velocities from Exploding Liquid Propellant Tanks,*" The Shock and Vibration Bulletin, Part 3: Shock Testing, Shock Analysis. Naval Research Laboratory, Aug. 1974, pp. 133-139.

[7] D. E. Taylor, C. F. Price, "*Velocity of Fragments From Bursting Gas Reservoirs,*" Journal of Engineering for Industry, Nov. 1971.

[8] R. H. Nichols, R. W. Tramel, and P. G. Buning, "*Solver and Turbulence Model Upgrades to OVERFLOW 2 for Unsteady and High-Speed Applications,*" AIAA Paper 2006-2824, June 2006.

[9] S. L. Lawrence, D. L. Mathias, K. Gee, and M. Olsen "*Simulation Assisted Risk Assessment: Blast Overpressure Modeling,*" PSAM-0197, 8th International Conference on Probabilistic Safety Assessment and Management (PSAM8), New Orleans, LA, May 2006.

[10] J. C. T. Wang, M. F. Werner, and D. R. Langley, "*Transient Flowfield from a Ruptured Pressurized Cylinder,*" AIAA Paper No. 1998-2658, 29th AIAA Fluid Dynamics Conference, Albuquerque, NM, June 1998.

[11] H. W. Liepmann, and A. Roshko, "*Elements of Gas Dynamics,*" John Wiley and Sons, 1957.

[12] The MathWorks, MATLAB vR2014a. http://www.mathworks.com/help/matlab/ref/ode45.html

[13] W. Chan, R. Gomez, S. Rogers, and P. Buning, "*Best Practices in Overset Grid Generation,*" AIAA-2002-3191, 32nd AIAA Fluid Dynamics Conference, St. Louis, MO, June 24-26, 2002.

A Failure Propagation Modeling Method for Launch Vehicle Safety Assessment

Scott Lawrence[*], Donovan Mathias, and Ken Gee
NASA Ames Research Center

Abstract: A method has been developed with the objective of making the potentially intractable problem of launch vehicle failure propagation somewhat less intractable. The approach taken is to essentially decouple the potentially multi-stepped propagation process into a series of bi-component transition probabilities. These probabilities are then used within a simple Monte Carlo simulation process through which the more complex behavior evolves. The process is described using a simple model problem and some discussion of enhancements for real-world applications is included. The role of the model within a broader analysis process for assessing abort effectiveness from launch vehicle failure modes is also described.

Keywords: Launch vehicles, Crew Safety, Failure Propagation, Explosions, Abort Effectiveness.

1 INTRODUCTION

Crewed launch vehicle ascent risk assessment requires consideration of two primary elements: the reliability of the launch vehicle and the effectiveness of the abort process should a failure of the launch vehicle occur. The reliability of the launch vehicle is typically provided in the form of failure scenario types with quantified probabilities. Characterization of launch system safety then builds upon the characterization of the system's reliability by developing an understanding of the consequences of the system's failure modes. Three factors that contribute significantly to crew safety in the face of a launch vehicle failure are: 1) the type of end state resulting from progression of the vehicle failure (e.g., confined explosion, vehicle breakup, vehicle dynamics, etc.), 2) the time required to reach that end state with respect to the time at which the failure is detected (i.e., the warning time), and 3) the severity of the resulting environments. The environment severity might, for example, be represented in terms of the time required for the crew module to reach a distance at which it can endure the environment created. The relative importance of each factor depends on the system's capabilities with respect to the other factors. For example, if failures are contained such that severe environments are rarely developed, then warning time and/or strong abort acceleration are less important. Conversely, if failures are detected early relative to the generation of severe environments, detailed knowledge regarding the failure propagation becomes less essential. The ability to detect catastrophic failures early enough to provide significant warning time is typically difficult to guarantee and, therefore, it is important to understand the potential failure propagation paths of the given launch vehicle when assessing the potential for successful abort.

Previous work within the Engineering Risk Assessment (ERA) team at NASA Ames Research Center [1,2] has focused on characterizing the end-state environments. Specifically, this work has integrated blast overpressure and debris modeling with abort system capabilities to provide failure probabilities as functions of mission time and warning time. The present work represents an effort to address the failure propagation aspect of the problem, i.e., determining the probability that failure environments will be created at all, given the occurrence of a specific launch vehicle failure mode. In the past, this propagation has been represented by relatively simple mappings, e.g., the probability of a stage explosion given an uncontained engine failure. These mappings have been based to a large extent on engineering judgment and expert opinion, supported occasionally with physics-based analysis. The ultimate objective of this work is to obtain this quantitative mapping information through a more systematic application of physics-based modeling and simulation. Further, it is envisioned that the

---

[*] Scott.L.Lawrence@nasa.gov

inherent complexity involved in the failure propagation process would be "automatically" reflected as the result of capturing relatively simple interactions between pairs of many components. Clearly, the failure propagation process is one with large uncertainties attached and, at some level, engineering judgment will be necessary in performing the analysis. Ideally, however, the analyses will be framed such that the judgment required pertains to parameters with which engineers are more familiar. For example, the question of the probability that stage explosion follows engine explosion is replaced with questions such as the speed and number of fragments generated by an uncontained engine event.

This paper will describe a process that the ERA team has developed to address the failure propagation problem, which is thought to represent a step toward the objectives described above. The process has been applied in support of NASA's Space Launch System (SLS) Program. For the SLS analysis, the propagation process is coupled with a characterization of the failure environments to produce data tables, which are then used to assess launch vehicle failure-related loss-of-crew (LOC) probabilities.

1.1 Definitions

Some critical components of the analysis process are defined below.

- **Loss-of-Mission Environments (LOMEs)** – Local conditions existing at the time of loss of mission, i.e., loss of critical functionality. These are the starting conditions for the analysis described in this document and are, therefore, occasionally referred to as "initiators." LOMEs generally do not pose a direct threat to the crew or crew module, but have the potential to propagate in such a way that larger energy releases are produced, either in the initiating element or adjacent elements of the architecture.

- **Element-Level Environments (ELEs)** – Environments generated by a single vehicle architecture element (stage) that have the potential to directly threaten the crew or crew module. For example, an ELE can be either a solid rocket booster (SRB) case burst or a liquid stage confined-by-missile (CBM) explosion. These are composed of one or more sub-environments, such as blast overpressure, explosion-generated debris or shrapnel, and/or radiant heating from a fireball.

- **Vehicle-Level Environments (VLEs)** – The complete environment resulting from the initial failure. The set of VLEs includes the null case in which no ELEs are created and generally are composed of zero or more ELEs occurring nearly simultaneously. These may result from the initiator generating multiple ELEs simultaneously or from propagation of one ELE to another.

- **Abortability** – The probability of successfully surviving the environments produced given a launch vehicle failure. Abortability can be assessed with respect to a specific failure type (e.g., LOME), or in an integrated sense, given a characterization of the relative likelihoods of various types of launch vehicle failures (failure probabilities). Because the analysis in this study is performed independently of the failure probabilities, it is only capable of providing abortability with respect to a specific failure type. Given the conditional loss-of-crew probability, P_{LOC}, for a specific failure type, abortability is given simply as $(1-P_{LOC})$

- **Warning Time** – The time interval between the moment that the abort vehicle has begun to separate from the launch vehicle and the moment that the explosive event is initiated. Note that other definitions of warning time exist in other contexts, but this is the definition used for the abortability table development.

- **Aft Skirt** – This term has been used to refer to an enclosed engine section between the aft propellant tank dome and a close-out located somewhere along the engine nozzle, aft of the engine combustion chamber and turbopumps.

2 PROPAGATION PROCESS

A schematic of the propagation process is shown in Figure 1. At the center of the process is a recursive algorithm that traverses what is called the propagation matrix (discussed below) and, in the process, creates a tree of failure events. The initial event tree generated through this process will then be "pruned" to ensure that the events in the tree are all compatible with one another such that the propagation is, in fact, realizable. Finally, end-state, or element-level, environments are identified. These steps are applied repetitively within a Monte Carlo process during which the results are binned according to the combination of end-state environments encountered.

Figure 1. Failure propagation process schematic.

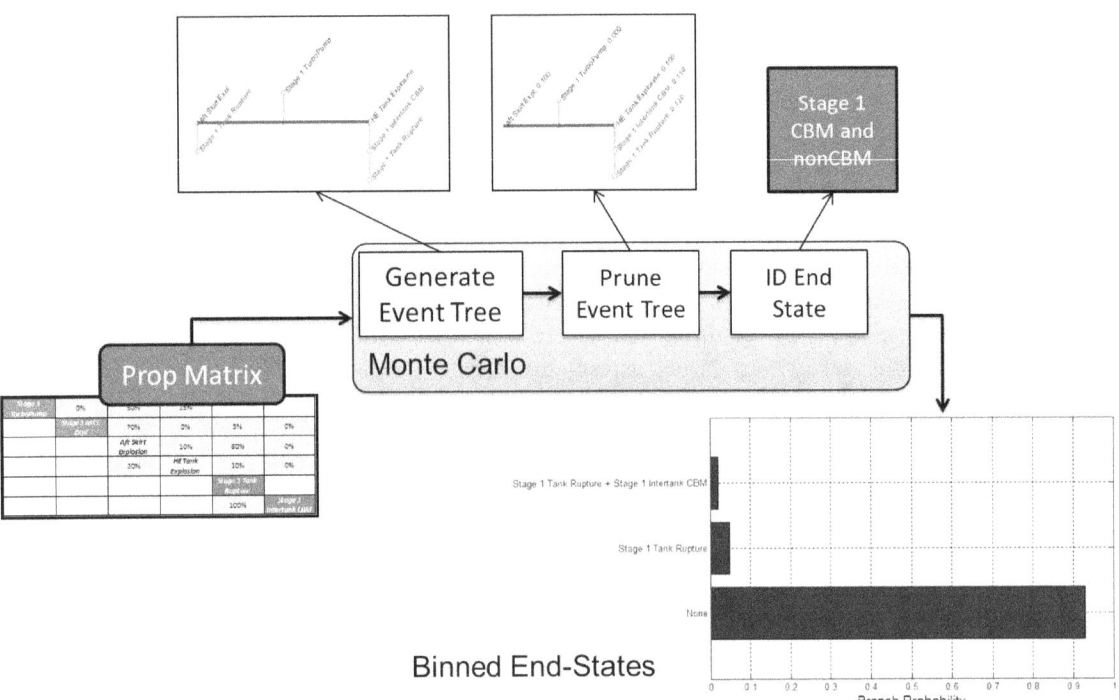

2.1 Propagation Matrix

The propagation matrix serves as the primary input to the propagation evolution algorithm, with diagonal cells representing the various environments that have been identified as potentially contributing to the failure propagation process and off-diagonal elements capturing the potential for transition between these environments. A simple example of a propagation matrix is shown in Figure 2.

The diagonal elements can be classified according to whether they are initial states (blue in Figure 2), intermediate states (yellow), or end states (red). Initial states in the context of propagation are the Loss-of-Mission Environments, i.e., the conditions that exist at the time the mission is lost (e.g., engine turbopump burst). Intermediate states are environments that are not considered to occur naturally without the occurrence of another failure environment, but that may contribute to the propagation process. For example, high-pressure tanks in the engine section may be thought to be so reliable that spontaneous explosion is not considered, but they may burst if struck by other failure-generated debris. Finally, the element-level environments described above are also included along the diagonal of the propagation matrix.

Figure 2. Example of a simple propagation matrix.

Stage 1 TurboPump	0%	50%	15%			
	Stage 1 MCC Expl	70%	0%	5%	0%	
		Aft Skirt Explosion	10%	80%	0%	
		20%	HE Tank Explosion	10%	0%	
				Stage 1 Tank Rupture		50%
				100%	Stage 1 Intertank CBM	
						Stage 2 Tank Rupture

One requirement of the propagation algorithm is that the results should be independent of the order in which the environments are placed along the diagonal. The result of re-ordering these elements should only be to cause some off-diagonal probabilities to be shifted from above the diagonal to below, or vice versa.

The transition probability in a given off-diagonal cell represents the probability that the environment lying horizontally at the diagonal (source environment) will generate the environment lying vertically at the diagonal (target environment). For example, the 15% transition probability in the 4th column of the 1st row indicates a 15% probability that the turbopump burst will cause a rupture of the high-pressure helium tank. Likewise, the 20% probability in the 3rd column of the 4th row indicates a 20% probability that a helium tank explosion will cause explosion of the aft skirt. Typically, engineering judgment is applied initially to rule out many of the potential transitions (blank cells). In other cases, the transitions are not initially ruled out, but may ultimately be set to zero because analysis shows the transition is either not ever credible or is not possible under certain conditions. For example, in early mission phases for which the aft propellant tank is relatively full, the possibility for fragments from aft explosions to penetrate the intertank walls enabling a confined explosion is considered virtually zero.

2.2 Propagation Simulation

The data in the propagation matrix is used to flesh out a particular propagation scenario by starting at a user-specified initial condition (one of the blue boxes) and "rolling the dice" against each of the transition probabilities in the associated row of the propagation matrix. Here, rolling the dice consists of generating a random number in Matlab [3] and evaluating it against the transition probability to determine whether a "hit" is observed. Given a hit, the same process is applied to the target environment, i.e., the target becomes the source for the next step. An environment is not allowed to recur along a given propagation path.

Figure 3 shows an illustration of one such path that might be generated in response to the inputs of Figure 2, starting with stage 1 turbopump burst. In this case, the turbopump burst releases shrapnel that impacts and ruptures the high-pressure helium tanks. Overpressure and/or shrapnel from the helium tank rupture then causes leakage of propellant, from either the fuel tank or feed-line or both, into the aft skirt volume. This flammable mass is assumed to be ignited, causing an explosion that ruptures the aft propellant tank, which is then driven into the upper stage causing rupture of the upper stage tanks.

Figure 3. Sample failure propagation path.

Stage 1 TurboPump	0%	50%	15%			
	Stage 1 MCC Expl	70%	0%	5%	0%	
		Aft Skirt Explosion	10%	80%	0%	
		20%	HE Tank Explosion	10%	5%	
				Stage 1 Tank Rupture		50%
				100%	Stage 1 Intertank CBM	
						Stage 2 Tank Rupture

This propagation scenario is somewhat unique in the sense that it is a single path. In fact, the nature of the propagation matrix shown allows for much more complexity because multiple hits could be observed from a given source environment. For example, shrapnel from the turbopump may create leakage as well as impact the helium tank. Each of those triggered target environments would then be used as source environments for subsequent propagation, and so on, with the final result being a potentially complex tree structure of environments. In the present model, each of the non-zero transition probabilities in the row associated with a given environment is queried independently. If no transitions are triggered, then the process ends for that path. The algorithm makes use of tree data structures and is coded in Matlab.

This approach is not dynamic in the sense that transition probabilities at a given point in the process might be influenced by preceding events. The present implementation assumes transition probabilities from a given state are independent of how that state was reached. This is a recognized limitation and is being studied to determine whether it can be removed or whether a substantially different approach is required.

2.3 Event Trees and Pruning

An example of an event tree that might be produced with this model is shown in Figure 4a. The simplicity of the example propagation matrix leads to relatively simple trees. However, one still observes branching in this case. In these plots, element-level environments are highlighted in red. One can also observe that, while no environments are repeated along either branch, stage 1 tank rupture occurs on both branches. In a sense, the branches are incompatible with each other. Furthermore, if the tank is ruptured by the aft skirt explosion, one might question whether the confined explosion (CBM) in the intertank region is possible.

In order to address these issues, a pruning process is applied to these raw event trees. The pruning depends on the introduction of some timing information, which may be physical time associated with the transition process if available, but in the simplest case may be a fixed time increment analogous to a computer clock cycle. All that is needed is to be able to identify which event happened first. The present model allows the user to specify an uncertain range of transition times, which are sampled and accumulated during the development of the raw trees. Then, all the events in a given propagation are sorted in chronological order of their occurrence, and repeat environments are removed along with any subsequent events on that "branch" of the tree.

Figure 4. Propagation event trees.

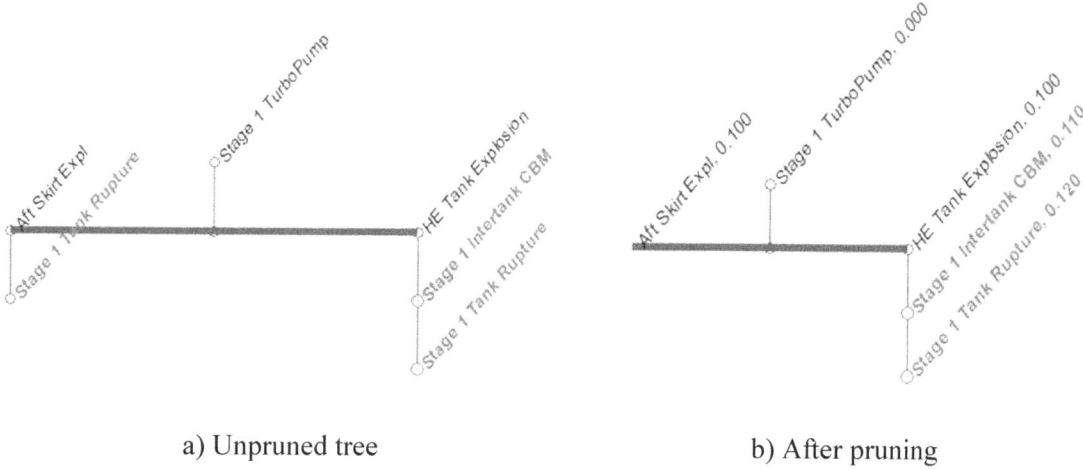

a) Unpruned tree   b) After pruning

The model also allows the user to specify groups of environments for which, if the first environment in the group has been triggered, all other subsequent environments in the group are thought to be impossible and are removed from the tree. In the present case, an example of an "exclusion group" would be {stage 1 tank rupture, stage 1 intertank CBM}. That is, rupture of the tanks precludes subsequent confined explosions. The reverse is not true, however; occurrence of a confined explosion does not prevent a subsequent unconfined explosion. In fact, it is highly likely that non-confined explosion(s) will follow a confined explosion. Thus, the exclusion is based only on the first environment in the group.

Figure 4b shows the effect of applying such a pruning process to the tree shown in Figure 4a. For clarity, two fixed time intervals are used in this case: 0.01 seconds and 0.1 seconds. The resulting tree indicates that, in this case, the intertank CBM occurs earlier than the tank rupture caused by the aft skirt explosion, and so the latter has been removed.

2.4 End States and Monte Carlo

The final step in analyzing a single instance of the propagation is to identify and log the final state. This simply involves collecting all of the element-level environments that have been triggered, e.g., the red environments in Figure 4. Since each of the N possible ELEs can be triggered or not, there are 2^N possible outcomes. These combinations of ELEs have been termed vehicle-level environments (VLEs). Note, one of the VLEs is the null environment, which is the case where none of the ELEs are generated.

The probabilities in the propagation matrix are effectively integrated using a Monte Carlo approach with the outcome of each instance categorized in terms of the particular combination (of the 2^N possible) of ELEs observed. The end result of the Monte Carlo process is a distribution among the possible VLEs, such as the example shown in
Figure 5. Note that only 3 of the 8 possible VLEs have been encountered, with the null environment the most likely in the current example.

This mapping information can be used directly within an integrated safety model such as that described in [4]. In this model, the VLEs are characterized using tables in which the conditional loss-of-crew probability is expressed in terms of functions of mission time and warning time. The integrated model then uses the mapping to determine which of these tables will be queried, and relative frequency with which they will be queried, given the occurrence of the initial failure type.

Figure 5. Example of Monte Carlo result for mapping Stage 1 turbopump failure.

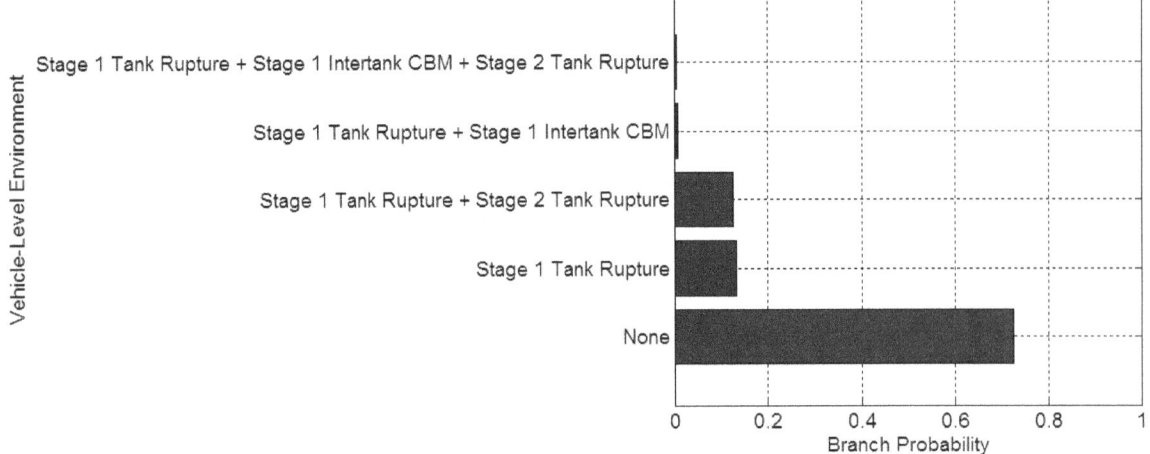

3 COMBINING PROPAGATION AND ENVIRONMENT CHARACTERIZATION: ABORTABILITY AND LOSS-OF-CREW

The mapping information resulting from the Monte Carlo simulations provides the probabilities of different classes of outcomes given the initial failure manifestation. What is ultimately needed is the probability of the crew surviving the environments generated as a result of the initial failure manifestation. To obtain these probabilities, the mapping information of the type shown above must be coupled with the failure environment information. Blast overpressure and debris environment characterization are discussed in [1] and [2], respectively. These environments have historically been produced in terms of tables of conditional failure probability as functions of mission time and warning time.

The ERA team characterizes failure environment severity using tables containing probabilities of failure from overpressure or debris as functions of mission time and available warning time. These tables include effects of the environment initiation, environment propagation and decay, abort trajectories, and crew module vulnerability to overpressure and debris. The environments can be generated using different assumptions regarding the initiation (blast location, blast yield, debris catalogs, etc.) to represent effects from explosions of different parts of the vehicle. The individual tables for overpressure and debris may be convolved into tables for the element-level environments, for example, stage 1 CBM. Under the assumption that the ELEs involved in a given VLE are triggered simultaneously, tables for each possible VLE can then be generated using a similar convolution of the associated ELE tables.

As mentioned in the previous section, the coupling of the propagation mapping with the environment characterization can be performed within an integrated mission safety model (see [4], for example). Alternatively, it can be performed outside such a model by generating what have been termed abortability tables. In this approach, tables for loss-of-crew probability given the occurrence of the initial failure manifestation type (LOME) are generated by combining the set of VLE tables using a simple weighted average in which the weightings are the mapping probabilities generated through the propagation analysis, e.g., those shown in
Figure 5. The results are expressed in terms of abortability by simply subtracting the failure probabilities from unity.

In this model, abortability for a specific failure scenario and mission time is obtained, given its initial manifestation (LOME), by querying the appropriate table at the appropriate mission time with an estimated warning time. The present form of the propagation modeling does not address the important problem of warning time estimations, but could potentially be extended to do so. In the SLS program, warning times are determined by the SLS Mission and Fault Management (M&FM) group based on knowledge of the available abort trigger sensor design information. See [5] for discussion of abort trigger assessment and selection.

4 IMPLEMENTATION ENHANCEMENTS

The approach used in the preceding sections is somewhat limited in the sense that maintaining the propagation matrices can quickly become unwieldy if one wants to account for variations of the transition probabilities with mission time or phase. This can be important if the transitions are effected through blast overpressure, for example, or another mechanism that depends upon ambient pressure. Further, these transition probabilities are often subject to large uncertainties, either because they are, at least initially, based on engineering judgment or because the process itself is inherently sensitive to small differences in the failure initiation. One may want to investigate and capture the effect of these uncertainties on the propagation. These issues have been addressed using a relatively compact input data format, maintained in an Excel spreadsheet, and a Matlab script that queries the Excel data and generates the propagation matrix on-the-fly.

A section of a sample input data table is shown in Figure 6. Each row represents one potentially non-zero element of the propagation matrix. In the present implementation, a static, qualitative propagation matrix is maintained in which transition probabilities in the off-diagonals are replaced by pointers to the appropriate row of the data table (the ID column in Figure 6 gives the cell index of the associated propagation matrix cell). The table allows the transition probabilities to be varied with mission phase, with each column representing a different phase. Within each cell, a range of values may be captured, separated by slashes, with different values capturing different parts of an uncertain distribution. Typically, the left-most value is considered a lower bound, the right-most is the upper bound, and the middle value is a "baseline" value. Higher values tend to allow the propagation to more easily reach the explosion outcomes so the left-most values are typically referred to as the pessimistic or worst-case set and the right-most values are then the optimistic or best-case set. Colors can be used to represent the level of confidence in the quantitative transition probabilities provided.

This format also allows for convenient documentation (not shown) for specific information regarding the source of the data and any other pertinent information. Also, with this data format, it is relatively easy to implement additional features that may require additional supporting data (e.g., the timing information, shown in the far right column as minimum/maximum ranges).

Figure 6. Sample input data for the propagation analysis.

ID	Pre-Launch w/ LAS PL	First Stage Burn FSB	Staging FSS	Upper Stage Burn, w/ LAS USL	Upper Stage Burn, no LAS USN	Spacecraft Staging USS	Source	Target	Timing
E6	0/ 0/ 0	0/ 0/ 0	0/ 0/ 0	0//	0//	0//	Stage 1 TurboPump	Stage 1 MCC Expl	0.01/0.01
F6	90/50/15	90/50/15	90/50/15	0//	0//	0//	Stage 1 TurboPump	Aft Skirt Expl	0.1/0.1
G6	25/15/5	25/15/5	25/15/5	0//	0//	0//	Stage 1 TurboPump	HE Tank Explosion	0.1/0.1
F7	100/70/20	100/70/20	100/70/20	0//	0//	0//	Stage 1 MCC Expl	Aft Skirt Expl	0.1/0.1
G7	5/0/0	5/0/0	5/0/0	0//	0//	0//	Stage 1 MCC Expl	HE Tank Explosion	0.01/0.01
H7	100/15/0	15/5/0	0//	0//	0//	0//	Stage 1 MCC Expl	Stage 1 Tank Rupture	0.01/0.01
I7	0//	0//	0//	0//	0//	0//	Stage 1 MCC Expl	Stage 1 Intertank CBM	0.1/0.1

Figure 7 shows an example of the effect of the launch phase-dependent transition probabilities on the VLE mappings. The launch phases are abbreviated as follows: PL is pre-launch, FSB is first stage boost, and FSS is first stage staging. In this case, the probability of an aft skirt explosion to cause rupture of the 1st stage propellant tank diminishes with altitude resulting in the trend toward fewer tank ruptures in later mission phases.

The most recent analysis of the SLS vehicle involves 24 environments, including 13 initiating environments and 10 ELEs. The analysis is further complicated by the need to divide the ascent trajectory into 15 separate mission phases. The present data format and pre-processing scripts allow relatively straightforward implementation of this complexity as well as a way to track progress in the maturation of the transition probability data.

Figure 7. Phase-dependent mapping results: Stage 1 turbopump failure.

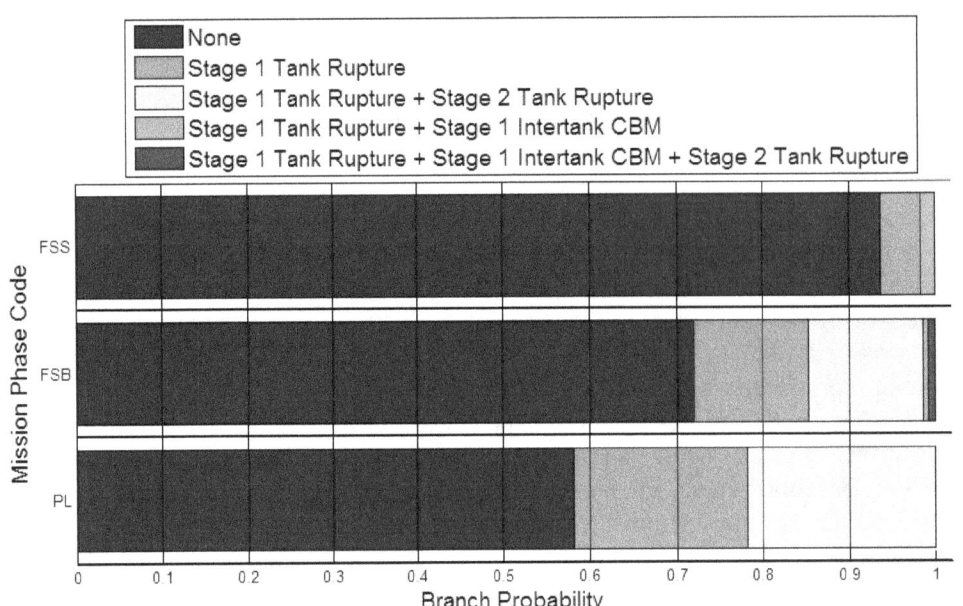

5 TRANSITION PROBABILITY SPECIFICATION

One feature of the current propagation model is that a simple representation can be created and applied relatively quickly. As the vehicle design matures and/or more detailed questions are asked, additional elements/environments can be introduced. The transition data upon which the analysis is based can also mature, starting with a set of transition probabilities that may be largely based on engineering judgment and evolving into a set that is supported predominantly by analysis and, possibly, test data. Even analysis-based transition probabilities will have uncertainties associated with them, but in general, the transition from judgment-based probabilities to analysis-based probabilities would be accompanied by a reduction in the data spread.

The current thought process for specifying the transition probabilities begins, for a given source and target environment, with an effort to identify the mode by which the source effects the failure of the target. Five general classifications have been identified for these modes:

1) energy transfer from the source, either by blast overpressure, kinetic energy (fragments), or heat transfer,
2) structural shock or vibration from failure of the source causes the failure of the target,
3) an environment created by a source failure, such as a flammable mass or pressure environment, cannot be withstood by the target,
4) the ability of the target to contain its energy may depend on the source functionality, e.g., loss of fuel to the engine combustion chamber caused by turbopump failure leads to combustion chamber failure, and
5) direct transfer.

Direct transfer captures cases in which the LOM environment presents a direct threat to the crew. For algorithmic convenience, these situations have been treated with a separate LOME and ELE within the propagation matrix, but with a direct transition between them (100% always). Solid rocket booster burst is an example of such a case.

Relatively simple analyses may be sufficient to provide significant insight into reasonable settings for transition probabilities. Figure 8 shows an example of a simple geometric assessment of the likelihood of debris from a given turbopump burst impacting a high-pressure pneumatic support tank. Here, the debris pattern is expected to disperse within a limited azimuthal range perpendicular to the rotation axis of the pump. The figure indicates a fragment probability distribution with respect to this angle, and the target tank lies within one tail of the distribution. A rough estimate of the debris strike probability would be given by:

$$P_{strike} = 1 - e^{-\lambda N}$$

where N is the number of fragments released and λ is given by

$$\lambda = FR / \pi d$$

Here, F represents the shaded area under the curve in Figure 8, R is the radius of the target tank, and d is the distance from the source to the target. This accounts for the distributed dispersion shown in Figure 8 as well as uniform dispersion in the circumferential direction.

Figure 8. Example of simple geometric strike probability analysis.

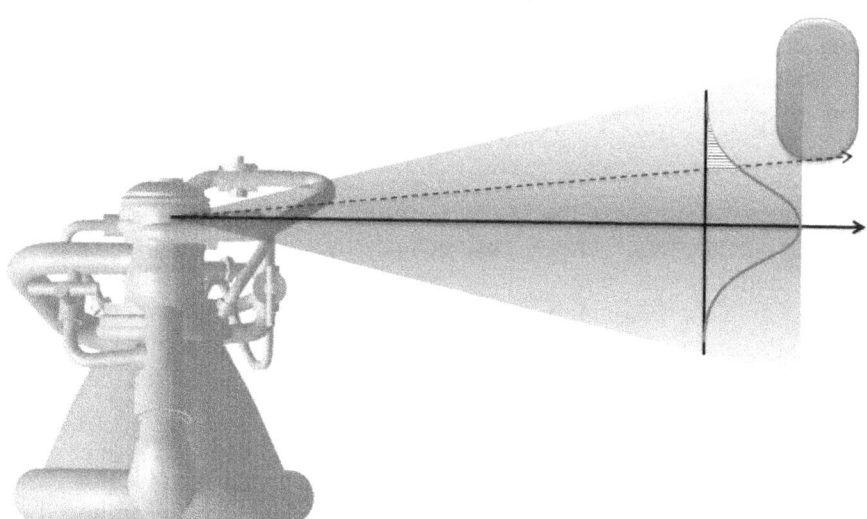

The failure probability given a strike would depend on the likely velocities of the debris, e.g., the spin velocity of the pump (likely conservative), as well as the vulnerability of the target tank, e.g., the ballistic limit velocity based on the material of the tank and some estimate of the fragment sizes. The transition probability from turbopump burst to high-pressure tank burst would then be set to the strike probability if the debris velocity exceeds the limit velocity for puncture, or would be set to zero otherwise.

The threat posed by the main combustion chamber could be similarly analyzed, except that the angular distribution and initial velocities of the debris would be different (see [6] for a discussion of modeling pressure vessel burst velocities). This simple example illustrates how analysis methods do not necessarily eliminate uncertainty—many unanswered questions remain regarding the number of fragments released, the initial velocity after escape from the pump walls, the true vulnerability of the tanks to debris, etc. However, the nature of the uncertainty is such that engineering judgment can perhaps be more reliable. Further, these questions are more amenable to engineering analysis and testing than is the question of the probability of stage explosion given an engine failure.

6 CONCLUSIONS AND FUTURE WORK

A relatively simple method to address the complex problem of failure propagation following launch vehicle failure has been presented. The method is extensible in the sense that one can generate a simple model for a given scenario and evolve it to account for additional factors as the vehicle design, or the need to understand the behavior, matures. Presently, the model is used only to capture the propagation of one energy-containing component to another in the cascade of events that might connect a local vehicle failure manifestation to crew-threatening explosion. The analysis has to date, therefore, been limited to determining the probability of explosion given various classes of vehicle failures.

Extension of the model to include realistic transition timing information as well as detection elements that might be activated in response to triggering certain environments, is currently under consideration. In this way, the model could potentially be applied to the problem of estimating abort warning time.

Ultimately, the approach envisioned is a system in which the components identified as potential contributors to the propagation process are defined in terms of engineering parameters such as contained pressure and temperature, wall thickness and material, etc. These engineering parameters would then be used to determine both potential threat of the components as sources (e.g., burst overpressure, fragment environments) as well as their vulnerabilities as targets. The off-diagonal terms in the propagation matrix would capture the relationships between the diagonal components: the distances, view factors, obstructions, etc. Given that these component and relationship properties are defined, transition probabilities could be generated automatically. This is in work by the present authors.

A separate effort is underway to apply the Monte Carlo process directly to uncertain elements in the component properties: the number of thrown fragments, the burst yield, component vulnerabilities, etc. Relationships in this model would be evaluated automatically through the direct use of CAD design data. This approach, while more complex to implement, would potentially provide a fully coupled simulation and would therefore relieve any limitations introduced by decoupling.

Acknowledgements

The authors would like to acknowledge the important contributions to this work by Louise Strutzenberg of the NASA Marshal Space Flight Center, the Sub-Discipline Lead for Abort Environments under the Space Launch System Structures and Environments Discipline. The authors would also like to acknowledge Hamed Nejad, presently of Liquid Robotics, Inc., for his important contributions, including cross-checking with Bayesian belief networks, during the early exploration of the approach.

References

[1] Lawrence, S., and Mathias, D., "Blast Overpressure Modeling Enhancements for Application to Risk-Informed Design of Human Space Flight Launch Abort Systems," RAMS 06B-3, 2008 Reliability and Maintainability Symposium, Las Vegas, NV, January, 2008.
[2] Gee, K. and Lawrence, S. L., "Launch Vehicle Debris Models and Crew Vehicle Ascent Abort Risk," Reliability and Maintainability Symposium (RAMS), Orlando, FL, January, 2013.
[3] The MathWorks, MATLAB vR2014a. http://www.mathworks.com/help/matlab/ref/ode45.html

[4] Go, S., Mathias, D., Mattenberger, C., Lawrence, S., and Gee, K., "An Integrated Reliability and Physics-Based Risk Modeling Approach for Assessing Human Space Launch Systems," 12th International Conference on Probabilistic Safety and Management (PSAM12), Honolulu, HI, June 2014.

[5] Y. Lo, S. B. Johnson, and J. T. Breckenridge, "Application of Fault Management Theory to the Quantitative Selection of a Launch Vehicle Abort Trigger Suite," 2014 IEEE International Conference on Prognostics and Health Management, Spokane, Washington, June 22-25, 2014.

[6] Manning, T. A., and Lawrence, S. L., "Physics-Based Fragment Acceleration Modeling for Pressurized Tank Burst Risk Assessments," 12th International Conference on Probabilistic Safety and Management (PSAM12), Honolulu, HI, June 2014.

An Integrated Reliability and Physics-based Risk Modeling Approach for Assessing Human Spaceflight Systems

Susie Go[*a], Donovan Mathias[a], Chris Mattenberger[b], Scott Lawrence[a], and Ken Gee[a]
[a] NASA Ames Research Center, Moffett Field, CA, USA
[b] Science and Technology Corp., Moffett Field, CA, USA

Abstract: This paper presents an integrated reliability and physics-based risk modeling approach for assessing human spaceflight systems. The approach is demonstrated using an example, end-to-end risk assessment of a generic crewed space transportation system during a reference mission to the International Space Station. The behavior of the system is modeled using analysis techniques from multiple disciplines in order to properly capture the dynamic time- and state- dependent consequences of failures encountered in different mission phases. We discuss how to combine traditional reliability analyses with Monte Carlo simulation methods and physics-based engineering models to produce loss-of-mission and loss-of-crew risk estimates supporting risk-based decision-making and requirement verification. This approach facilitates risk-informed design by providing more realistic representation of system failures and interactions; identifying key risk-driving sensitivities, dependencies, and assumptions; and tracking multiple figures of merit within a single, responsive assessment framework that can readily incorporate evolving design information throughout system development.

Keywords: PRA, simulation, physical modeling, reliability modeling, risk-informed design.

1. INTRODUCTION

Over the years, NASA has developed a working knowledge and body of standards to guide both the design and the evaluation of safe human space flight systems. These include specific requirements on procedures and best practices in addition to quantitative mission risk requirements. As an example, NASA's Commercial Crew Program (CCP) has identified quantitative requirements for loss-of-mission (LOM) and loss-of-crew (LOC) probabilities that establish safety and mission success metrics. To analytically verify these requirements before a system is operational, NASA programs often depend on probabilistic safety analysis (PSA).

In addition to requirement verification, PSA techniques can also be incorporated into the design process to optimize safety and mitigate critical risk drivers throughout system development. The quantitative safety and reliability information from PSA enables a risk-informed design process that includes risk as a design metric along with mass, performance, and cost. By injecting such information into the design process, the post-facto assessment of a system's LOC/LOM probabilities is avoided, and actionable insights are provided at a point in the program where changes can still be impactful.

Traditionally, PSA, or probabilistic risk assessment (PRA), is performed using logical probability models [1]. Most commonly used are the event tree/fault tree (ET/FT) models that have been employed for systems such as the Space Shuttle and International Space Station (ISS). These methods are effective at combining failure probabilities into a series of top events or specified end-states. More recently, simulation methods have been employed to extend the traditional methods such that physical factors and system dynamics are inserted directly into the risk models [2]. The goal of these approaches is to more accurately represent the interactions among on-board systems and between the system and its operating environment. Whereas classical ET/FT approaches strive to generate a very precise solution to an approximate, static problem, simulation methods may be able to better represent a system's fully interactive dynamics by producing an approximate solution to a more realistic problem. The ultimate choice of the "best" analysis approach is determined by the problem specifics, questions being addressed by the analysis, resources available, and desired level of precision.

---

[*] Susie.Go@nasa.gov

This paper presents an example of a simulation-based risk model for a human spaceflight mission. The model blends engineering analyses commonly used during the design process with traditional reliability approaches to produce LOM and LOC estimates that can be used for risk-based decision-making and requirement verification. The example illustrates how the simulation model can produce top-level risk estimates, such as LOC/LOM, while also representing the physical aspects of the system that often drive the risk and are meaningful to system designers. In addition, the model can track multiple figures of merit (FOMs), time-based risk intensities, and system interactions within the same assessment. This paper utilizes a generic space transportation system to illustrate these capabilities.

2. EXAMPLE PROBLEM

An example mission for the end-to-end risk model was developed with similar mission requirements as those defined for the CCP [3]. The baseline mission includes delivering four astronaut crewmembers and their equipment to the ISS and returning them to Earth at least twice a year. A fictitious, generic human spaceflight launch vehicle (LV) and spacecraft (SC), shown in Figure 1, were used for the reference mission in order to illustrate the risk modeling process with representative design and performance data [4]. The launch vehicle is a two-stage, liquid-propellant rocket with four liquid oxygen LOX/RP-1 engines on the first stage and two LOX/hydrogen engines on the upper stage. The spacecraft is a capsule with a five-meter diameter and a tractor-type ascent launch abort system, sized to support the crew for seven days in free-flight or to dock with ISS for a six-month stay. All risk data and physical models in this paper are based on this generic space architecture, which is presented not as a closed architecture, but rather as an example architecture used to demonstrate the risk assessment process. The baseline mission timeline for this architecture was assumed to include a 9.25-minute ascent phase before separation of the spacecraft from the upper stage, a one-day flight to ISS, a six-month stay docked to the ISS with weekly one-hour spacecraft checkouts, and a four-hour entry, descent, and landing (EDL) phase, as shown in Figure 2.

The top-level LOC/LOM metrics were computed for this example mission and the leading risk drivers called out. In addition, the time-based risk intensity was produced for the system. Key interactive examples where dynamics and physics are first-order risk drivers are shown. All of this information was produced by a single model, removing the need for multiple risk models to track specific FOMs.

Figure 1: Launch vehicle and spacecraft specifications [4].

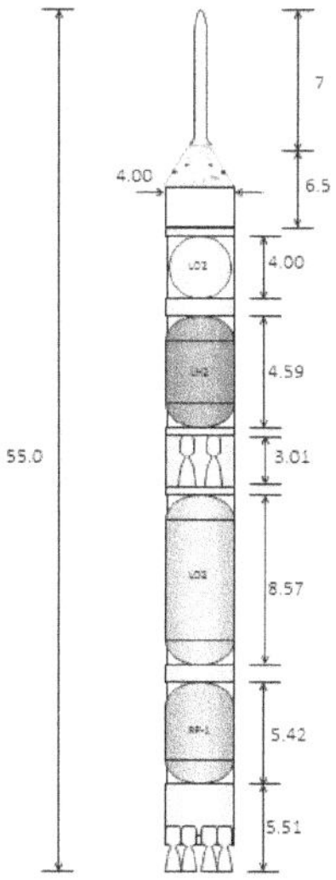

TSTO LV Design Summary
- Payload (spacecraft) mass: 8436 kg
- LAS mass: 5367
- LAS ejection 30 sec after 2nd stage ignition

Item	Stage 1	Stage 2
Prop	LO$_2$/RP-1	LO$_2$/LH$_2$
T	3280800 N	393369 N
P_c	20680 kPa	17240 kPa
MR	2.47:1	6.0:1
I_{sp}	350 s	450 s
ε	0.11	0.16
π	0.177	0.147
m_p	172954 kg	33331 kg
m_0	194330 kg	39680 kg
m_{GLOM}	247814 kg	53483 kg
T/W	1.35	0.75
ΔV	4290 m/s	5209 m/s
# engines	4	2

*Units are in meters

Figure 2: Nominal mission timeline for sample reference mission.

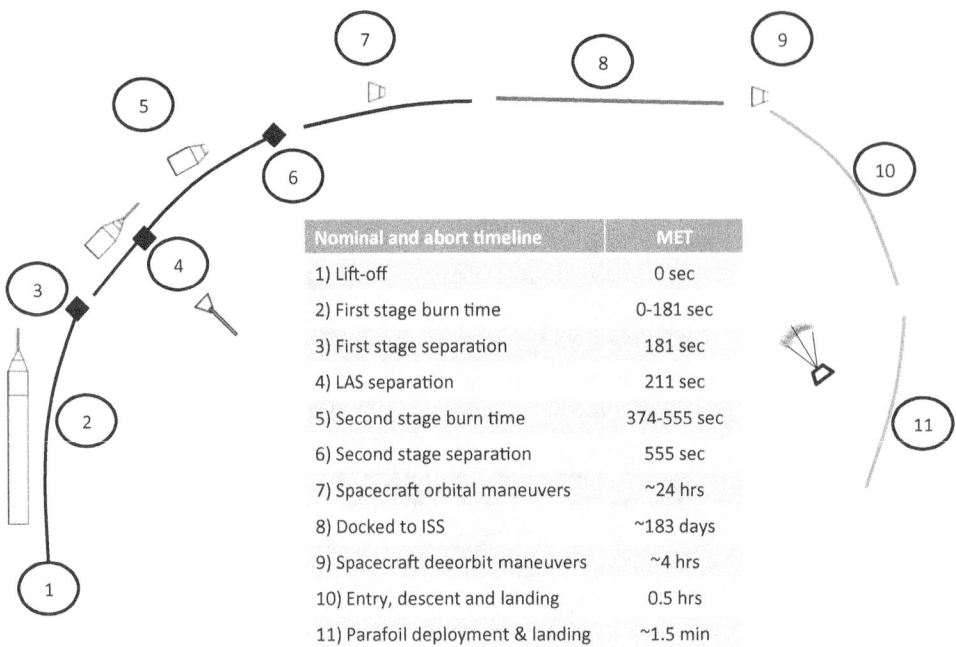

3. DYNAMIC MISSION RISK MODEL

3.1. Risk models and required data

A space launch system travels through multiple environments with vastly different timescales, and the integrated risk model was developed with these variations in mind. Table 1 summarizes some of the key data required for a typical mission risk assessment model. The data covers multiple science and engineering disciplines, including vehicle design, blast physics, trajectory analysis, propulsion system performance, aerodynamic and aerothermal analyses, environment characterization, material responses, and reliability engineering. Traditionally, these modeling methods are used in the design process with an emphasis on optimizing system performance or defining a feasible system. However, the outputs from each of these analysis disciplines can also be focused on understanding system performance during off-nominal conditions or degraded states to provide valuable risk information for both designers and decision makers. In this way, success-based design models can be used to populate the risk model.

Table 1: Summary list of data requirements for a typical mission risk model.

	Ascent Phase	On-orbit Phase	EDL Phase
Launcher design and reliability	X		
Spacecraft design and reliability	X	X	X
Integrated stack design and reliability	X		
Trajectory data	X		X
Aerodynamic environments	X		X
Aerothermal heating environments			X
Space environments		X	
Orbital and deorbit maneuvers		X	
Abort conditions	X		
Material response	X	X	X
Concept of operations and contingencies	X	X	X

In order to best represent the space exploration system, the dynamic mission risk model uses a hybrid approach that integrates the outputs from each of these disciplines to calculate estimates of key system metrics. Figure 3 contains an overview of the model's data structure. At the top of the model are the elements that represent the dynamic behavior of the system and track system-level parameters of interests. Examples of such dynamics include physical failure propagation, system failure response, and the ultimate evolution of hazardous failure environments. The dynamic simulation elements are only incorporated to the level required to resolve dynamics that drive the overall assessment. Below these dynamic levels are static reliability elements that seed the dynamic simulation with likelihood information. These probabilities are often generated by representing components using traditional FT-like approaches. As a result, the simulation model handles the system dynamics and the full FT cut-sets while the static aspects live below the dynamics and can be incorporated back into the mission results through post-processing.

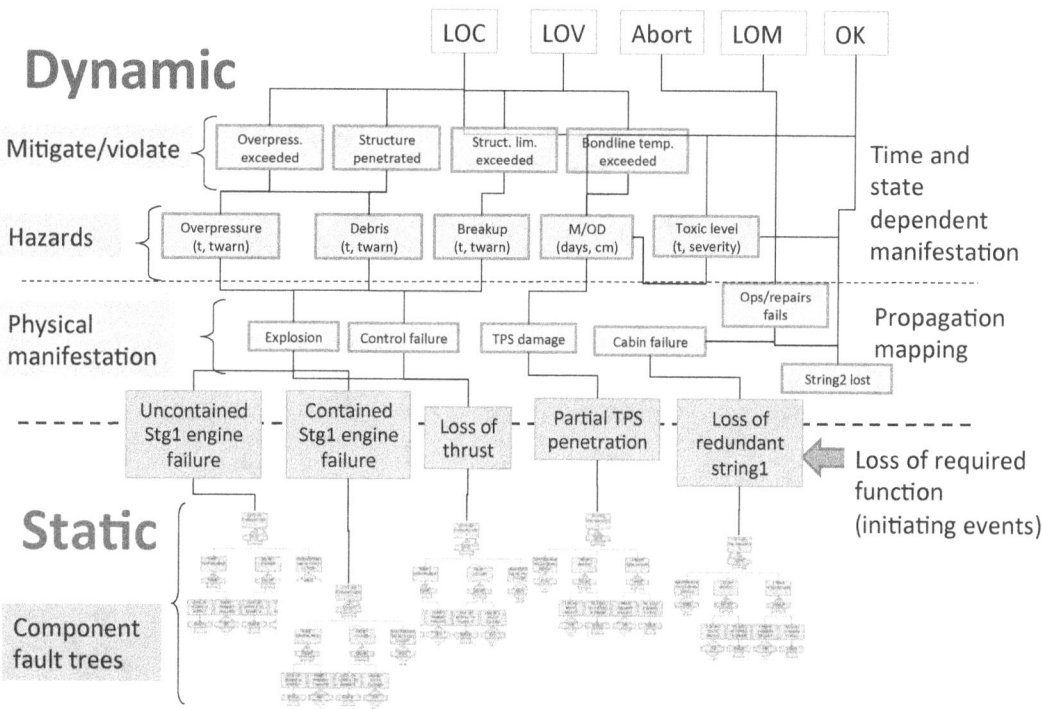

Figure 3: Example of multi-disciplinary, hierarchical failure data.

Starting at the bottom of Figure 3, representative fault trees can be seen to represent the component reliabilities. FT math is utilized when there are no important system dynamics or when a collective set of failures has a similar effect on the simulation, i.e. the simulation can't tell the difference between failure types in the tree. This can occur when the system itself is accurately represented using static methods, when the data used to characterize the subsystem does not support dynamics, or when design fidelity is limited and a static representation serves as a reasonable proxy until additional information becomes available. Though the data in the trees may be static, the simulation can easily incorporate such data in a time-dependent way. For example, the simulation can apply a time-based duty-cycle that queries the tree data only during certain phases in the mission. Also, though FTs are often used to describe component reliability, the basic simulation elements can represent physical elements as well. For example, the current model incorporates offline EDL assessments, such as the one described in [5], at the basic simulation element level. This does not mean the EDL results are imported as a single probability, but that the results can be represented parametrically by a single element. Figure 4 shows a sample sub-element tree for the spaceraft's pressure control system (PCS), based on component data from the master equipment list.

Figure 4: Pressure control system sub-element fault tree.

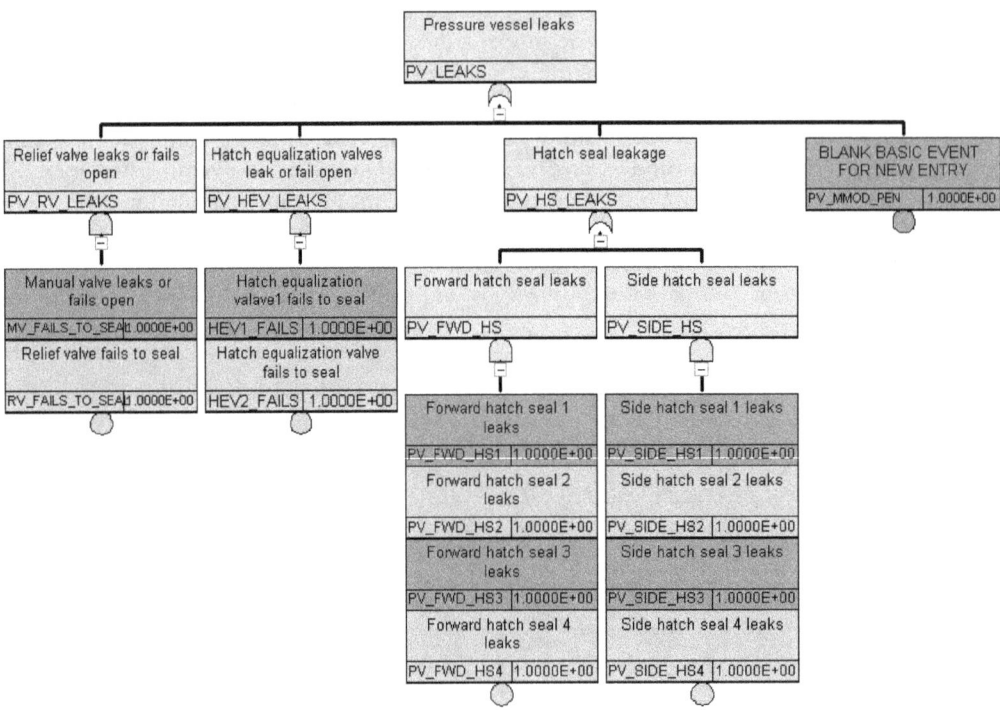

Moving upward through Figure 3, the level above the fault trees and basic simulation elements represents failure propagation through the system. The lowest levels, described above, determine the probability of a failure initiating.[†] The middle levels of the model represent the translation of the initiators into cases that threaten the mission or crew. One such translation is performed by mapping the initiators to subsequent failure environments. This approach is relatively easy to implement and can be very effective in trade studies where a range of failure paths are to be considered. Consider the possible failure modes of the launch vehicle as shown in Figure 5 (from Ref. [6]). Each failure can progress down a number of paths, leading to a range of failure environments. The propagation table allows these "branches" to be assessed using offline models whose results can be directly incorporated into the mission risk model. Should the propagation analysis be updated, or have different assumptions applied, the values in the table can be updated without any change to the mission risk model. For more details on this approach, see Ref. [6].

Figure 5: Sample propagation table for launch vehicle failure modes [6].

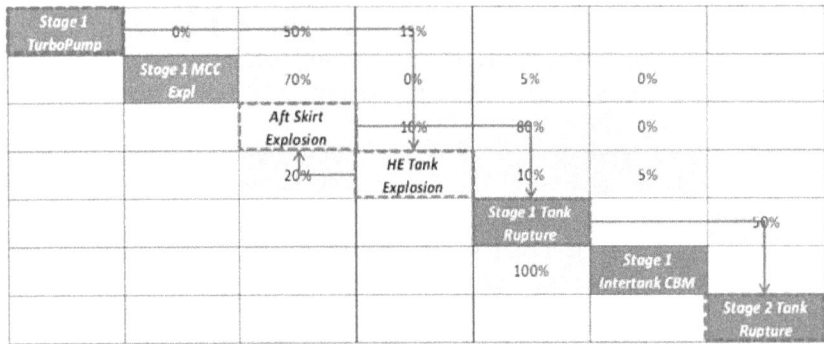

[†] Though the example considers the elements at the bottom of the model to represent failures that trigger LOM, they can just as easily represent system failures that represent a loss of redundancy. Loss of redundancy can be mitigated by the design of the system, can trigger an abort in cases where the degraded system is considered too risky for mission continuance, or can represent a true loss of system function. In all cases, the dynamic simulation model can manage the occurrence and associated response.

While the previous example represents different discrete propagation paths resulting from a launch vehicle failure, continuous failure propagation models can also be incorporated. Consider a seal leak due to one of the PCS hatches shown in Figure 4. The simple, worst-case assumption would be that any PCS failure would lead to LOC immediately, but this assumption would be excessively conservative and drive the risk of the subsystem. To refine this assumption, the actual cabin environment can be modeled to allow the spacecraft to abort from orbit and return the crew safely. Reference [7] describes such a model, which has been incorporated in this assessment. With this additional modeling incorporated, the different leakage failure modes can now map to specific physical responses. For example, a valve that has failed open resulting in a leak of cabin pressure will not lead to immediate LOC. Instead, the time-to-LOC is predicted based on various mission parameters such as the size of the leak, available pressurant (itself a function of mission time and state), leak detection capability, and time required to safely depart ISS and return to Earth. The curves from [7] are shown in Figure 6 and represent parameterized responses to cabin leakage. The mission model selects from the curves, based on the appropriate pre-failure mission state, and determines a resulting time-to-LOC given a cabin leak. Again, because the physics of the failure propagation are imbedded in the dynamic mission risk model, changes to the design, abort policy, crew protection using pressure suits, or any number of other factors can be directly reflected in the analysis without re-architecting the model.

Figure 6: Example LOC hazard data for various cabin leakage scenarios [7].

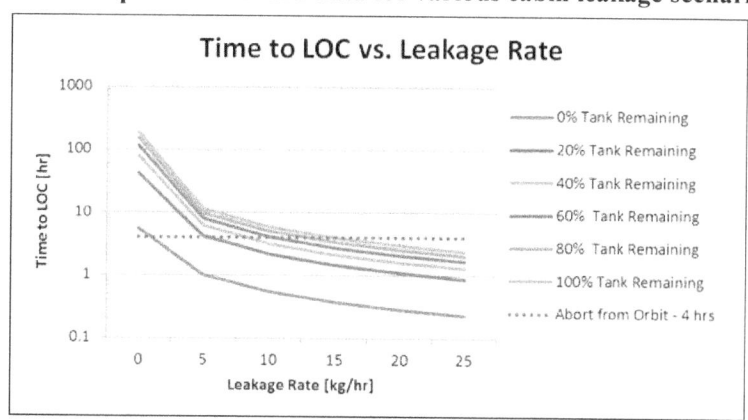

Sometimes the physics itself is responsible for the initiator. For example, consider micrometeoroid orbital debris (MMOD) threats to the spacecraft in orbit. An analysis of the MMOD threat was performed using the BUMPER II code developed at NASA Johnson Space Center [8]. BUMPER II predicts the probability of MMOD penetrating the spacecraft in orbit, which is used to populate a basic simulation element in the model. If penetration occurs, then one would assess the damage done to underlying systems to determine the criticality of the impact. If the pressure vessel was compromised, the cabin physics model could be employed to determine the time-to-LOC, or the time the spacecraft could remain pressurized such that alternate action could be taken. In the current model, any full spacecraft penetration when docked to the ISS is assumed to result in LOM if unoccupied, or LOC if occupied. Of more interest is the case where an MMOD strike does not fully penetrate but causes damage to the thermal protection system (TPS). In this case, the impact damage is passed to the EDL element, described above, and the dynamic simulation model computes the LOC probability based on the heating environment, location of the damage, TPS material response, and temperature limits of the spacecraft structure. All of this is handled within the mission model automatically.

All of these examples begin with an off-nominal situation that initiates a potentially fatal chain reaction. The estimation of LOC depends on the ability of the space transportation system to respond in a manner that a) inhibits the failure propagation so that no threatening failure environment develops, b) protects the crew in spite of the failure environment, or c) can move the crew safely away from the hazard. In the second and third cases, the evolution of the failure environment must be characterized.

Much work has been done to estimate the blast overpressure, fireball, and debris environments resulting from an explosive system failure [9,10,11]. The current risk model incorporates these hazardous environments using parameterized tables of results computed outside of the risk model. In this way, the best estimates of the hazardous environments' impact on the system are incorporated into the model and the resulting failure probabilities are computed using the state of the launch vehicle and spacecraft at the time of the failure.

A short list of the physic-based models needed to represent failure and propagation conditions encountered in a space mission risk model includes:
- LOC probability due to blast overpressure as a function of mission elapsed time (MET) and amount of abort warning time available.
- LOC probability due to debris striking the crewed cabin as a function of MET and amount of abort warning time available.
- Time to structural breakup of the launch vehicle in a loss-of-control situation, as a function of MET and failure mode (rapid versus slow control loss).
- Probability of penetration and partial penetration from MMOD as a function of TPS thickness and time in orbit.
- LOC probabilities due to TPS bondline temperature exceedance as a function of TPS thickness and TPS stack-up.
- Time-to-LOC cabin environment conditions as a function of failure severity, MET, and abort from orbit duration.

At the top of the model hierarchy shown in Figure 3 are the elements that track the failures, mitigations, and responses of the system. These top elements determine whether LOC, LOM, or any other simulation parameter of interest occurs. An attribute of simulation techniques is that multiple FOMs can be tracked within a single simulation. For example, one model can estimate LOC, LOM, failure time/risk intensity, failures that did not lead to abort, etc. These are the multiple top events shown in Figure 3. The parameter estimation is accomplished by simulating a large number of mission realizations, storing the end-state of each, and reporting the statistical results at the end of the simulation. Figure 7 shows a diagram of how the data items combine to produce these statistics.

Figure 7: Flow chard of mission risk data.

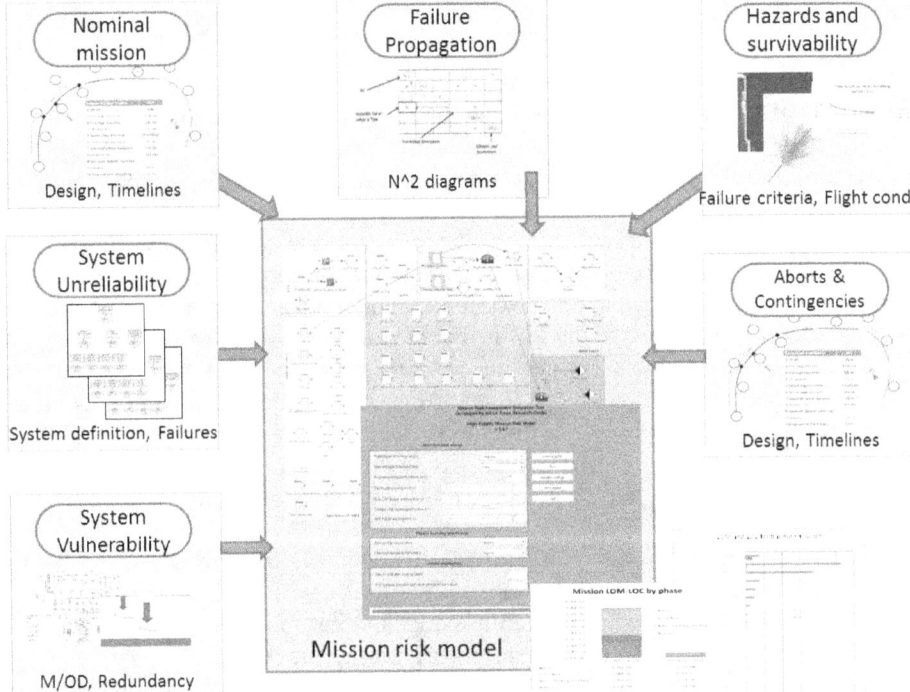

4. RESULTS

All results in this paper were run using a commercial Monte Carlo simulation code called GoldSim (www.goldsim.com/). A Monte Carlo simulation framework was chosen for its flexibility in incorporating probabilistic data from multiple sources and linking them together as event-triggered responses in a time-stepped model. The physics models were run offline in advance to generate response surface failure probability tables that are used as inputs to the simulation. A lookup table is created for each model and the tables are linked to different failure events along the mission timeline, as prescribed by the propagation maps. If a failure event is triggered during a simulation realization, the time of occurrence is captured and the appropriate value(s) are pulled from the relevant hazard lookup table(s). Each mission in a realization is "flown" to completion until a LOC event occurs. Unless otherwise noted, all simulations were run with 40,000 Monte Carlo realizations using Latin Hypercube sampling.

4.1. LOM and LOC for Baseline Design

First, consider the integrated mission LOC/LOM assessment typically associated with PSA. This type of analysis would normally be used for requirement verification. Figure 8 shows the mean LOM and LOC probability breakdowns by mission phase for a 183-day mission to ISS. The mean probability of LOM for this space system is 1 in 26 and the mean probability of LOC is 1 in 244 for a single Monte Carlo run. The stacked bar chart shows most of the LOM contribution occurring during the long orbital phase, followed by LOM contributions of ~1:80 during the ascent phase. The LOC contributions show most of the failures occurring during the ascent phase and the long orbital stay.

Figure 8: LOM and LOC estimates for generic spaceflight system mission to ISS.

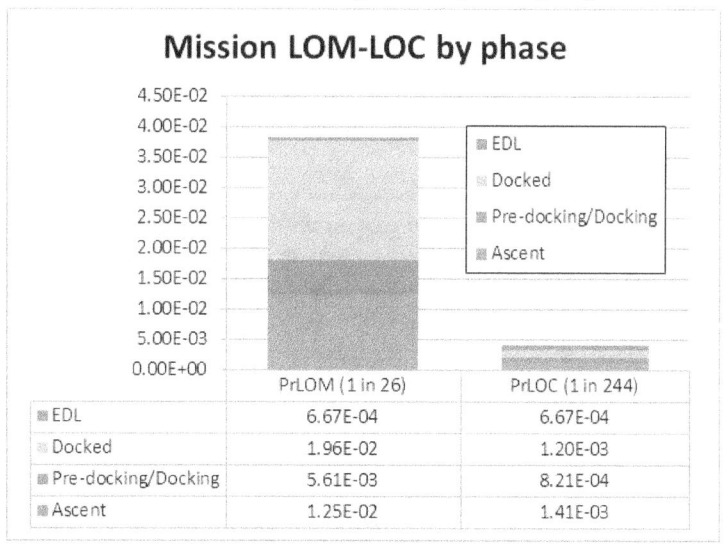

Figure 9 shows a chart of the top 20 LOM and LOC risk drivers for this example spaceflight system, sorted by descending contribution to LOC and then LOM. The driver labels are the loss of critical function "top events" in the risk model data hierarchy and are coded "[LV-xx]" for launch vehicle and "[SC-xx]" for spacecraft. Here, LOM is further differentiated as either a mission abort or a loss of vehicle (LOV). The top-20 list shows that the largest contributor to LOC comes from pressure vessel leaks and catastrophic first stage engine failures. The top LOM driver during the on-orbit phase stems from critical MMOD strikes to the spacecraft: ~1% occurrence for a 183-day stay while docked to the ISS. Since the spacecraft is largely unoccupied during this time (checkouts with crew occur weekly and for only 1 hour while docked), these critical penetration failures specifically lead to an LOV and would require a rescue mission to return the crew safely back to Earth, unless some repair capability was defined.

Figure 9: Top 20 LOM and LOC risk drivers for a generic spaceflight mission to ISS.

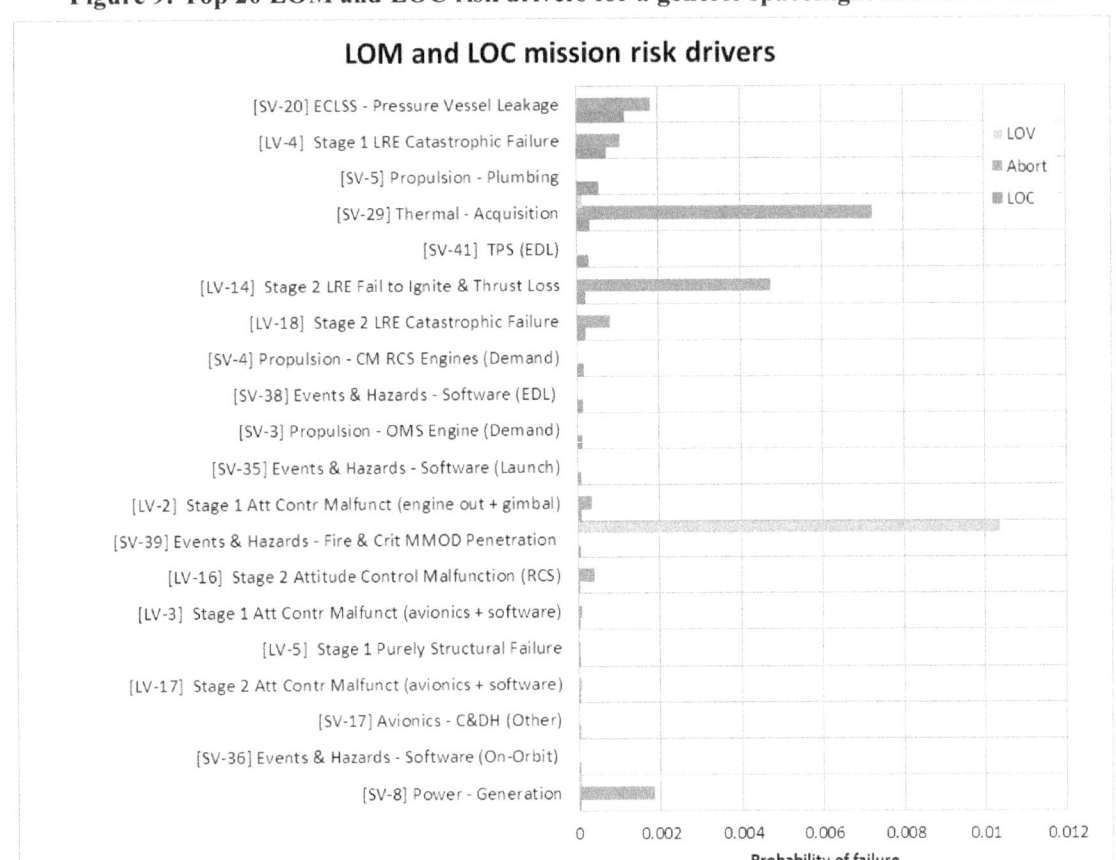

4.2. Trade Studies and Uncertainty Studies

While a LOC/LOM assessment has value, the requirement verification aspect of such assessments often occurs fairly late in a program, after the design has frozen to the point where a complete analysis can be performed. This is a bit of a catch-22, since the risk information is most impactful while the design is still progressing. A model such as the one described above, however, can provide additional insight throughout the design process. While the results will not necessarily characterize the final system to a large degree of precision, they offer tremendous value by providing context for design choices. The model's modular design allows sensitivity and trade studies to be easily performed by sampling the corresponding rows or columns in the lookup tables.

For example, consider a system trade study looking at ascent abort triggers that detect launch vehicle failures and recommend when an abort should be attempted. Suppose a large number of abort trigger options are being considered, but the system impact (e.g., cost, schedule, or complexity) is significant if the sensor suite and software must be re-engineered. Assume three options have been proposed for the abort trigger thresholds: the baseline option, a moderate option, and a sensitive option, as shown in Table 2. The baseline system has little or no abort capabilities (a large vehicle deviation rate trigger, no vehicle attitude trigger, and no early warning systems) and is compared against the two alternate abort systems.

Table 2: Ascent abort modeling assumptions used for abort trigger trade study.

	Trigger	Baseline	Moderate	Sensitive
Abort system	Launch vehicle rate trigger (deg/s)	4	3	2
	Launch vehicle attitude trigger (deg)	90	8	4
	Abort system latency (ms)	300	300	300
	First stage warning time (s)	0	1	5
	Second stage warning time (s)	0	1	5

The risk model was run with a single failure environment assumption set and the three different abort trigger options. Figure 10, shows the 90% confidence intervals for LOC generated from the Monte Carlo simulations of each abort system. An abort system with the moderate design thresholds offers a large benefit to LOC, reducing the mean LOC probability from 1 in 660 missions to 1 in 1,220 missions compared to the baseline abort system. The sensitive abort system with even more responsive thresholds further reduces the occurrence of LOC to a mean of 1 in 2,830 missions. This does not imply that such systems are desirable or even feasible, but it allows designers and decision-makers to perform what-if studies and make risk-informed decisions about the specific design threshold values that are needed to achieve a certain level of protection.

Figure 10: Abort trigger trade study ascent LOC results (5th, mean, and 95th percentile confidence bounds).

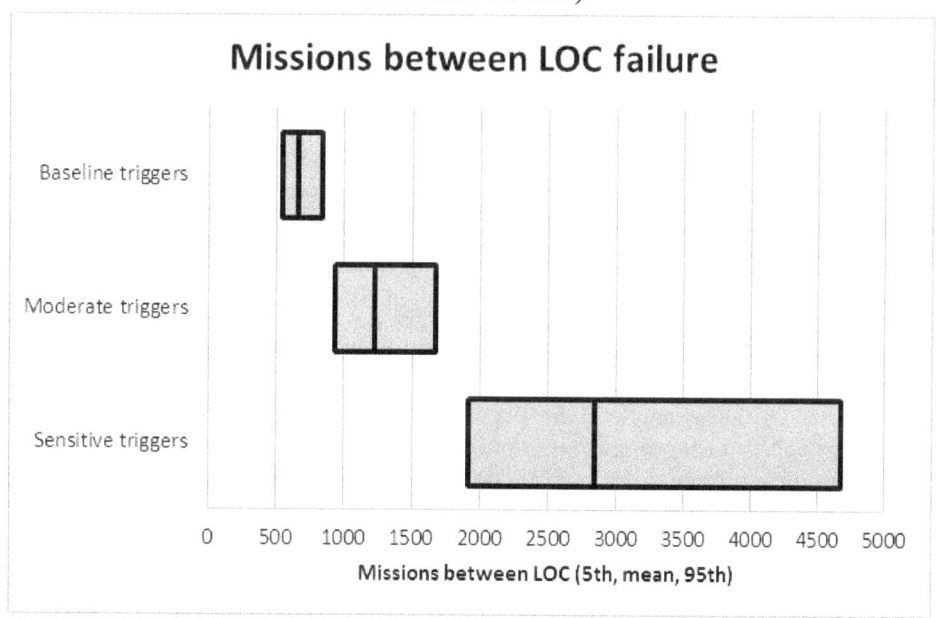

While the impact of the abort trigger options could be very significant in meeting a LOC requirement, it is also important to consider whether the results reflect the best knowledge of the system. For example, assume previous studies have shown that the assumptions about the blast and debris environments are very impactful to ascent LOC results. The project team would benefit from insight into how effective different the trigger results would be under different explosion scenario assumptions. Uncertainty distributions could be assigned to the likelihood of the range of potential blast environments, but this would yield the same mean results shown above, assuming the mean of the distribution was used for the sensitivity study. The confidence intervals would likely be different, but it is unlikely that they would characterize the differences in a way that could meaningfully impact trigger selections. Alternately, the trigger trade study could be re-assessed under specific blast assumption cases. This would provide a direct connection between the trigger results and the blast assumptions and would enable a decision based on the belief of those assumptions. This takes traditional uncertainty quantification to another level and highlights whether a result is *sensitive to the uncertainty*. In this way model uncertainty is included in the assessment, not in a combined way that masks its importance, but in a manner that explicitly quantifies the impact. In addition, this model now allows a user to understand the range of possible conditions following a failure rather than choosing one specific set of assumptions.

Incorporating the risk assessment into the design process can also produce significant value by iteratively identifying where overly conservative or insufficiently defined assumptions may need to be refined. For example, Figure 9 shows the top LOC driver to be on-orbit seal leakage. This seems intuitively wrong, so this risk driver should be examined further. It turns out that the con-ops stated that an abort would be initiated in the event of any PCS failure, so the model had the crew enter a "leaky" spacecraft and return to earth. While this may sound ridiculous, is it the result of a bad model? Maybe not, as it is not unusual to have blanket policies such as these defined as placeholders during the early design phases.‡ At this point, the hypothetical abort policy could be raised as an issue. While the LOC impact was certainly over-stated, the risk model highlighted an incomplete con-ops definition. By re-visiting and refining this assumption, the LOC became LOV, and further definition of mission operations and repair capabilities could remove these LOV even further.

Similarly, the top LOM driver is MMOD penetration on orbit. As mentioned, this is LOV, not LOC, because a penetration can be isolated and the crew protected while the spacecraft is docked. However, the model allows simple trades to reduce the potential for these LOM (and LOV) scenarios. Changing the TPS design definition by increasing the TPS insulation thickness from 1 inch to 1.5 inches reduces the probability of a LOV from 1.04e-2 ($\pm$ 8.33e-4 at 90% confidence) to a probability of LOV 4.08e-3 ($\pm$ 5.24e-4 at 90% confidence).

5. CONCLUSION

A multi-disciplinary risk modeling approach was developed to merge the best information from each engineering field into a cohesive view of a space transportation system. A generic space architecture and mission were used to demonstrate how such a modeling approach can provide valuable risk-based insights during the design cycle rather than just at the end.

For the risk model to impact the design, including con-ops, it must tie to meaningful design parameters and be responsive to the design process. It is suggested that the real value of PSA for space system design is not as much from the LOC/LOM verification but from the insight it can provide throughout the development process. The example in this paper illustrated aspect of dynamics modeling, inclusion of physics, trade support, model uncertainty, and using the PSA to drive discussions that would be easy to discount and delete from a risk model. The example also showed how a model can highlight a gap, not necessarily in the design or data, but in the operation assumptions themselves. Assumptions causing high variations in the outputs of interest merit further exploration and refinement of the underlying inputs, while assumptions that do not drive the key results may be suitable to use as-is. As design data matures, an iterative feedback loop between designers and risk analysts can be established early, when trade studies are still being performed and hold the most promise of being implemented. In order for such value to be realized, the model needs to be responsive so that improved data, assumptions, policy, or design features can be readily incorporated. If the analysis cannot be updated frequently, then such things as the seal leaks causing LOC will either encourage the design team to discount the PSA altogether, or even worse, put design resources into fixing a non-problem. Responsiveness is a key aspect of the modeling approach presented in this paper.

In these ways, the risk model's responsiveness and ability to represent meaningful parameters enable it to provide actionable, risk-informed design data. In addition to simply verifying requirements, the risk model provides an intelligent roadmap for a spaceflight program to plan future analysis studies and optimize design improvements throughout the system development lifecycle.

---

‡ The architecture used for this study, as mentioned, was conceptual only and the abort policies were defined such that an abort was triggered on a loss of redundancy or function.

Acknowledgements

The authors wish to thank Samira Motiwala for developing the design and reliability data for the generic space launch system and spacecraft used in this paper.

References

[1] *"Probabilistic Risk Assessment Procedures Guide for NASA Managers and Practitioners,"* NASA/SP-2011-3421, Second Edition, December 2011.
[2] S. Go, D. L. Mathias, H. Nejad. *"Human Space Mission Architecture Risk Analysis,"* Reliability and Maintainability Symposium (RAMS), Orlando, FL, January, 2013.
[3] "Commercial Crew Program." NASA Facts. National Aeronautics and Space Administration, 2011. <http://www.nasa.gov/pdf/609181main_12.08.11_CCP.pdf>.
[4] S. A. Motiwala and D. L. Mathias. *"Conceptual Launch Vehicle and Spacecraft Design for Risk Assessment,"* NASA USRP – Internship Final Report, December 2013.
[5] K. Gee, K., L. Hunyh and T. Manning. *"Physics-based Entry, Descent and Landing Risk Model,"* Probabilistic Safety Assessment and Management (PSAM12) Conference, Honolulu, HI, June, 2014.
[6] Lawrence, S. L., Mathias, D.L., and Gee, K.,*"A Failure Propagation Modeling Method for Launch Vehicle Safety Assessment,"* 12th International Conference on Probabilistic Safety and Management (PSAM12), Honolulu, HI, June 2014.
[7] C. J. Mattenberger and D. L. Mathias. *"Cabin Environment Physics Risk Model,"* Probabilistic Safety Assessment and Management (PSAM12) Conference, Honolulu, HI, June, 2014.
[8] E. L. Christiansen, et.al. *"Handbook for Designing MMOD Protection,"* JSC-64399 Version: A, January 28, 2009.
[9] S. L. Lawrence, S., D. L. Mathias, K. Gee and M. Olsen. *"Simulation Assisted Risk Assessment: Blast Overpressure Modeling,"* PSAM8-0197, Probabilistic Safety Analysis and Maintenance #8, New Orleans, LA, May, 2007.
[10] S. L. Lawrence and D. L. Mathias. *"Blast Overpressure Modeling Enhancements for Application to Risk-Informed Design of Human Space Flight Launch Abort Systems,"* RAMS 06B-3, 2008 Reliability and Maintainability Symposium, Las Vegas, NV, January, 2008.
[11] K. Gee and S. L. Lawrence. *"Launch Vehicle Debris Models and Crew Vehicle Ascent Abort Risk,"* Reliability and Maintainability Symposium (RAMS), Orlando, FL, January, 2013.

Apportioning Transient Combustible Fire Frequency via Areal Factors: More Complicated Than It May Seem

Raymond H.V. Gallucci, Ph.D., P.E.
U.S. Nuclear Regulatory Commission (USNRC), MS O-10C15, Washington, D.C. 20555
Ray.Gallucci@nrc.gov

Abstract: Apportioning the frequency of transient combustible fires to vary <u>within</u> a physical analysis unit for a fire probabilistic risk assessment (PRA) has been discussed and attempted by various analysts to date with limited success. The technique presented here illustrates the complexity involved in such a calculation, considering the constraints on preserving transient fire ignition frequencies within the Fire PRA, which may lend insight into why this has proven a difficult process. While the approach offered can be used, the goal is more to provide "food for thought" that may lead to a more straightforward, even if approximate, technique that would reasonably represent the reality of the situation without being overly complex.

Key Words: Fire Frequency, Transient Combustibles, Apportionment, Fire PRA

1. INTRODUCTION[1]

NUREG/CR-6850 / EPRI (Electric Power Research Institute) 1011989, as enhanced by National Fire Protection Association (NFPA) 805 Frequently Asked Question (FAQ) 12-0064 [2], addresses "ignition source weighting factors" for transient combustibles located in the "physical analysis units (PAUs)" of a nuclear power plant to guide the allocation of transient fire ignition frequencies for Fire PRAs. Implicit in this modeling is an assumption that the weighting, or "influence," factors remain the same throughout the PAU, which is typically a "spatial subdivision of the plant ... generally defined in terms of fire areas and/or fire compartments," such as a fully enclosed room. [3] In the process of applying this guidance for Fire PRAs, especially ones developed to support transitions to 10CFR50.48(c), analysts have sought ways to account for variations within certain PAUs to represent particular sub-locations where reduced fire ignition frequency due to stricter transient combustible controls than generally present in the rest of the PAU can be credited, e.g., G. Zucal and R. White, "The Influence of Spatial Geometry on Transient Fire Likelihood." [4] However, due to the need to preserve the PAU frequency regardless of any redistribution within the PAU itself, this author shows that the process can be quite complicated, as demonstrated through an illustrative example.

2. ANALYSIS

Assume a plant has a total area of $10^2 = 100$ square units, divided into the seven PAUs shown in Figure 1. Each PAU is rated "average" for transient combustible fires for all four influence factors, i.e., 3 each for Hot Work Maintenance, Electrical/Mechanical (E/M) Maintenance, Occupancy and Storage. [1-2] Thus, for Hot Work maintenance, the total plant hot work ignition frequency, λ_{hw}, is apportioned equally among all seven PAUs, i.e., $\lambda_{hw,PAU-x} = \lambda_{hw}/7$. Likewise for E/M Maintenance, $\lambda_{e/m,PAU-x} = \lambda_{e/m}/7$. The fact that the PAUs have different areas does not affect the apportioning of the transient combustible fire frequency.

---

[1] This paper was prepared by an employee of the U.S. NRC. The views presented do not represent an official staff position.

Figure 1: Layout for Physical Analysis Units

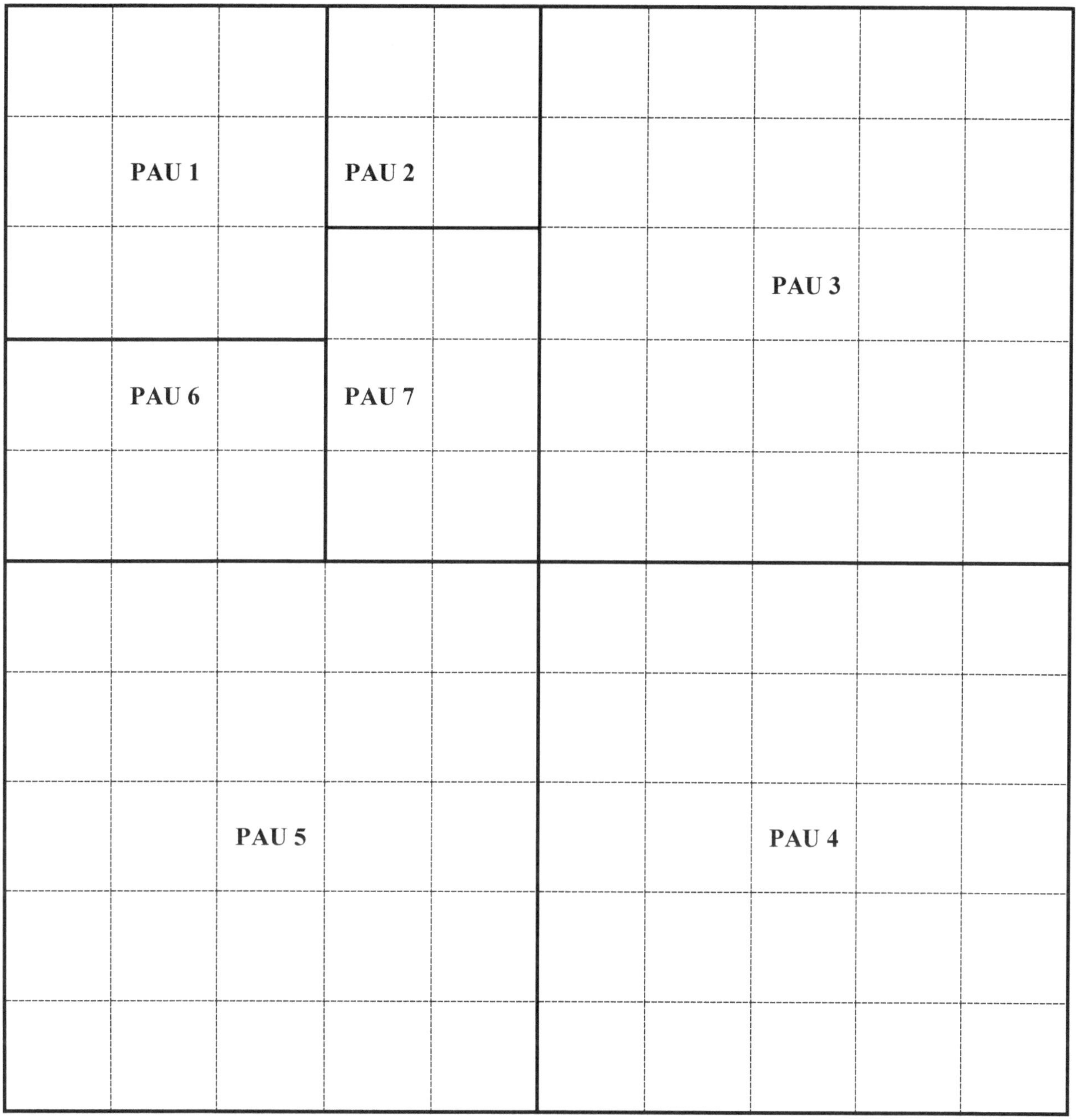

Now, assume the only zones of influence (ZOIs) where fire damage can occur to risk-significant components at the plant are shown by the two single-hatched areas along the top and left side of the layout, with their intersection occurring only in one location in PAU 1 shown by the cross-hatching (Figure 2). The single-hatched areas could represent each of two redundant trains, with an associated conditional core damage probability (CCDP) = C_1. The cross-hatched area then represents the "pinch point" where both trains can be damaged, with CCDP = $C_2 > C_1$. We need to apportion the transient combustible fire frequencies such that they correspond to pairings of ZOI and CCDP, i.e., into three pairings: (1) no ZOI (no risk-significant components), shown non-hatched; (2) ZOI for single-train, i.e., pairing with CCDP = C_1 (single-hatch); (3) ZOI for "pinch point," i.e., pairing with CCDP = C_2 (cross-hatch).

Figure 2: Layout for Physical Analysis Units Showing Sub-locations where Potentially Risk Significant Fire Damage Can Occur

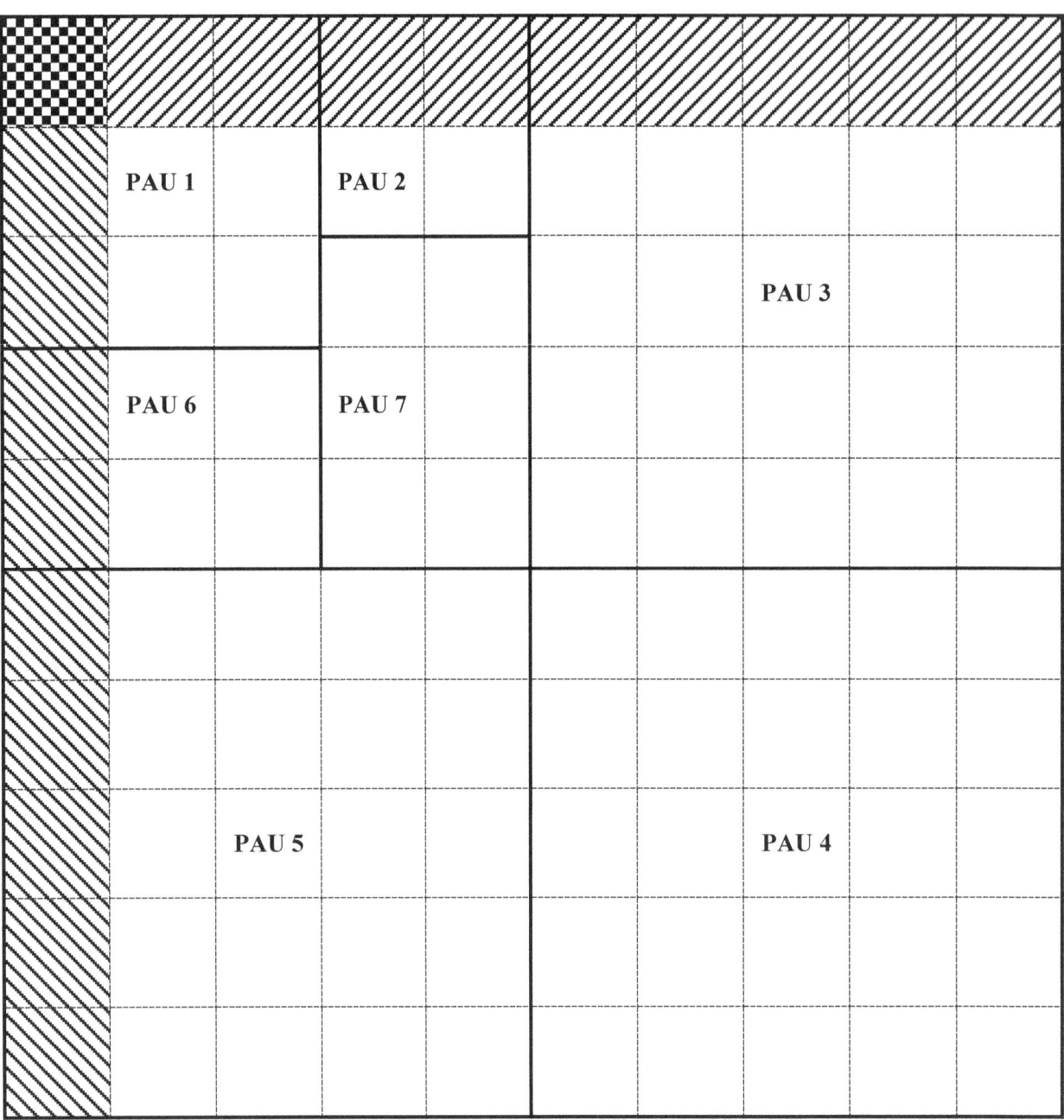

Where a PAU contains no ZOI (PAUs 4 and 7), there is no risk contribution, ($CDF_{PAU-4} = CDF_{PAU-7} = 0$), since there is no risk-significant component ZOI present. Transient combustible fires occurring here cannot contribute to risk. For the remaining PAUs, containing either or both a ZOI paired with C_1 or C_2, the apportioning of the transient combustible fire frequency within each depends on the likelihood of the fire being located within vs. without the ZOI.

<u>Simple Case</u>

The simplest case occurs if the fire is equally likely to be located anywhere within the PAU. For all but PAU 1, the corresponding CDFs are as follows (ignoring suppression, etc., i.e., considering only ignition and CCDP), dropping the subscript for Hot Work or E/M maintenance since the formulas would be the same for each:

$$CDF_{PAU-3} = CDF_{PAU-5} = (\lambda_{PAU-3}/5) \times C_1 = (\lambda_{PAU-5}/5) \times C_1 = (\lambda/7)/5 \times C_1 = \lambda/35 \times C_1, \text{ since the ZOI covers 1/5 of each PAU.} \tag{1}$$

$$CDF_{PAU-2} = (\lambda_{PAU-2}/2) \times C_1 = (\lambda/7)/2 \times C_1 = \lambda/14 \times C_1, \text{ since the ZOI covers 1/2 of PAU 2.} \tag{2}$$

$$CDF_{PAU-6} = (\lambda_{PAU-6}/3) \times C_1 = (\lambda/7)/3 \times C_1 = \lambda/21 \times C_1, \text{ since the ZOI covers 1/3 of PAU 6.} \tag{3}$$

For PAU 1, the analysis has an additional complication in that, within the ZOI, there are different CCDPs: $C_1 < C_2$. Therefore,

$$CDF_{PAU-1} = (\lambda_{PAU-1} \times 4/9) \times C_1 + (\lambda_{PAU-1}/9) \times C_2 = \lambda_{PAU-1}/9 \times (4 \times C_1 + C_2) = (\lambda/7)/9 \times (4 \times C_1 + C_2) = \lambda/63 \times (4 \times C_1 + C_2), \text{ since the ZOI for a either single train covers 4/9 of PAU 1, while the ZOI for both trains ("pinch point") covers 1/9 of PAU 1.} \tag{4}$$

Another way of viewing this, which will prove convenient, is to assign the probabilities of the transient combustible fire being located in each "unit square" of the plant as shown below. One can then express these CDFs as follows:

$$CDF_{PAU} = \lambda/7 \times \Sigma_{ZOI} (\pi [P_u \times C_u]) \tag{5}$$

Where Σ = sum, π = product, P_u = probability of transient combustible fire in unit square, C_u = CCDP for transient combustible fire in unit square.

The previous equations are recalculated below to give the same results (see Figure 3):

$$CDF_{PAU-3} = CDF_{PAU-5} = \lambda/7 \times (5 \times 1/25 \times C_1) = \lambda/35 \times C_1. \tag{6}$$

$$CDF_{PAU-2} = \lambda/7 \times (2 \times 1/4 \times C_1) = \lambda/14 \times C_1. \tag{7}$$

$$CDF_{PAU-6} = \lambda/7 \times (2 \times 1/6 \times C_1) = \lambda/21 \times C_1. \tag{8}$$

$$CDF_{PAU-1} = \lambda/7 \times (4 \times 1/9 \times C_1 + 1/9 \times C_2) = \lambda/63 \times (4 \times C_1 + C_2). \tag{9}$$

Adding Complexity

Let's consider some additional complexity. Assume that, in PAU 2, 3, 6 or 7 (PAU 1 is more complicated, and will be addressed subsequently), the analyst credits an effort to limit the potential for transient combustible fires within the risk-significant ZOI, perhaps by erecting at least a gated, cross-link fence around the ZOI with "combustible exclusion zone" signage. Further, assume as the result of this modification, the analyst can justify the likelihood of a transient combustible fire within the ZOI now being only 1/10 of that outside the ZOI within the PAU. That is,

Probability of transient fire within ZOI (P_{in}) = Probability of transient fire outside ZOI (P_{out})/10.

Total probability of transient fire within PAU = $P_{in} + P_{out} = P_{out}/10 + P_{out} = P_{out} \times 11/10 = 1$.
Therefore, $P_{out} = 10/11$ and $P_{in} = 1/11$.

Figure 3: Layout for Physical Analysis Units Showing Equal Likelihoods among Sub-locations where Potentially Risk Significant Fire Damage Can Occur

1/9	1/9	1/9	1/4	1/4	1/25	1/25	1/25	1/25	1/25
1/9	1/9	1/9	1/4	1/4	1/25	1/25	1/25	1/25	1/25
1/9	1/9	1/9	1/6	1/6	1/25	1/25	1/25	1/25	1/25
1/6	1/6	1/6	1/6	1/6	1/25	1/25	1/25	1/25	1/25
1/6	1/6	1/6	1/6	1/6	1/25	1/25	1/25	1/25	1/25
1/25	1/25	1/25	1/25	1/25	1/25	1/25	1/25	1/25	1/25
1/25	1/25	1/25	1/25	1/25	1/25	1/25	1/25	1/25	1/25
1/25	1/25	1/25	1/25	1/25	1/25	1/25	1/25	1/25	1/25
1/25	1/25	1/25	1/25	1/25	1/25	1/25	1/25	1/25	1/25
1/25	1/25	1/25	1/25	1/25	1/25	1/25	1/25	1/25	1/25

These probabilities are then apportioned among the unit squares, which effectively reduces the CDF within the PAU, readily observable as shown below (e.g., in PAU 3, there are five unit squares within the ZOI whose total probability = 1/11, so each receives a 1/5 contribution, or 1/55; similarly for the 20 unit squares outside the ZOI, whose total probability = 10/11, each receives a 1/20 contribution, or 1/22) (see Figure 4):

Figure 4: Layout for Physical Analysis Units Showing Physically-Limited Access to Sub-locations where Potentially Risk Significant Fire Damage Can Occur

1/55	1/55	1/55	1/22	1/22	1/55	1/55	1/55	1/55	1/55
1/55	5/22	5/22	5/11	5/11	1/22	1/22	1/22	1/22	1/22
1/55	5/22	5/22	1/6	1/6	1/22	1/22	1/22	1/22	1/22
1/22	5/22	5/22	1/6	1/6	1/22	1/22	1/22	1/22	1/22
1/22	5/22	5/22	1/6	1/6	1/22	1/22	1/22	1/22	1/22
1/55	1/22	1/22	1/22	1/22	1/25	1/25	1/25	1/25	1/25
1/55	1/22	1/22	1/22	1/22	1/25	1/25	1/25	1/25	1/25
1/55	1/22	1/22	1/22	1/22	1/25	1/25	1/25	1/25	1/25
1/55	1/22	1/22	1/22	1/22	1/25	1/25	1/25	1/25	1/25
1/55	1/22	1/22	1/22	1/22	1/25	1/25	1/25	1/25	1/25

As is evident, the hatched squares (ZOI) now have reduced probabilities for transient fires while the unhatched (no ZOI, excluding PAUs 4 and 7) have increased probabilities. Nonetheless, the sum of all probabilities within each PAU equals 1, as it must since we are still requiring the transient combustible fire to be located with the PAU.

Using our summation technique, the reduced CDFs in each PAU are readily calculated, as follows:

$$CDF_{PAU-3} = CDF_{PAU-5} = \lambda/7 \times (5 \times 1/55 \times C_1) = \lambda/77 \times C_1. \tag{10}$$

$$CDF_{PAU-2} = CDF_{PAU-6} = \lambda/7 \times (2 \times 1/22 \times C_1) = \lambda/77 \times C_1. \tag{11}$$

$$CDF_{PAU-1} = \lambda/7 \times (4 \times 1/55 \times C_1 + 1/55 \times C_2) = \lambda/385 \times (4 \times C_1 + C_2). \tag{12}$$

These results may not, at first, seem intuitively obvious, as one might have expected each of the previously calculated PAU CDFs to decrease by a factor of 11. In fact, each decreases by a factor equal to 11 divided by the ratio of the total number of unit squares to the number just for the ZOI (i.e., the ratio of the total PAU area to the area covered only by the ZOI). The latter ratios for PAUs 1, 2, 3, 5 and 6 were originally 9/5, 4/2 = 2, 25/5 =5, 25/5 =5, and 6/2 = 3, respectively. As a result, we see the CDF in each of these PAUs decreasing by factors of 11/(9/5) = 6.11, 11/2 = 5.5, 11/5 = 2.2, 11/5 = 2.2, and 11/3 = 3.67, respectively. This is evident from the denominators in the three equations above. For PAU 1, the denominator has increased from 63 to 385, a factor of 6.11; for PAU 2, from 14 to 77, a factor of 5.5; for both PAUs 3 and 5, from 35 to 77, a factor of 2.2; for PAU 6, from 21 to 77, a factor of 3.67.

Most Complex

As a final level of complexity, assume that within the gated fence in PAU 1, the analyst takes additional measures to ensure no combustible fires within the pinch-point ZOI (e.g., maybe a chain link with additional signage), such that the likelihood of a transient combustible fire is justified to be further reduced by a factor of 5. The analysis extends from the previous as follows:

Probability of transient fire within pinch-point zone ($P_{in,pinch}$) = Probability of transient fire in single-train zone (P'_{in})/5.

Probability of transient fire in single train zone (P'_{in}) = Probability of transient fire outside ZOI (P'_{out})/10.

Given the transient fire occurs within the ZOI, $P_{in,pinch} + P'_{in} + P'_{out} = P'_{in}/5 + P'_{out}/10 + P'_{out} = (P'_{out}/10)/5 + P'_{out}/10 + P'_{out} = P'_{out} \times 56/50 = 1$. Therefore, $P'_{out} = 50/56 = 25/28$, $P'_{in} = 5/56$ and $P_{in,pinch} = 1/56$.

As above, these probabilities are then apportioned among the unit squares, which effectively further reduces the CDF within PAU-1, readily observable in Figure 5 [for PAU 1 only; the "pinch point" unit square receives the full probability of 1/56; each of the four single-train ZOI unit squares receives a probability of (5/56)/4 = 5/224; each of the four non-ZOI unit squares receives a probability of (25/28)/4 = 25/112]:

Again using our summation technique, the new CDF in PAU-1 is readily calculated, as follows:

$$CDF_{PAU-1} = \lambda/7 \times (4 \times 5/224 \times C_1 + 1/56 \times C_2) = \lambda/7 \times (5/56 \times C_1 + C_2/56) = \lambda/392 \times (5 \times C_1 + C_2). \tag{13}$$

Depending how much C_2 exceeds C_1, this CDF may or may not be lower than the previous, even though we are further restricting the presence of transient combustible fires within the "pinch point." This is a consequence of the reallocation of likelihood of transient combustible fire location throughout the PAU. In fact, there will not be a decrease until $C_2 > 50 \times C_1$, as shown here. Assuming $C_2 = \alpha \times C_1$:

$$CDF_{PAU-1} \text{ (1:10 ratio)} = \lambda/385 \times (4 \times C_1 + C_2) = \lambda/385 \times (4 \times C_1 + \alpha \times C_1) = \lambda/385 \times (4 + \alpha) \times C_1 \tag{14}$$

$$\text{CDF}_{\text{PAU-1}} \text{ (1:5:50 ratio)} = \lambda/392 \times (5 \times C_1 + C_2) = \lambda/392 \times (5 \times C_1 + \alpha \times C_1) = \lambda/392 \times (5 + \alpha) \times C_1. \tag{15}$$

$\text{CDF}_{\text{PAU-1}}$ (1:5:50 ratio) < $\text{CDF}_{\text{PAU-1}}$ (1:10 ratio) when $\lambda/392 \times (5 + \alpha) \times C_1 < \lambda/385 \times (4 + \alpha) \times C_1$,
i.e., $\alpha > 51$
(16)

Figure 5: Layout for Physical Analysis Unit 1 Only, Showing Additional Physically-Limited Access to Sub-locations where Potentially Risk Significant Fire Damage Can Occur

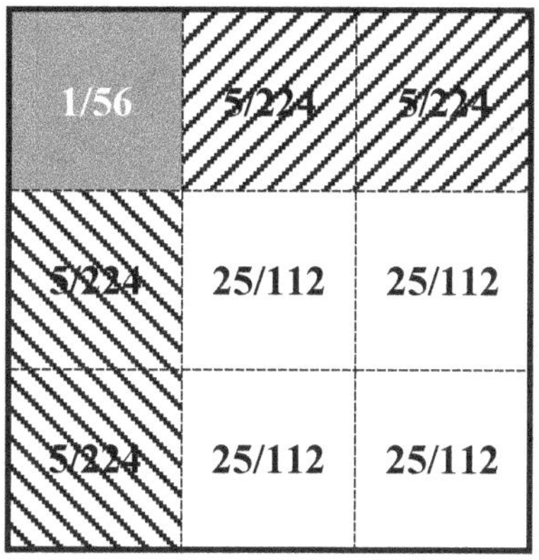

3. SUMMARY

As demonstrated here, the technique by which an analyst may wish to vary transient combustible fire frequency within a PAU to account for physical limitations on access, etc., to reduce the potential for fire damage to risk-significant components can be quite complex. The example presented offers one approach to accommodate this, albeit it is not intuitively obvious or calculationally simple. Nonetheless, it is offered as a starting point toward developing a technique that may involve approximations, but ones which are sufficiently representative of reality to become practical tools.

References

[1] USNRC/EPRI, *EPRI/NRC-RES (Office of Nuclear Regulatory Research) Fire PRA Methodology for Nuclear Power Facilities*, NUREG/CR-6850 / EPRI 1011989 (2005).
[2] USNRC, "Close-out of NFPA Standard 805 FAQ 12-0064, 'Hot Work/Transient Fire Frequency Influence Factors'" (2013), ADAMS (Agency-wide Documents Access and Management System) Accession No. ML12346A488.
[3] 10CFR50.48, "Fire Protection," Federal Register.
[4] G. Zucal and R. White, "The Influence of Spatial Geometry on Transient Fire Likelihood," *ANS PSA Topical Meeting on Probabilistic Safety Assessment and Analysis*, September 22-26, 2013, Columbia, SC.

Characterizing Fire PRA Quantitative Models: An Evaluation of the Implications of Fire PRA Conservatisms

M.B. Saunders[*], E.T. Burns
ERIN Engineering and Research, Inc., Walnut Creek, California, USA

Abstract: Conservative bias may be present in fire PRAs due to limitations in data or methodologies. An evaluation was performed to characterize the current situation with fire PRA models and the implications regarding perceived risk associated with the degree of conservative bias. The principal areas of the fire PRA data and modeling that may be subject to such biases were identified and the impacts these biases have on the reported point estimate CDF and the contributors were quantified. These biases were assessed using a number of sensitivity studies where in a set of modeling approaches or assumptions were varied from the NUREG/CR-6850 guidance that are considered to be conservatively biased. Three point estimates were developed using NUREG/CR-6850 guidance and by incrementally removing biases by crediting more realistic approaches supported in part by revised guidance or in-progress industry and NRC efforts. The conclusion from the evaluation is that reasonable (realistic) approaches to the assessment of the fire hazard will result in a reduced estimate of the fire risk, will likely change the primary risk insights, and could greatly influence the priority that is assigned to possible plant changes resulting from a re-characterization of the causes of risk significant fires and fire zones.

Keywords: Probabilistic Risk Assessment, PRA, Fire PRA.

1. INTRODUCTION

The development of external hazard Probabilistic Risk Assessment (PRA) models has generally lagged behind the development of internal events PRAs. This situation has been exacerbated because of a lack of adequate data and methods to allow a reasonably accurate quantification, i.e., to provide a quantification that is not subject to very large uncertainty bounds. The Nuclear Regulatory Commission (NRC) PRA Policy Statement states that "PRA evaluations in support of regulatory decisions should be as realistic as practicable..." [1]. During the maturation of fire PRAs, the NRC and the Electric Power Research Institute (EPRI) have established a set of data, rules, and methods that are agreed to be bounding for use in PRA applications. This set of data, rules, and methods have been codified in NUREG/CR-6850/EPRI 1011989 [2] (heretofore referred to as NUREG/CR-6850). Although the original intent was for the NUREG/CR-6850 methods to be piloted and revised prior to implementation in the industry as a whole, this was not fully achieved [3]. The resulting premature adoption of NUREG/CR-6850 has introduced a significant number of potentially conservative or bounding biases being incorporated in the data, rules, and methods. These biases generally result from a lack of knowledge. It is noted that the industry and NRC continue to make progress in developing the fire PRA methods; however, this work is not reflected in most current fire PRAs that are being used for risk informed decision-making. Given that the primary inputs to a risk-informed decision are the insights gleaned from the PRA, not a bottom line number, an undue conservative bias could confound good decision-making.

An evaluation was performed to characterize the current situation with fire PRA models and the implications regarding the degree conservative bias that may be present because of limitations in data or methodologies. For the purposes of this evaluation, the approach was to identify the principal areas of the fire PRA data and modeling that may be subject to such biases and to attempt to quantify the impact these biases have on the reported point estimate Core Damage Frequency (CDF) and the dominant contributors.

---
[*] mbsaunders@erineng.com

Example areas of potential bias include the treatment of fire initiating event frequencies, transient combustible fires, electrical cabinet fires, spurious operations, and fire effects on operator actions. Potential areas for investigation were identified and assessed using a number of sensitivity studies where a set of modeling approaches or assumptions were varied from the NUREG/CR-6850 guidance that are considered to be conservatively biased.

2. METHODOLOGY

In the effort to quantify the impact that key assumptions may have on the reported point estimate CDF and the dominant risk contributors, those areas in fire PRAs that generally contribute the most to the fire risk and have potential conservative biases were identified. These areas are categorized into types.

The first type (Type 1) includes areas of the fire PRA that have matured since the guidance in NUREG/CR-6850 was adopted into fire PRAs that are currently being used for decision making. There are three processes for which new or revised guidance has been provided. The first process is the National Fire Protection Association (NFPA) Standard 805 Frequently Asked Questions (FAQ) Program which is a mechanism to provide clarifications or enhancements related to fire PRA methods and applications [4]. The second process is an EPRI led Fire PRA Methods Review Panel. Lastly, additional testing performed by the NRC and industry has been documented in a series of subsequent NUREGs. Through these three processes several clarifications and enhancements have provided additional guidance for use in fire PRAs. Four areas of the fire PRA that may have applied NUREG/CR-6850 guidance were identified where potential conservative bias may be removed by applying additional guidance provided through one of the three processes identified above. These include fire ignition frequencies, transient administrative controls, circuit failure likelihood probability and spurious operation duration probability, and treatment of electrical cabinet fires.

The second type (Type 2) includes areas of the fire PRA where application of NUREG/CR-6850 guidance is believed to result in conservative biases; however, to date there is no supplemental guidance. Testing may be in progress to support revised guidance for these areas; however, the testing is in the early stages and conclusions cannot be made to date. Two areas of the fire PRA that may have applied NUREG/CR-6850 guidance were identified where potential conservative biases are believed to exist. These include fire ignition source heat release rates and fire growth and suppression modeling.

The third type (Type 3) includes areas of the fire PRA where assumptions are made that may introduce potential conservative biases into the fire PRA because of resource constraints, lack of knowledge, or a particular area has yet to be explored in detail. Four areas of the fire PRA where assumptions may introduce conservative biases were identified. These include fire induced initiating events, fire human reliability analysis (HRA), recovery of fire induced loss of offsite power, and mitigation of adverse environmental conditions in the Main Control Room (MCR).

To explore the individual and overall extent of conservative bias that may exist in these areas of the fire PRA, a sensitivity study is performed for each identified area. Three point estimate CDFs were developed to quantify the impact that conservative biases may have on the reported point estimate CDF and the corresponding dominant contributors used in decision making. An "upper bound" point estimate was developed using the guidance in NUREG/CR-6850. A "nominal" point estimate was developed using the guidance in NUREG/CR-6850 supplemented by clarifications and enhancements from FAQs, the industry review panel, and subsequent NUREGs. The "nominal" point estimate includes modifications to the four areas identified above categorized as Type 1. A "lower bound" point estimate was developed that includes modifications to each of the areas discussed above. Comparison of the three point estimates were made to identify the extent conservative biases may influence decision making.

Fire PRAs were selected for use from three operating U.S. nuclear power plants. These plants were selected because the plants are of differing designs such that certain areas of the fire PRA may influence the plant risk in substantially different ways. Therefore, each area explored can be quantified and the extent to which a conservative bias may influence fire PRAs can be more broadly investigated.

It should be noted that the introduction of multiple, significant conservative biases may not only obscure the important insights from the study, but the cumulative impact of these biases can lead to computed core damage frequencies that are overstated, in some cases even approaching or exceeding the subsidiary safety goals for total CDF. Such a result may mask areas where more realistically determined risk contributors could benefit from increased scrutiny.

3. SENSITVITY STUDIES

3.1. Fire Ignition Frequency

Fire ignition frequency (FIF) is an area of the fire PRA categorized as Type 1 where industry efforts have resulted in updated FIFs which are documented in EPRI 1016735 [5]. The NUREG/CR-6850 FIFs are based on a review of all of the industry fire events from 1968 through 2000. In EPRI 1016735 the FIFs were revised based on the trend of fire events in the industry and include industry fire events from 1991 through 2000. The NRC position on the use of the updated FIFs is documented in FAQ 08-0048 which is included in Supplement 1 of NUREG/CR-6850 [4]. The interim NRC position is that the NRC accepts the use of the revised FIFs with the provision that the sensitivity of the risk must also be evaluated using the NUREG/CR-6850 FIFs. Any situations in which the sensitivity of the FIFs changes the risk significance of elements of the fire PRA must be addressed.

An industry and NRC effort has been in progress to update fire FIFs based on a more complete data set and a re-classification of fire events for the years 1990 through 2009. Nevertheless, current fire PRA results used in decision-making are based on generic FIFs from NUREG/CR-6850 and EPRI 1016735. Table 1 presents the sensitivity of fire risk to FIFs used in decision-making when EPRI 1016735 FIFs were used instead of NUREG/CR-6850 FIFs. Approximately a 20-25% decrease in fire induced CDF is calculated when the revised EPRI 1016735 FIFs are applied.

Table 1: Fire PRA Sensitivity to Fire Ignition Frequency

Fire PRA	Change in CDF
Plant 1	-21%
Plant 2	-25%
Plant 3	-21%

3.2. Transient Combustible and Ignition Source Administrative Controls

The apportioning of transient FIF is an area of the fire PRA categorized as Type 1 where industry efforts have resulted in updated guidance. NUREG/CR-6850 guidance includes the use of weighting factors to apportion the FIFs to plant areas. Fire PRAs developed for the NFPA 805 application identified the need for a process that gives more credit to plant areas that have stricter transient controls in place. Through NFPA 805 RAIs, the NRC found the deviations from NUREG/CR-6850 unacceptable. However, with the understanding that the guidance in NUREG/CR-6850 had been interpreted differently and had limitations, the NRC proposed additional guidance and refinements to the NUREG/CR-6850 guidance to address the application of transient weighting factors in FAQ 12-0064 [6]. The guidance in FAQ 12-0064 did address some of the deviations; however, others were still ultimately found unacceptable. For example, the guidance in NUREG/CR-6850 and FAQ 12-0064 only address transient weighting factors on an area basis. Specifically, when additional credit was provided for individual postulated transients where transient controls would be stricter but do not apply to the area as a whole, the NRC found the deviation unacceptable.

The fire PRA results being used for decision-making generally do not include or fully incorporate the guidance included in FAQ 12-0064. Additionally, processes that refine the apportioning of transient FIFs continue to be explored and presented at industry conferences. Table 2 presents the results of a sensitivity study exploring the potential benefit of additional credit for administrative controls using the guidance in FAQ 12-0064 and considering additional potential refinements to the treatment of administrative controls. Additional transient administrative control credit may reduce the fire induced CDF between 3% and 11% based on the sample plants examined.

Table 2: Fire PRA Sensitivity to Transient Administrative Control Credit

Fire PRA	Change in CDF
Plant 1	-3%
Plant 2	-11%
Plant 3	-7%

3.3. Circuit Failure Likelihood and Duration Probability

Circuit failure likelihood and duration probabilities are areas of the fire PRA categorized as Type 1 where industry efforts have resulted in updated guidance even though some of the guidance may be in draft form. NUREG/CR-6850 provides guidance for applying circuit failure likelihood probabilities. NUREG/CR-6850 guidance includes probabilities for circuits with and without a control power transformer (CPT). The guidance includes a reduction of a factor of two for circuits with a CPT. This guidance was generally applied in fire PRAs. Subsequent testing in a joint effort by the NRC and EPRI resulted in the conclusion that the reduction of a factor of two does not accurately reflect the test data. The conclusion is based on a Phenomena Identification and Ranking Table (PIRT) panel and is documented in NUREG/CR-7150 [7]. In addition, the PIRT panel is in process of developing new circuit failure likelihood probabilities and spurious operation durations based on the test data. The preliminary results were presented at the 2013 NEI Fire Protection Forum and the PSA conference.

In decision-making, the NRC issued interim guidance in June 2013 that did not allow the use of the NUREG/CR-6850 circuit failure likelihood failure probabilities for circuits with CPT [8]. However, the interim guidance did not include spurious operation probabilities for motor operated valves (MOVs) or spurious operation duration probabilities. Therefore, fire PRA results being used in decision-making do not include circuit failure likelihood probabilities for MOVs and spurious operation duration probabilities have not been included. Table 3 presents the results of a sensitivity study that explores the potential reduction when spurious operation and duration probabilities are applied. These results indicate at least a 22% reduction in fire induced CDF was identified.

Table 3: Fire PRA Sensitivity to Spurious Operation Probability and Duration

Fire PRA	Change in CDF
Plant 1	-41%
Plant 2	-22%
Plant 3	-30%

3.4. Alignment Factor for Electrical Cabinet Fires

ERIN Engineering Research, Inc. developed a method that treated electrical cabinets based on a detailed review of the industry fire events which included distinguishing electrical cabinets by type (i.e., switchgears, motor control centers (MCCs), and low voltage cabinets). The application of alignment factors for electrical cabinet fires is an area of the fire PRA categorized as Type 1 because industry efforts have resulted in acceptance of the method [9]. On the other hand, the NUREG/CR-6850 guidance groups electrical cabinets together with a single fire ignition frequency and recommends heat release rates (HRRs) for the different types based on typical configurations. The revised method derived alignment factors for electrical cabinets based on the fire events for each type of electrical cabinet.

The method was reviewed by an industry panel which accepted the use of the method in certain circumstances with a dissenting opinion. Subsequently, the NRC sent a letter to the Nuclear Energy Institute (NEI) that did not endorse the method for use in risk informed regulatory applications [10]. Part of the basis for the dissent was that applying NUREG/CR-6850 guidance and detailed fire modeling would provide comparable results. Based on this dissent, NRC RAIs have requested risk results without the alignment factor for use in decision-making. In response to RAIs, significant resources have been expended to replace the alignment factors in an effort to achieve comparable results. A sensitivity study was performed to compare the fire risk with and without the alignment factors. In one case a significant reduction in plant fire induced CDF (65% reduction) was calculated when the electrical cabinet alignment factors were applied. Additionally, in each case plant configurations exist where fire modeling does not provide comparable results as suggested by the dissent.

Table 4: Fire PRA Sensitivity to Use of Panel Factors

Fire PRA	Change in CDF
Plant 1	-65%
Plant 2	-8%
Plant 3	-12%

3.5. Fire Growth and Suppression

The modeling of fire growth and suppression is an area of the fire PRA categorized as Type 2 because the available guidance is believed to be conservative; however, no updated guidance has been provided for use in risk informed applications. The part of fire growth in fire PRA that is believed to include the most significant bias is the evaluation of fire duration. NUREG/CR-6850 Appendix P provides guidance on fire non-suppression probabilities. NUREG/CR-6850 Table P-3 recommends a floor non-suppression probability of 1E-3 be applied. In NFPA 805 RAIs, this guidance has been interpreted by the NRC that a fire may continue to grow indefinitely and ultimately may result in full room damage and multi-compartment interactions. In contrast, NUREG/CR-6850 Section 12.5.3.6 suggests that the analysis of operator actions required after the first hour can assume the fire to be extinguished and therefore would not cause late scenario complications.

The recommended floor non-suppression probability is not realistic. For example, using this guidance the Main Control Room (MCR) fire non-suppression probability would be 1E-3 from 20 minutes until the end of the postulated fire scenario. Additionally, NUREG/CR-6850 Appendix P provides no guidance with respect to fire response practical application. Specifically, while a fire may not be declared suppressed, the plant operators and fire brigade may have engaged in sufficient activity to control the fire. A classic example would be the prescribed action to de-energize an electrical cabinet fire and let the fire self-extinguish.

The guidance in NUREG/CR-6850 Section 12.5.3.6 indicates that a sense of realism should be considered. However, in application RAIs, the NRC has interpreted the guidance that fires will continue to burn with a non-suppression probability of 1E-3. Table 5 presents the results of a sensitivity study which explored the fire risk assuming that fire growth is limited and fires do not continue to grow and result in full room damage if the fire brigade and operator response prevents fire growth. The results vary significantly among the sample plants based on plant configuration. A large reduction is calculated for certain plant configurations.

Table 5: Fire PRA Sensitivity to Fire Growth

Fire PRA	Change in CDF
Plant 1	-64%
Plant 2	-1%
Plant 3	-4%

3.6. Heat Release Rate

The NUREG/CR-6850 recommended HRRs is an area of the fire PRA categorized as Type 2 because the available guidance is believed to be conservative; however, no updated guidance has been provided for use in risk informed applications. NUREG/CR-6850 Appendix E and G provide guidance for ignition source HRRs. The HRRs are based on a set of generic tests that are largely considered to not reflect equipment or practices in a nuclear power plant. The recommended HRRs have been considered conservative by the industry for application in nuclear plant applications.

Industry efforts to revise the NUREG/CR-6850 HRRs include the work published in EPRI 1022993 [11] regarding electrical cabinet HRRs and clarifications regarding transient HRRs resulting from the industry review panel [12]. Subsequently, the NRC endorsed the transient HRR review panel decision, but did not endorse the EPRI 1022993 electrical cabinet HRRs for risk informed regulatory applications [10]. The NRC did not endorse the EPRI 1022993 HRRs because of planned additional fire testing of electrical cabinets by the NRC to develop improved guidance. The electrical cabinet fire tests are in progress. Nevertheless, the fire PRA results used in decision-making are based on the recommended NUREG/CR-6850 HRRs. A sensitivity study explored the potential reduction in risk given a smaller electrical cabinet HRR that did not result in damage to target cables. A fire induced CDF decrease of up to nearly a factor of 4 (up to a 72% reduction) was calculated for the sample fire PRAs examined.

Table 6: Fire PRA Sensitivity to Heat Release Rate

Fire PRA	Change in CDF
Plant 1	-69%
Plant 2	-72%
Plant 3	-34%

3.7. Fire PRA Human Reliability Analysis

Fire PRA human reliability analysis (HRA) is an area of the fire PRA categorized as Type 3 because the available guidance is believed adequate; however, the guidance has been difficult to implement in full. NUREG/CR-6850 Section 12 provides guidance on fire PRA HRA and the determination of human error probabilities (HEPs). Subsequently, NUREG-1921 [13] was published and provides additional guidance on fire PRA HRA. One of the greatest obstacles in fire PRA HRA is not necessarily related to the available guidance but the required resources to adequately apply the guidance. Two areas of significant resources that have generally been treated conservatively are the treatment of available cues and scenario specific influences on operator actions.

Fire PRA HRA introduces the need to explicitly identify primary and secondary cues and address the potential for degraded cues because of fire damage. A large set of primary and secondary instruments may be available to include in the full set of cues available to operators. In general, primary cues include instrumentation on the plant's safe shutdown equipment list. However, secondary cues are generally not included and require circuit analysis and cable location identification to explicitly credit in fire PRA. Therefore, generally the fire PRA HRA incorporates a conservative bias that assesses the secondary cues as degraded or provides no credit for secondary cues because the lack of data.

Fire HRA generally builds on the internal events HRA which includes sequence timing and available personnel. The timing involved in fire scenarios may be significantly different because of fire growth, time to target cable damage, and suppression activities. Additionally, there is the potential that some operators may participate in fire brigade activities reducing the number of available operators. Each of these items is specific to the postulated fire scenario. However, the fire PRA typically includes hundreds to thousands of fire scenarios. Therefore, scenario specific HRA may be deferred to the most risk significant scenarios if performed at all.

A sensitivity study explored the potential fire risk reduction if the fire HEPs were no worse than the internal events HEPs. A fire induced CDF reduction of up to 46% was calculated for the sample fire PRAs examined.

Table 7: Fire PRA Sensitivity to Human Error Probability

Fire PRA	Change in CDF
Plant 1	-46%
Plant 2	-13%
Plant 3	-13%

3.8. Assumed Plant Trip

Fire PRA plant response is an area of the fire PRA categorized as Type 3 because the available guidance is believed adequate; however, the guidance has been difficult to implement in full. One of the more difficult elements of a fire PRA is determining how an ignition source and subsequent damage to cables will affect plant operation. NUREG/CR-6850 equipment selection task provides guidance for identifying equipment which if lost may result in a fire induced initiating event. Fire damage to equipment and cables that may result in a more challenging initiating event typically has the necessary cable and routing data available. For example, damage to cables that may cause spurious opening of valves that could result in an inventory loss would be identified and the cables location in the plant would typically be known. Therefore, it is easy to identify if a postulated fire scenario will result in the fire induced initiating event. However, cable data for all equipment in the plant is not typically identified. Therefore, for many scenarios it is uncertain what may be the effect of the fire if any at all. NUREG/CR-6850 guidance acknowledges that for these cases a fire induced initiating event may have to be assumed. Typically, at a minimum a plant trip will be assumed.

A sensitivity study was performed in which a plant trip was not assumed. In the sensitivity study, a fire induced initiator was considered only if the consequence of the fire damage was known. A reduction of up to 21% in fire induced CDF was calculated for the sample fire PRAs examined.

Table 8: Fire PRA Sensitivity to Assumed Plant Trip

Fire PRA	Change in CDF
Plant 1	-21%
Plant 2	-4%
Plant 3	-13%

3.9. Main Control Room Abandonment Modeling

Modeling MCR abandonment because of environmental conditions is an area of the fire PRA categorized as Type 3 because there is available guidance; however, the guidance may not fully assess the potential operator or fire brigade response actions. The MCR abandonment criteria guidance is included in NUREG/CR-6850 Section 11.5.2. The guidance recommends the use of zone and field fire modeling tools to determine the time the habitability criteria are exceeded. When the models predict these conditions, then MCR abandonment is assumed. A factor that has not been explored is operator actions and fire brigade response activity in preventing abandonment conditions. For example, operators may open MCR doors or use portable fans. There are complications in crediting such actions because these types of actions may interfere with plant response actions and may not be proceduralized or trained. However, when MCR abandonment is identified as a risk significant fire scenario, then conservative biases should be explored.

A sensitivity study resulted in a reduction of up to 10% in the fire induced CDF when the MCR abandonment scenarios were removed from the sample fire PRAs examined.

Table 9: Fire PRA Sensitivity to MCR Abandonment

Fire PRA	Change in CDF
Plant 1	-10%
Plant 2	0%
Plant 3	-2%

3.10. Fire Induced Offsite Power Recovery

Modeling fire induced offsite power recovery is an area of the fire PRA categorized as Type 3 because the available guidance addresses equipment recovery; however, the lack of knowledge may prevent explicit credit. Fire scenarios that result in a fire induced loss of offsite power typically do not credit the recovery of offsite power in the fire PRA. It is often considered that the cause of the loss of offsite power may not be easily diagnosed. Also, the timing associated with identifying the cause for the loss of offsite power and taking actions to restore it may be difficult to assess. Another factor is that the necessary actions may not be proceduralized or may require special skill of craft knowledge.

A sensitivity study was performed in which offsite power recovery was credited. Table 10 presents the results which indicate up to an 18% reduction in fire induced CDF for the sample fire PRAs examined.

Table 10: Fire PRA Sensitivity to Offsite Power Recovery

Fire PRA	Change in CDF
Plant 1	-10%
Plant 2	-18%
Plant 3	-1%

4. RESULTS

4.1. Overall Fire CDF Point Estimates

Three fire PRA point estimate CDFs were developed for each of the three plants. The "upper bound" was developed using NUREG/CR-6850 guidance. The "upper bound" point estimate CDF may be representative of fire PRA results that were developed and used for risk informed applications prior to NUREG/CR-6850 supplemental guidance. Using this guidance, a "high" fire CDF could be calculated (i.e., fire CDF > 1E-4/yr.) when the NUREG/6850 guidance is strictly applied. The "nominal" point estimate CDF may be representative of fire PRA results developed more recently and used for risk informed applications that implement NUREG/CR-6850 supplemental guidance or methods from FAQs, industry review panel, or subsequent NUREGs. It is noted that not all methods accepted by the industry review panel is endorsed by the NRC for use in risk informed applications. Using these sets of guidance, it was found that the calculated fire CDF could be at least a factor of two less than the "upper bound" fire induced CDF. Additionally, for each plant the "nominal" fire CDF was calculated to be less than 1E-4/yr. The "lower bound" point estimate CDF may represent fire PRA results that could be expected as fire PRAs mature to the level of internal event PRAs. In fact, the "lower bound" fire CDF estimates could be comparable to the internal events calculated CDFs.

Figure 1 presents the results for each plant fire PRA. Each fire CDF point estimate is presented as a percentage of the "upper bound" point estimate CDF.

Figure 1: Comparison of Fire CDF to the Upper Bound Estimate

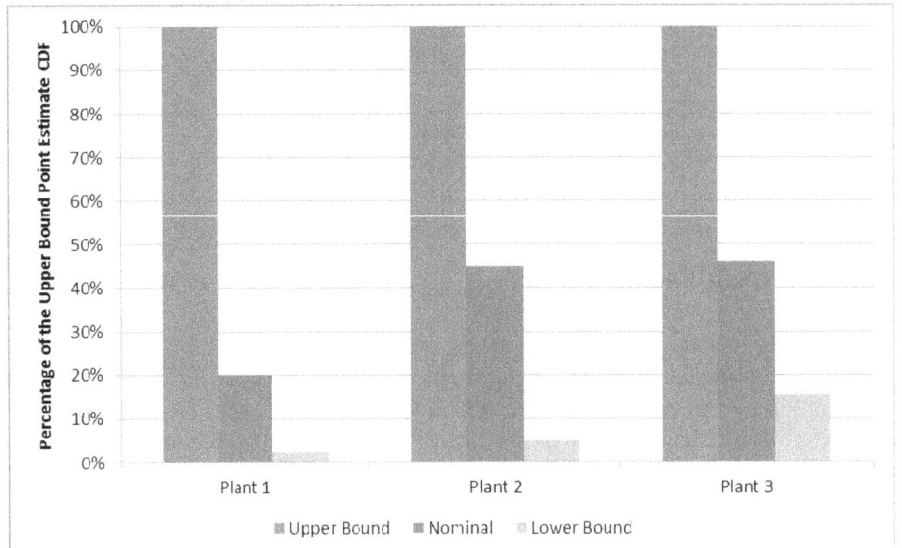

4.2. Contributors to Fire Related CDF

The value of the fire PRA is the ability to provide the plant with insights so that risk informed decisions can be made in support of plant operations. Conservative biases can result in insights that may lead the plant to make decisions that do not provide real safety benefit, and could introduce additional risks. At the very least, biases could result in not prioritizing the most realistic sources of risk sufficiently high to be treated by decision makers. Three examples of the use of the fire PRA results insights is the understanding of the risk by plant location, the risk by ignition source type, and the risk by fire scenarios.

Fire related risk by plant location may be useful for identifying ways to reduce the plant fire risk. For example, a high risk area may not have an automatic detection or suppression system. A plant modification to install a system may be a reasonable solution to reduce plant fire risk. Figure 2 presents one of the plant's fire risks by location. For this plant, the Auxiliary Electric Equipment Room (AEER) contributes approximately 50% to the fire risk when looking at the "upper bound" CDF. However, as conservative biases are removed other plant locations become larger contributors which may not necessarily have been properly prioritized in making safety decisions regarding resource allocation.

Fire related risk by ignition source type may be useful for identifying ways to reduce the plant fire risk. For example, a high risk ignition source type may be vented panels that damage or ignite risk significant cables. A plant modification to seal the panels or install tray covers may be a reasonable solution to reduce plant fire risk. Figure 3 presents one plant's fire risk by ignition source type. For this plant, switchgear fires contribute approximately 50% to the fire risk when looking at the "upper bound" CDF. However, as conservative biases are removed high energy arching fault (HEAF) events become more significant contributors to the plant risk. Also, when looking at the "lower bound" CDF it is noticed that the importance of low voltage panels and motor control centers (MCCs) may be masked by conservative biases.

Fire related risk ranked by fire scenario contribution may be useful for identifying ways to reduce the plant fire risk. For example, plant resources could be focused on the most significant fire scenarios to reduce the risk. Changes in administrative controls, procedures, circuits, or plant operation could reduce the risk associated with a particular fire scenario. Figure 4 presents one plant's fire risk by the top 10 fire scenarios. For this plant, a transient fire scenario contributes approximately 5% to the fire risk when looking at the "upper bound" CDF. However, as conservative biases are removed the contribution of the transient fire scenario increases to approximately 40% when looking at the "lower bound" CDF.

Figure 2: Example CDF Point Estimates by Plant Location

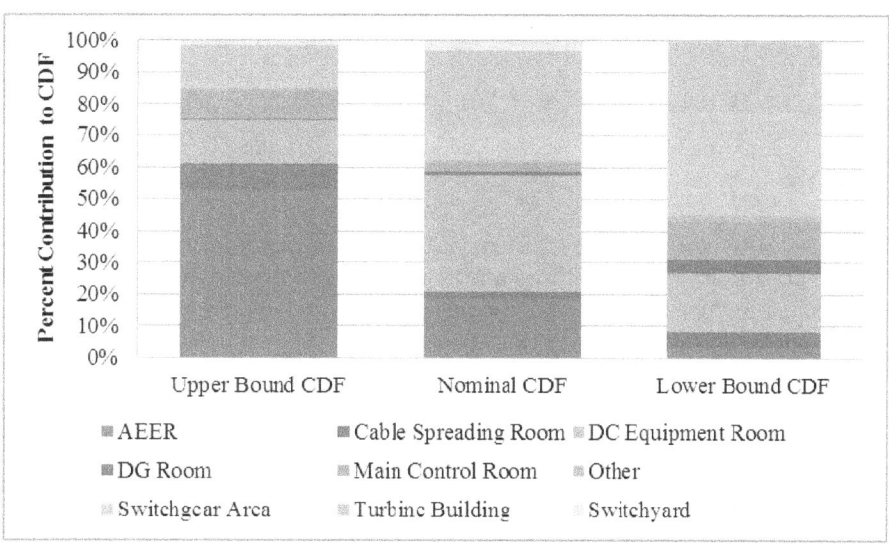

Figure 3: Example CDF Point Estimates by Ignition Source

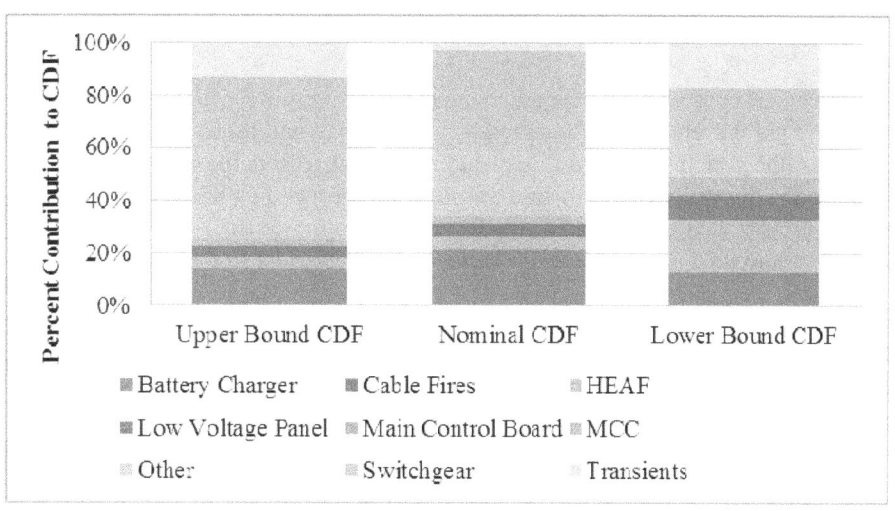

Figure 4: Example CDF Point Estimates Top 10 Scenarios

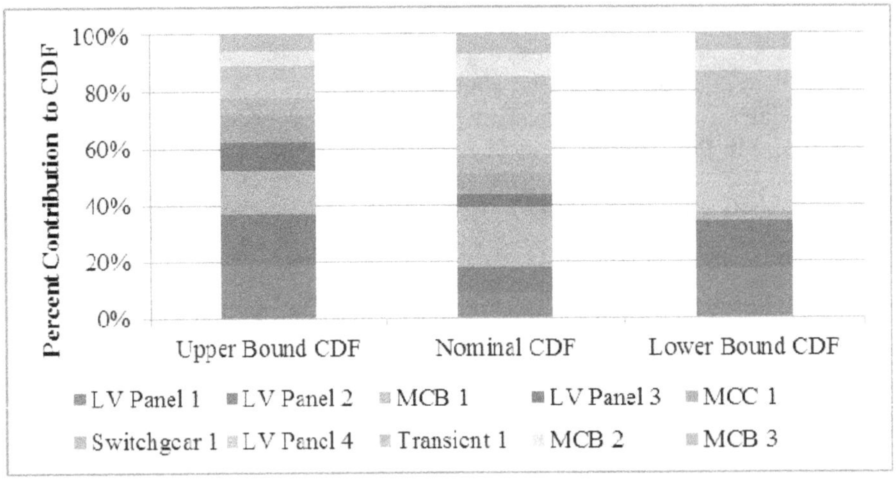

6. CONCLUSION

As identified in the NRC's PRA Policy Statement, PRA is most useful when it presents a realistic assessment of the conditions and then allows a realistic portrayal of the uncertainties. The evaluation reported here resulted in "upper bound" CDF point estimates greater than 1E-4/yr. and reductions in calculated CDF up to approximately one order of magnitude. The reductions were obtained by using realistic modeling approaches for only a limited number of issues. The reductions were obtained without fully addressing other potential biases not identified in this paper. The results were summarized from different perspectives to determine the impact on the dominant contributors. A comparison of the results showed that the relative plant location contribution to the risk profile could vary dramatically depending on the assumptions or biases imposed on the modeling. For example, a comparison of the results identified that the application of the same treatment regardless of the types of electrical cabinets (i.e., switchgears, MCCs, and low voltage panels) could mask the relative importance of a specific electrical cabinet. A comparison of the results also identified that the fire scenario contribution to the risk profile could vary depending on the assumptions imposed on the modeling.

There are interface and competing issues associated with identifying which of these uncertainties or conservative biases are currently being pursued as viable today and those that might be a "bridge too far". Nevertheless, the conclusion from the evaluation is that reasonable (realistic) approaches to the assessment of the fire hazard will result in a substantially reduced estimate of the fire risk and will likely change the priority that is assigned to the fire zones and the causes of risk significant fires. This change in insights could then lead to reprioritization of possible plant or procedural changes.

In addition to the methodological differences that may affect the fire risk profile, there are real plant or procedure changes that may influence the fire risk profile. The risk significance of these may dramatically change based on the degree of conservative biases that are imposed on the fire PRA. These might include protecting, replacing or rerouting cables, changes in system design or normal configuration, circuit modifications, or procedural changes to mitigate multiple spurious operations. Therefore, the risk reduction because of plant changes may be over stated due to conservative biases, and there may be risk beneficial plant changes being masked. Again, the conclusion from the evaluation is that reasonable (realistic) approaches to the assessment of the fire hazard will result in a reduced estimate of the fire risk and will likely change the priority that is assigned to plant changes.

Lastly, conservative bias fire PRAs may not adequately support risk input for other risk-informed applications. The possibility of mis-prioritizing plant modifications is a critical issue for an industry that is under severe economic pressure. The need to perform those safety enhancements that are most beneficial is critical to public safety. In addition, point estimate CDFs and computed delta CDFs input into the Significance Determination Processes (SDPs), Notice of Enforcement Discretions (NOEDs), or License Amendment Requests (LARs) may result in greater findings or administrative controls than are realistic. Alternatively, conservative bias fire PRAs may result in masking effects in computed delta CDFs in applications such as technical specification completion time evaluations. EPRI 1026511 Section 5.3.3.2 addresses potential issues related to conservative biases because of large uncertainty [14]. While this evaluation did not explore the potential issues related to conservative bias effects on computed delta CDF input into applications, an insight from this evaluation is that the comparison between the point estimate CDFs shows that the models could lead to different conclusions when used in applications and compared to specific acceptance guidelines or risk metric thresholds.

References

[1] "Use of Probabilistic Risk Assessment Methods in Nuclear Regulatory Activities; Final Policy Statement", U.S. Nuclear Regulatory Commission. (1995). 60 FR 42622.

[2] EPRI/NRC-RES Fire PRA Methodology for Nuclear Power Facilities: Volume 2: Detailed Methodology. Electric Power Research Institute (EPRI), Palo Alto, CA, and U.S. Nuclear Regulatory Commission, Office of Nuclear Regulatory Research (RES), Rockville, MD: 2005. EPRI-1011989 and NUREG/CR-6850.

[3] *EPRI/NRC-RES Fire PRA Methodology for Nuclear Power Facilities: Volume 1: Summary and Overview*. Electric Power Research Institute (EPRI), Palo Alto, CA, and U.S. Nuclear Regulatory Commission, Office of Nuclear Regulatory Research (RES), Rockville, MD: 2005. EPRI-1011989 and NUREG/CR-6850.

[4] *Fire Probabilistic Risk Assessment Methods Enhancements: Supplement 1 to NUREG/CR-6850 and EPRI 1011989.* EPRI, Palo Alto, CA, and NRC, Washington, D.C.: December 2009. 1019259.

[5] *Fire PRA Methods Enhancements: Additions, Clarifications, and Refinements to EPRI 1019189.* EPRI, Palo Alto, CA: 2008. 1016735.

[6] "Close-Out of National Fire Protection Association 805 Frequently Asked Question 12-0064 on Hot Work/Transient Fire Frequency Influence Factors", U.S. Nuclear Regulatory Commission. (2013). ADAMS Accession Number ML12346A488.

[7] *Joint Assessment of Cable Damage and Quantification of Effects from Fire (JACQUE-FIRE), Volume 1: Phenomena Identification and Ranking Table (PIRT) Exercise for Nuclear Power Plant Fire-Induced Electrical Circuit Failure,* U.S. Nuclear Regulatory Commission, Office of Nuclear Regulatory Research (RES), Rockville, MD, 2012, and Electric Power Research Institute (EPRI), Palo Alto, CA, NUREG/CR-7150, BNL-NUREG-98204-2012, and EPRI 1026424.

[8] "Interim Guidance Pending Publication of Expert Elicitation Results", U.S. Nuclear Regulatory Commission. (2013). ADAMS Accession Number ML13165A214.

[9] "Recent Fire PRA Methods Review Panel Decision: Treatment of Electrical Cabinets", Nuclear Energy Institute. (2012).

[10] "Recent Fire PRA Methods Review Panel Decisions and EPRI 1022993, Evaluation of Peak Heat Release Rates in Electrical Cabinet Fires", U.S. Nuclear Regulatory Commission. (2012). ADAMS Accession Number ML12171A583.

[11] *Evaluation of Peak Heat Release Rates (HRRs) in Electrical Cabinet Fires*, EPRI, Palo Alto, CA: 2012. 1022993.

[12] "Recent Fire PRA Methods Review Panel Decision: Clarification for Transient Fires and Alignment Factor for Pump Oil Fires", Nuclear Energy Institute. (2011).

[13] *EPRI/NRC-RES Fire Human Reliability Analysis Guidelines.* EPRI, Palo Alto, CA, and U.S. Nuclear Regulatory Commission, Washington, D.C.: 2012. 1023001/NUREG-1921.

[14] *Practical Guidance on the Use of Probabilistic Risk Assessment in Risk-Informed Applications with a Focus on the Treatment of Uncertainty.* EPRI, Palo Alto, CA: 2012. 1026511.

Approach for Integration of Initiating Events into External Event Models

Nicholas Lovelace[a]*, Matt Johnson[a], and Michael Lloyd[b]
[a] Hughes Associates, Lincoln, NE, USA
[b] Risk Informed Solutions Consulting Services, Ball Ground, GA, USA
* E-mail: nlovelace@haifire.com

Abstract: Probabilistic Risk Assessments (PRAs) are increasingly being used as a tool to support risk informed applications. As a result, the scope and quality of these PRA models has expanded to account for the risk associated with external events, such as fires, seismic events, external floods, and high winds. Improved methods developed to create a PRA model for one of these external events can often be used to improve the process used to develop PRA models for other external events. This paper explores one such method. It describes an improvement in the method to incorporate fire initiating events into a Fire PRA model. The improved method reduces the potential for mapping external event failures to multiple induced initiating events. Such mapping can have the undesirable effect of generating duplicate cutsets, subsuming issues, etc. during the quantification process. This is especially an issue when the quantification engine applies the "rare event" approximation and cutset basic events probabilities are relatively high (i.e., greater than 0.1). The improved method allows multiple external event induced initiating event mapping, while addressing the limitations of the quantification engine.

Keywords: PRA, PSA, Initiating Event, External Events, Quantification

1. INTRODUCTION

Improved methods have been developed to utilize the existing PRA event trees to model external event initiators without creating quantification and subsuming issues utilizing new programs such as Fire Risk Analysis (FRANX). This methodology has been developed using lessons learned in Fire PRAs and other external event models. This methodology meets the requirements of the American Society of Mechanical Engineers (ASME) PRA Standard [1].

This paper presents the methodology for the approach utilizing Computer Aided Fault Tree Analysis (CAFTA) and FRANX. However, this approach is applicable to any software. Section 2 of this paper provides an overview of the major steps required to implement the methodology. Section 3 provides the conclusion.

2. METHODOLOGY

The following approach involves the identification of the external event initiating events. A review needs to be performed of the model to identify the minimum set of induced initiating events (IEs) that can be used to fully represent all accident scenarios applicable for each external event scenario. The example approach shown is from a Fire PRA model but has been successfully utilized in other external event applications such as seismic and high winds.

The starting point in implementing the methodology is to review the current set of event trees in the model to ensure they bound the hazards, failures, and plant response to the external event. Generally, the internal events model event trees in the PRA model will bound the external event initiators. Once this has been determined it's simply a matter of mapping the external event initiators to the internal event initiators that result in the same plant response to the event.

For simplicity, it's initially assumed that any fire will cause a plant turbine trip. If this assumption leads to unnecessary conservatism in the analysis, the assumption should be revisited, as necessary, for the affected fire compartments in order to avoid potentially inappropriate conclusions in future Fire PRA applications.

The simplest way to set the turbine trip Initiating Event (IE) to TRUE for a Fire PRA is to set the internal IE to TRUE. However, to avoid quantification problems (i.e., automatic exclusion of valid cutsets by the quantifier when multiple initiators occur in the same cutset), each of these turbine trip IEs should be "generically" applied in the model by replacing each by an OR gate with two contributors: (1) the random turbine trip IE and (2) a gate representing the fire initiated turbine trip. An example of this logic is shown in Figure 1, below. The fire induced IE (%1FIRE) is set to TRUE for all fire scenarios and, consistent with the assumption that any fire in any compartment will induce a turbine trip, the fire induced turbine trip flag basic event (FIRE1TT) is generically set to TRUE for all scenarios. Gate FIRE_COMPARTMENTS is an OR gate of basic events (BEs) each representing each and every fire compartment. This gate was AND'd with %1FIRE to indicate the compartment associated with each fire quantification cutset.

Figure 1: Gate Representing Fire Initiated Turbine Trip

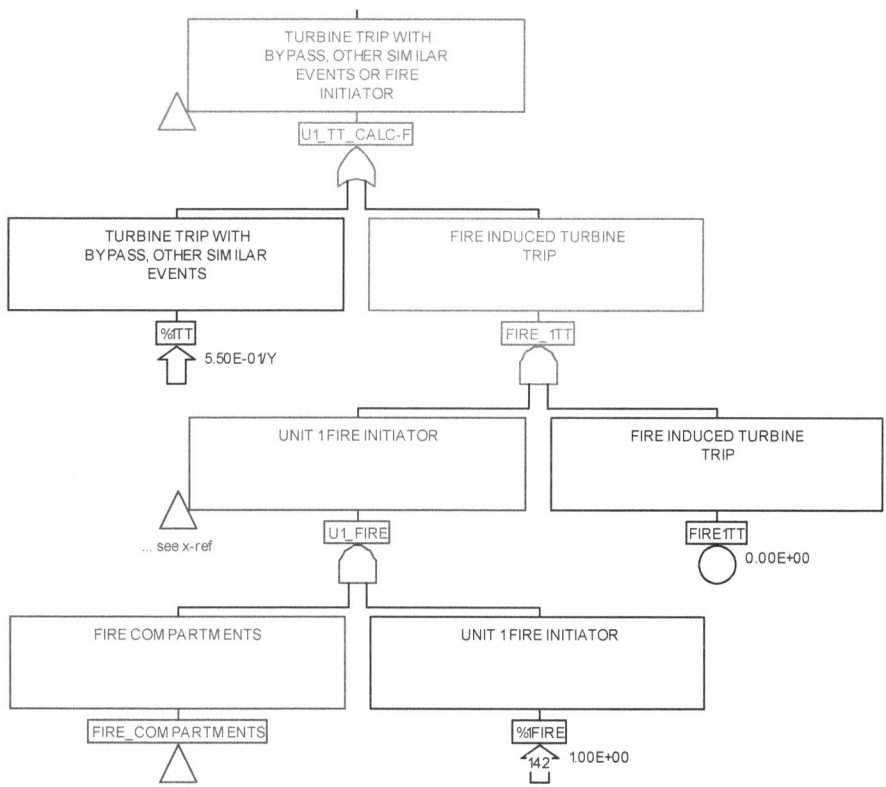

Even though all components damaged by the fire are failed, use of the turbine trip (TT) as the sole generic fire induced IE is not expected to fully account for all effects of a fire, since this initiator only applies the general transient event tree accident sequence fault tree logic. Some fires should invoke other accident sequences. For example, use of the TT as the sole fire induced IE will not fully account for fires that induce LOCAs due to pressure boundary equipment impacts and need to be modelled using the applicable internal model LOCA event tree.

To ensure that every IE that could be caused by a fire is incorporated into the quantification results, each of the candidate fire induced IEs should be reviewed for unique model impact. This review involves two steps. The first step is to find the minimum set of internal events initiating events that are needed in the fire model and the second step is to identify which of these minimal IEs should be set to TRUE for each fire scenario. These two steps are described below.

Step 1: Identify Minimal Set of Internal Event IEs Needed in the Fire PRA

In this first step, every gate that each internal event IE reports to in the Fire PRA model was traced upwards (towards the top event) in the fault tree. A candidate fire induced IE can be eliminated, if every gate to which the IE reports are also failed by either one or a combination of the following criteria:

Criterion 1: a generically applied fire induced IE fails the gate or
Criterion 2: component failures associated with the IE that are failed by the same fire fail the gate.

As noted above, initially, the only "generic" fire induced IE was the turbine trip IE.

The search for which of the candidate fire induced IEs are "subsumed" by generically assumed IEs involves a CAFTA fault tree review for each candidate IE. The process involves the creation of a special unit specific Core Damage Frequency CDF top gate extracted from the internal events PRA fault tree model, excluding mutually exclusive logic. This logic contains all of the candidate fire induced IEs uncomplicated by the presence of Large Early Release Frequency (LERF), Equipment Out Of Service (EOOS), and event tree success and mutually exclusive logic. The basis for this approach is that a large early release requires core damage, that not all EOOS logic reports to the CDF top gate, and that removal of success and mutually exclusive logic only maximizes the potential for a fault tree Basic Event (BE) to result in core damage.

Using the applicable unit specific fault tree, the gates to which each candidate fire induced IE reports were identified, and gates to which these gates report are identified, etc. The resulting "IE impact" table is a combination of the CAFTA BE and gate cross reference reports extending to multiple gate levels. This table is used to assess the impact of each candidate IE and to determine whether this impact is addressed by either assumed by generic fire induced IEs, by fire induced system failures, or by a combination of these.

For each IE, the fire impact review on a gate path can be terminated if the gate is TRUE'd by a generic fire induced IE (Criterion 1) or if the gate is TRUE'd by BEs that are failed by the same fire (Criterion 2). To apply Criterion 1, each unit's generic IE (e.g., %1TT) is set to TRUE. As can be seen in Figure 1, this is equivalent to setting the following unit specific BEs and gates to TRUE: (1) the generic fire induced IE flags (e.g., FIRE1TT), (2) the fire initiator (e.g., %1FIRE),), and (3) gate FIRE_COMPARTMENTS.

This type of review, looking upwards into the fault tree from the IE, should be performed on all candidate fire induced IEs. The results of these reviews should be documented. As noted above, initially, only one generically applied fire induced IE flag FIRE1TT is set to TRUE for all fire scenarios. This event should be found to subsume a number of other candidate fire induced IEs. During this review you may find that setting a second fire induced IE flag such as the Main Steam Isolation Valve (MSIV) closure flags to TRUE for all fire scenarios could cause additional IEs to be subsumed. Setting multiple "generic" fire induced IE to TRUE for all scenarios eliminates the need to include many additional candidate fire induced IEs in the model.

If there are candidate fire-induced IEs that have pre-initiator support system fault trees (SSFTs) in the PRA model, they can be explicitly addressed via minor model revisions. Essentially, the revision is to replace the existing IE flag event with an OR gate with two contributors: the IE flag and the existing gate U1_FIRE. Figures 2 and 3 provide an example of this revision. These revisions ensure that the effects of the initiator propagate into the model if a fire causes a component contributing to the initiator in its SSFT to fail. Thus, these IEs need not be explicitly addressed in the Fire PRA model.

In addition, because many cutsets that come out of the SSFTs will have a 1 year, 1.5 year, or a 2 year exposure interval, these terms must be divided by 1/365, 1/(1.5*365), and 1/(2*365), respectively to convert to a 24 hour exposure. These corrections are BE specific and convert these BE "frequency" events to "probability" events, i.e., probability of failure during the 24 hour mission time. The subject BEs can be identified via review of the SSFTs contributing to each of the SSFT initiating events.

Figure 2: %1VI Logic before SSFT Logic Revision

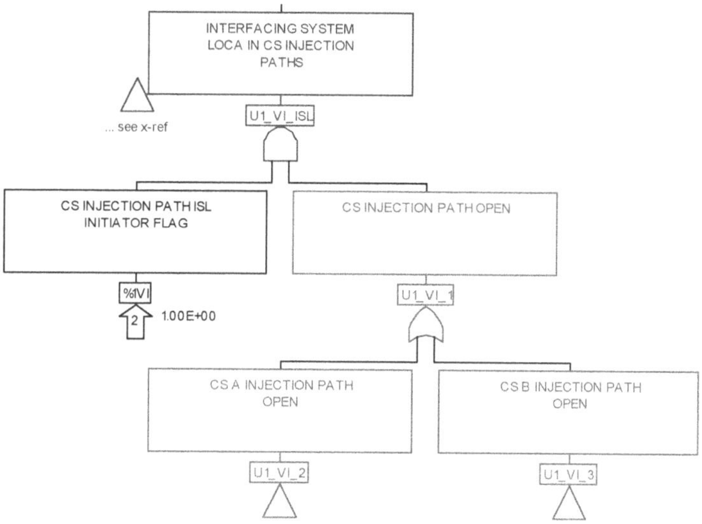

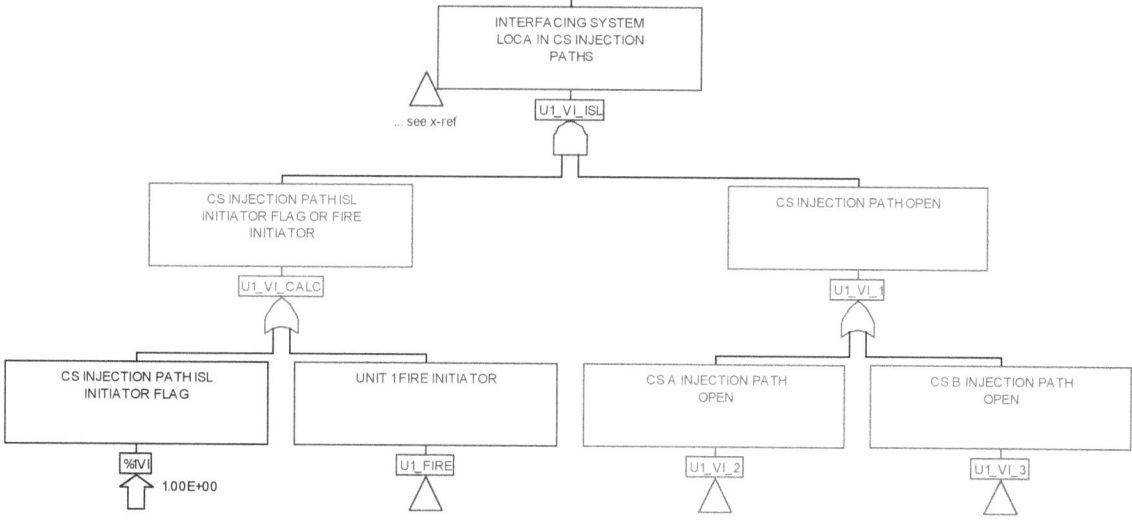

Figure 3: %1VI Logic after SSFT Logic Revision

Since they are not subsumed by other internal event IEs that are assumed to be TRUE in the Fire PRA model (i.e., TT) and are not explicitly modelled as SSFTs, the remaining candidate fire IEs that are considered a set of minimal IEs should be explicitly accounted in the Fire PRA model. Additional insights and assumptions must be made to determine which fire scenarios that would result in each of these IEs. This determination is part of Step 2 in this assessment.

Step 2: Identify Scenarios in which Minimal IEs Should be TRUE in the Fire PRA

It is reasonable to assume that only certain fires cause certain effects. For example, it is reasonable to assume that a fire in the plant switchyard will result in a loss of off-site power IE. Thus, the corresponding loss of off-site power fire initiator flag event will be set to TRUE for all fire scenarios in the switchyard. Similarly, since off-site power cables are routed from the switchyard to the plant's electrical buses through the Turbine Building, it may be assumed that a fire in either the switchyard or the Turbine Building will result in a loss of off-site power IE. Thus, the loss of off-site power fire initiator flag will be set to TRUE for all fire scenarios in both the switchyard and in the Turbine Building.

The remaining candidate fire induced IEs that are not subsumed by a "generically" applied IE and also are not subsumed by their post-initiator support logic failures must be taken into account. However, unlike the loss of off-site power IEs, these IE may not be easily correlated with any single plant physical location. Determining the list of applicable fire scenarios for which they apply requires special consideration.

Fortunately, there are generally specific gates in the model which, should it occur, can be used to identify whether a fire in a given scenario could result in each of these initiators. Initiator Inadvertent Open Relief Valve (IOOV) will be used to demonstrate this process. The effect of all of these initiators is addressed in a like manner.

Although in this case, only one gate is of particular interest, in the general case, as many panels as gates of interest can be used in this process. Once the gates of interest are identified, the FRANX model, specifically the FRANX Status Panel feature, can be used to determine which fire scenarios TRUE which initiating event related gates of interest.

The effort involves creating a new status panel configuration file (*.cfg) containing a panel for each of the gates of interest. The easiest way to create the new FRANX status panel configuration file is to edit the existing panel configuration file for the Fire PRA model, replacing the existing model gate associated with each panel with an initiating event gate of interest and renaming the panel title to the initiating event gate name or description. The editing can be performed in FRANX by viewing an existing status panel file (via FRANX command: View, Show Status Panel, and selecting the panel configuration file), revising it (via FRANX command: Edit, Configure, Edit Status Panel), saving the new panel configuration file, and then creating a status grid for each fire scenario (via FRANX command: Tools, Show Status Grid). Figure 4, below, provides an example of a status panel developed to show the status of initiating event gates, including a panel labelled %1IOOV, which represents the status of the spurious opening of a relief valve.

Figure 4: Example Status Panel for a Set of Gates of Interest

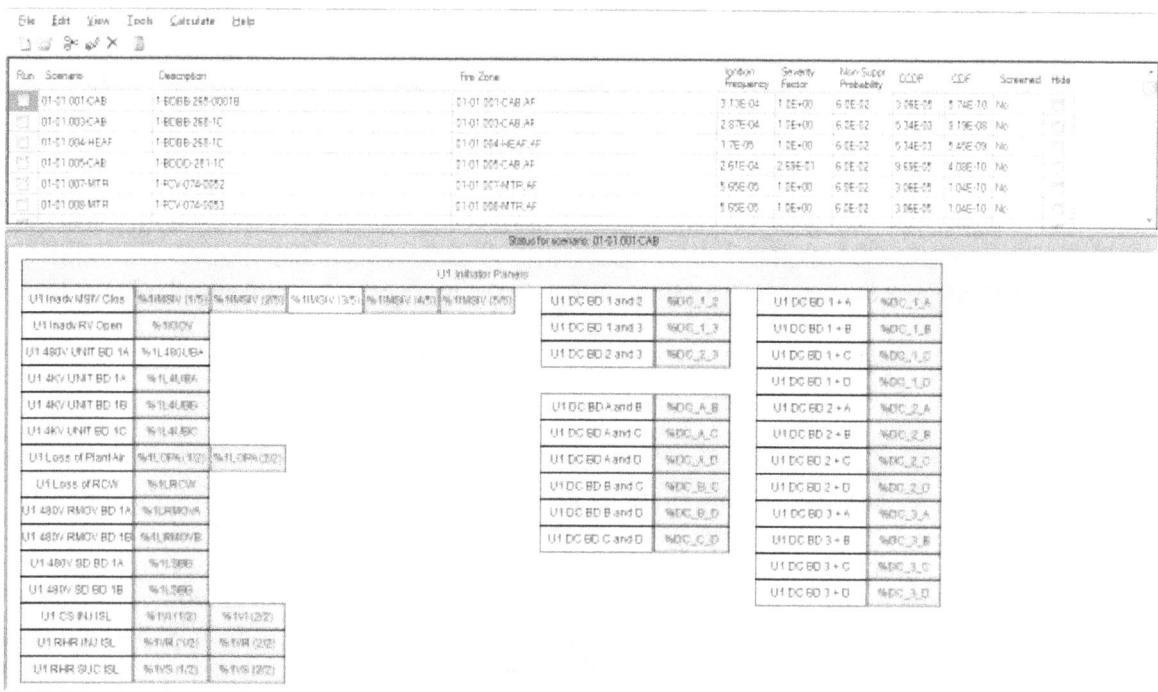

Depending on the number of fire scenarios, the complexity of the PRA model logic, and the number of gates of interest, it may take several hours to create the new status panel grid. The grid is stored in FRANX table SystemStatus. It can be copied to a MS EXCEL file by opening the FRANX model using MS ACCESS. Note that attempting to open this table within FRANX (via FRANX command: Edit, Configure, Edit Database in MS Access) can result in the deletion of all records in this table. Further processing can be performed using the MS EXCEL file to identify which fire scenarios fail which gates of interest.

As a result of running the FRANX status panel for %1IOOV, the list of fire scenarios can be provided that are identified to cause the spurious relief valve opening gate to be TRUE. The FRANX Zone_to_Raceway table can be used to set event FIRE1IOOS in the model to TRUE for the scenarios identified.

A similar process using the FRANX status panel can be performed for the other initiators to identify which scenarios their corresponding FIRE1* events should be set to TRUE.

3. CONCLUSION

As discussed, this proposed methodology can be utilized to identify which internal model initiating events and event trees should be utilized in modeling the various external event scenarios in the PRA model. This approach limits quantification issues and avoids the need to create additional external event specific event trees with accurate results. This example presented is for a Fire PRA, however, the methodology has been applied to other external events applications successfully.

References

[1] ASME/ANS RA-Sa-2009 Addenda to ASME/ANS RA-S-2008 Standard for Level 1/Large Early Release Frequency Probabilistic Risk Assessment for Nuclear Power Plant Applications, February, 2009.

Development of Margin Assessment Methodology of Decay Heat Removal Function Against External Hazards
− Project Overview and Preliminary Risk Assessment Against Snow −

Hidemasa Yamano[a,*], Hiroyuki Nishino[a], Kenichi Kurisaka[a], Takaaki Sakai[a], Takahiro Yamamoto[b], Yoshihiro Ishizuka[b], Nobuo Geshi[b], Ryuta Furukawa[b], Futoshi Nanayama[b], and Takashi Takata[c]

[a] Japan Atomic Energy Agency; 4002 Narita, Oarai, Ibaraki, 311-1393 Japan
* Tel: +81-29-267-4141, Fax: +81-29-266-3675; E-mail: yamano.hidemasa@jaea.go.jp
[b] National Institute of Advanced Industrial Science and Technology; 1-1-1 Higashi, Tsukuba, Ibaraki, 305-8567 Japan
[c] Osaka University; 2-1 Yamada-oka, Suita, Osaka, 565-0871 Japan

Abstract: This paper describes mainly preliminary risk assessment against snow in addition to the project overview. The snow hazard indexes are the annual maximum snow depth and the annual maximum daily snowfall depth. Snow hazard curves for the two indexes were developed using 50-year weather data at a typical sodium-cooled fast reactor site in Japan. Snow hazard categories were obtained from a combination of the daily snowfall depth (snowfall speed) and snowfall duration that can be calculated by dividing the snow depth by the snowfall speed. For each snow hazard category, accident sequences were evaluated by producing event trees that consist of several headings representing the loss of the decay heat removal. Snow removal operation and manual operation of the air cooler dampers were introduced into the event trees as the accident managements. In this paper, a snow risk assessment showed less than 10^{-6}/reactor-year of core damage frequency. A dominant snow hazard category was a combination of 1−2 m/day of snowfall speed and 0.75−1.0 day of snowfall duration. Sensitivity analyses indicated important human actions, which were improvement of the speed of snow removal and awareness of snow removal necessity.

Keywords: PRA, External Hazard, Sodium-Cooled Fast Reactor.

1. INTRODUCTION

External hazard risk is increasingly being recognized as important for nuclear power plant safety after the Fukushima Daiichi nuclear power station accident. To improve the nuclear plant safety, risk assessment methodologies are necessary against various external hazards, although a probabilistic risk assessment (PRA) methodology against earthquake has been developed as a priority because of the importance of consequences by the earthquake. The Atomic Energy Society of Japan published a seismic PRA standard [1] in 2007 and a tsunami PRA standard [2] in 2012 which was vigorously developed as an important issue after the Fukushima Daiichi accident caused by the tsunami. Except for the two external hazards, there are no PRA standards against various external hazards in Japan. An alternative methodology different from the PRA was developed in Europe after the accident for complementary safety assessments, so called stress tests [3]. The stress test methodology is useful to show a margin to core damage against earthquakes and floods. Since challenging tasks in external PRA methodologies are quantitative external hazard evaluation, the stress test methodology would be useful and effective to suggest safety measures and accident managements that extend margins to core damage against external hazards. To improve the plant safety against various external hazards, it is necessary to develop risk assessment methodologies, such as the PRA or stress test methodology.

A four-year research project has started since 2012 to develop a margin assessment methodology of decay heat removal function against external hazards. In this project, only the decay heat removal function was taken into account assuming no loss of reactor shutdown function because the reactor trip was successful in the Fukushima Daiichi accident. Although this accident lessons suggested the importance of a spent fuel pool, this study focuses on event sequences resulting in core damage as a first step. The developed methodology is applied mainly to sodium-cooled fast reactors (SFRs), while it would be applicable basically to light water reactors (LWRs). A typical SFR heat sink is air, which is different from a heat sink in LWRs. Therefore, it is important external hazards that influence to air coolers (ACs) which are located at high elevation. This project addresses extreme weathers (snow, tornado, wind and rainfall), volcanic eruption and forest fire as representative external hazards. In this study, the external hazard evaluation, the event sequence and the margin assessment methodologies are developed for each external hazard.

This paper describes the project overview, followed by a presentation mainly of preliminary risk assessment methodology against extreme snow which is one of outcomes from this project in Japanese fiscal year 2012.

2. PROJECT OVERVIEW

2.1. Scope of External Hazards

The external hazards are roughly categorized into three groups: underground, ground-surface, and above-ground hazards. One of the representative underground hazards is earthquake which would have a structural impact on the nuclear power plant. Since significant boundary/component failures might lead to core damage, seismic design with an appropriate design margin to component failure has been preferentially implemented. The ground-surface hazards consist of tsunami (sea), flood (river), etc. The tsunami in the Fukushima Daiichi site in Japan and the flood in the Blayais site in France [4] have given full recognition to the significance of their hazard potential. From this background, nuclear regulatory authorities in many countries strongly require some actions and/or measures against their external hazards. This study aims mainly at a contribution to the risk assessment and safety improvement of the typical SFR in Japan. As shown in Fig. 1, the scope of external hazards in this study is above-ground hazards which might influence the decay heat removal system of the SFR. Air is usually taken not only into the decay heat removal system but also ventilation and air-conditioning system, emergency power supply system, etc. It should be noted that the PRAs against earthquakes and tsunami would be performed separately based on the regulatory requirement for the typical SFR in a similar way to risk assessments in LWRs.

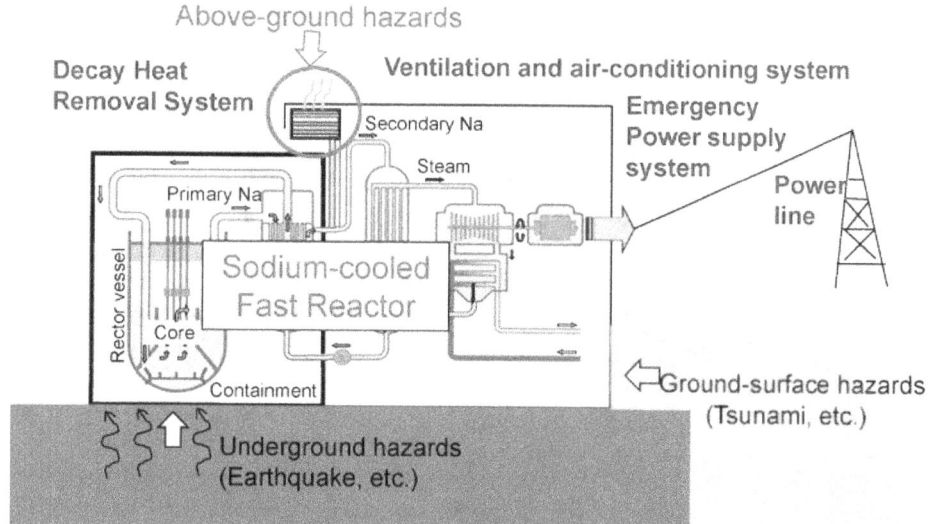

Figure 1: Scope of external hazards in this project

2.2. Selection of Representative External Hazards

In this section, a screening process is described to select the representative external hazards in this project: extreme weather, volcanic eruption and forest fire. At first, all foreseeable external hazards shall be exhaustively identified, including the potential for human-induced events directly or indirectly affect the safety of the nuclear power plant. There are a wide variety of external events by referring the International Atomic Energy Agency (IAEA) reports and so on [5-8]. Figure 2 shows the screening process for the typical SFR site in this study. As an initial step, a wide variety of external events are screened out in terms of site conditions, impact on plant, progression speed, envelop and frequency, in a similar manner to the NUREG/CR-4550 report [9-10]. For example, a drought can be precluded because it is less significant in Japan since nuclear plants are usually located near sea coast. In the second screening process, the external hazards are selected on a basis of the scope of this project, which are performed in view of natural hazards and above-ground hazards. Similar hazards are merged; e.g., hail can be enveloped by tornado-induced missiles. Through this screening process, this project selected extreme weathers (snow, tornado, wind and rainfall), volcanic phenomena and forest fire as representative external hazards.

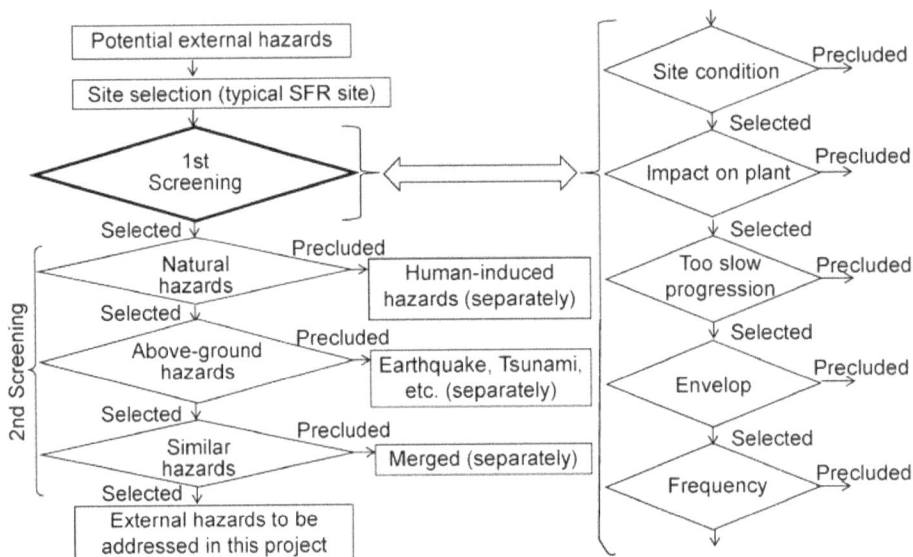

Figure 2: Screening process to select representative external hazards

2.3. Concept of Assessment Methodology

In general, an external hazard evaluation has a large uncertainty to quantify an occurrence frequency. Based on such background, the term "likelihood" is used in this paper. As with the stress test, an advantage of margin assessment is un-necessity of quantitative external hazard evaluation. Only an index is necessary to specify hazard intensity; e.g. peak ground acceleration in seismic margin assessment. On the contrary, the PRA requires a hazard curve that creates a relation between the likelihood and the hazard intensity. Since the event sequence evaluation is needed both for the margin assessment and PRA, a difference between them is quantification of external hazards. As illustrated in Fig. 3, both the margin assessment and PRA methodologies are developed because this project makes an attempt to develop the external hazard curve. The PRA would indicate a core damage frequency (CDF), whereas the margin assessment would show the extension of a margin to the core damage by introducing several measures including accident management.

The snow PRA methodology has been developed in the first year. Next, the PRA methodologies against tornado and wind in the second year, and against rainfall and volcanic eruption in the third year are scheduled to be developed. Finally, the PRA methodology development against forest fire and combination events is planned.

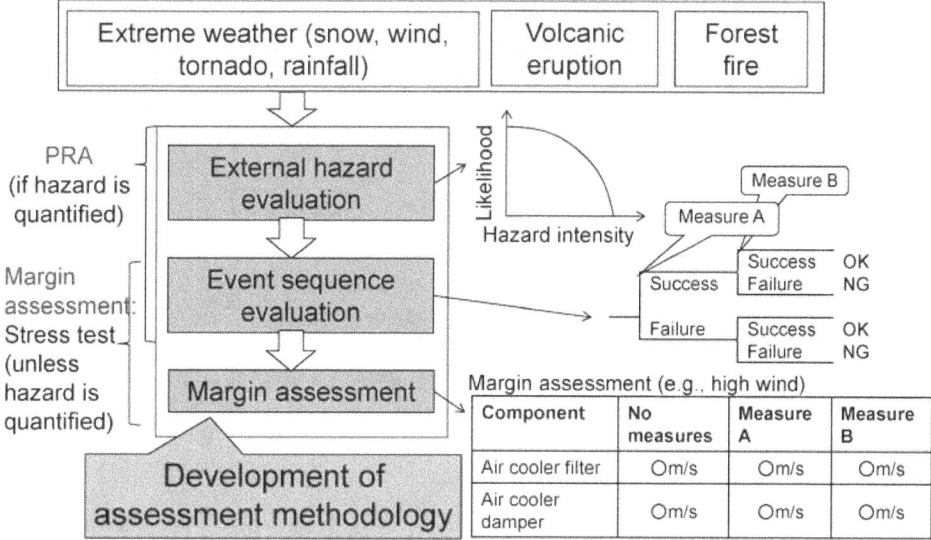

Figure 3: Concept of assessment methodology

3. SNOW PRA METHODOLOGY

3.1. Snow Depth Leading to Component Failure

Initiating events can be identified because components would be failed by a certain large level of earthquake and/or tsunami in a very short time without personnel actions. On the other hand, the personnel (except the one in the control room) can remove snow which is accumulated very slowly as not to have a serious impact on the plant. Therefore, the snow removal can be expected in the snow PRA. In this study, the initiating events were simplified and the event sequence was evaluated considering the time dependence, for which snow hazard category was used as mentioned in Section 3.2.

In this study, no snow removal capability for higher than six meters of snow depth was assumed for conservativeness and simplification, as explained in Section 3.2. Emergency power supply function would be lost by filter clogging at the air inlet of the emergency diesel generator room. In addition, the emergency diesel generator (EDG) could be failed by heating in the EDG room in case that the snow clogs the air inlets of ventilation and air-conditioning system. Thus, it was assumed that the loss of EDG power supply function enveloped the loss of the ventilation and air-conditioning system. The main decay heat removal system is an auxiliary cooling system (ACS) consisting of three loops in the typical SFR. This ACS is normally operated in a forced circulation mode with pony motors, but a natural circulation mode in this system is also available even in station blackout. Moreover, a maintenance cooling system (MCS) in the forced circulation mode with one loop using an electromagnetic pump is also available even if the decay heat removal capability by the ACS is lost. To put it all together, important components of the decay heat removal function that are affected by the extreme snow are representatively regarded as the ACS, MCS and EDG power supply system.

In fact, the heights are different between the air inlet and outlet of each important component. Since the snow would be melted due to hot exhausted air at the outlet, their important components are actually available for the decay heat removal if air is taken into the inlet. To conservatively evaluate in this study, however, the important components were assumed to be failed when the snow reached a lower elevation between the inlet and outlet. In this paper, the snow depths leading to the failure of the important components are specified as follows: 1.5 m for ACS, 2.0 m for MCS and 1.2 m for EDG.

The loss of the decay heat removal function is roughly divided into two types: its functional failure and the structural failure of the system. The failure of the important components stated above is categorized to the functional failure which is caused by isolating air ventilation due to snow. The structural failure could arise from a heavy snow load exceeding the proof strength of a building or component. A reactor auxiliary building can withstand higher than ten meters of snow depth in the typical SFR in Japan. Other components also keep their integrity in a view of structural strength under a deeper snow condition, compared to the functional failure. Therefore, only the functional failure is addressed in this study.

In this paper, the lowest height leading to the failure of the important components is 1.2 m, at lower than which in turn no core damage sequence appears in the snow PRA. Sometimes, offsite power is lost at several ten centimeters of snow depth by disconnecting the power line. Therefore, the loss of offsite power can be regarded as an initiating event in the snow PRA.

3.2. Snow Hazard Category

3.2.1. Historical Records of Snow

In Japan, snow data is recorded at representative local offices of the Japan Meteorological Agency (JMA). Near the typical SFR site, a local weather observatory measures and collects various weather data including snow at the Japan Sea side central area in Japan. This study used snow data of 50-year from 1961 to 2010 based on the JMA database [11]. Historical records are plotted in terms of the annual maximum snow depth and the annual maximum daily snowfall depth in Fig. 4. At maximum, the annual maximum snow depth and the annual maximum daily snowfall depth are 1.96 m and 0.78 m/day, respectively. The snow depth has tended to decrease since 1980. As shown in Fig. 4 (b), the heavier the daily snowfall is, the deeper the snow depth is. Scattering, however, is large in deeper regions. In other words, duration of heavy daily snowfall is not always continuously long, so that a snowfall duration is important in the hazard evaluation.

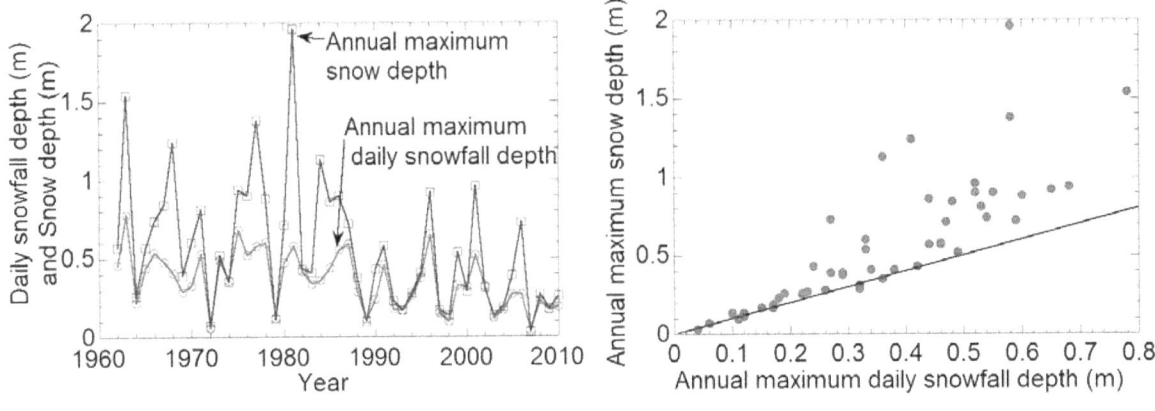

(a) Historical records of snow   (b) Correlation of daily snowfall and snow depth

Figure 4: Annual maximum data of daily snowfall depth and snow depth

3.2.2. Snow Hazard Evaluation Methodology

In this study, a snow hazard evaluation methodology was developed as described in Fig. 5 based on a probabilistic precipitation estimation methodology proposed by the JMA [12]. A basic concept of this methodology is a generalized estimation way. This is characterized by obtaining appropriate probability distribution through the conformance and stability evaluations.

The annual maximum data of the snow depth and daily snowfall depth was collected in Fig. 4. At first, using these data, the annual exceedance probability can be evaluated by plotting position formula: Weibull, Hazen and Cunnane for general use. Of the three formula, it is said that the Cunnane is the best suitable and applicable to all probability distributions. Next, the parameters of Gumbel or Weibull cumulative probability distributions are determined by a least square method. Using the annual exceedance probability, the snow hazard curves can be obtained after checking the conformance and stability evaluations.

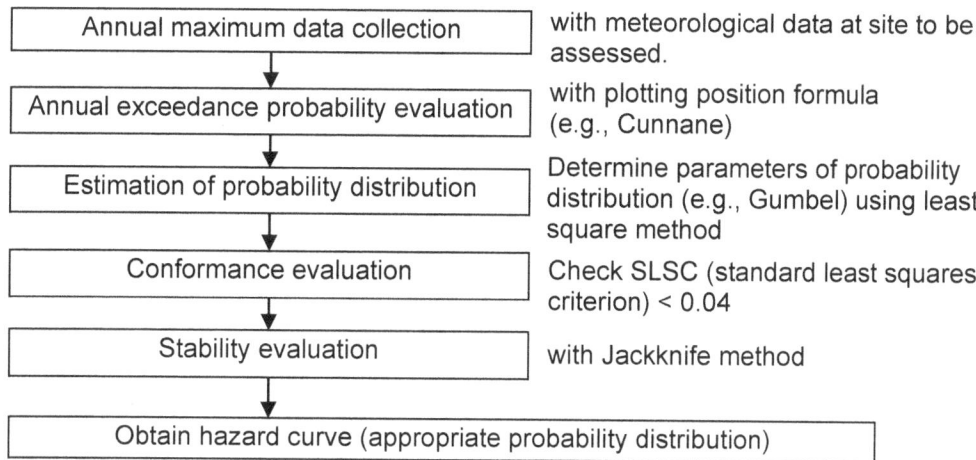

Figure 5: Snow Hazard Evaluation Methodology

3.2.3. Snow Hazard Category for PRA

Based on the snow hazard evaluation methodology in Section 3.2.2, the snow hazard curves were obtained in terms of the annual maximum snow depth and the annual maximum daily snowfall depth, as presented in Fig. 6. Given that the snowfall is time dependent, the snowfall speed (daily snowfall depth) and snowfall duration are important in the PRA for the personnel snow removal action. Using them, we have categorized the snow hazard to evaluate event sequences with the time dependence of snowfall. The snow hazard categories are obtained as a combination of the snowfall speed and snowfall duration that is defined as the snow depth divided by the snowfall speed.

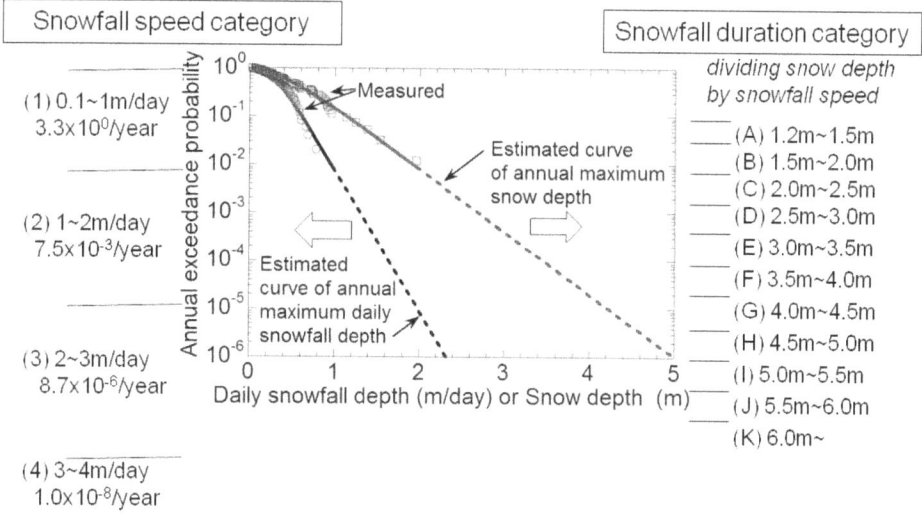

Figure 6: Snow Hazard Category

The snowfall speed lower than 0.1 m/day was precluded in this study because it was expected very low failure probability of snow removal, which needed 12 days to reach 1.2 m that could affect the plant. The present PRA also precluded the snowfall duration corresponding to the snow depth lower than 1.2 m because of no core damage. The snow hazard categories were represented as 44 combinations of four snowfall speed and eleven snowfall duration categories. The snow hazard category higher than 4 m/day of snowfall speed was regarded as the core damage because the annual exceedance probability was estimated less than 10^{-11}/year. For conservativeness, the present PRA used the maximum value in each snowfall speed category; e.g., 1 m/day in the category of 0.1−1 m/day.

The snowfall duration can be calculated by dividing the snow depth by the snowfall speed. For example, the snowfall duration category for 1.2−1.5 m becomes 0.4 day (9.6 hrs) − 0.5 day (12 hrs) of snowfall in case of 3 m/day. In this paper, the snowfall duration category corresponding to higher than 6 m of snow depth was also assumed as the core damage because the probability was estimated less than 10^{-7}/year by the snow depth curve. It should be noted that the CDF assessed in this paper is not lower than 1.7×10^{-7}/year, corresponding to the snow hazard category of a combination of the snowfall speed of 0.1−1 m/day and the snowfall duration corresponding to higher than 6 m.

3.3. Event Tree

As described in Section 3.1, air ventilation channels shall be secured for the important components in this PRA: EDG, ACS and MCS. The natural circulation decay heat removal is expected in the SFR, so that manual operation of the ACS-AC dampers is required in a total blackout situation (the loss of direct current powered equipment). They became headings of an event tree developed in the present PRA (see Fig. 7). Although personnel are usually aware of snowfall as much as necessary snow removal by weather forecast, it was assumed in this study that the awareness of snow removal was required when snowfall started. If there is no awareness of snow removal necessity, the event sequence would results in core damage. Several opportunities were actually expected for the awareness, but the present PRA assumed three chances for the awareness of snow removal necessity. This was incorporated into the first heading in the event tree.

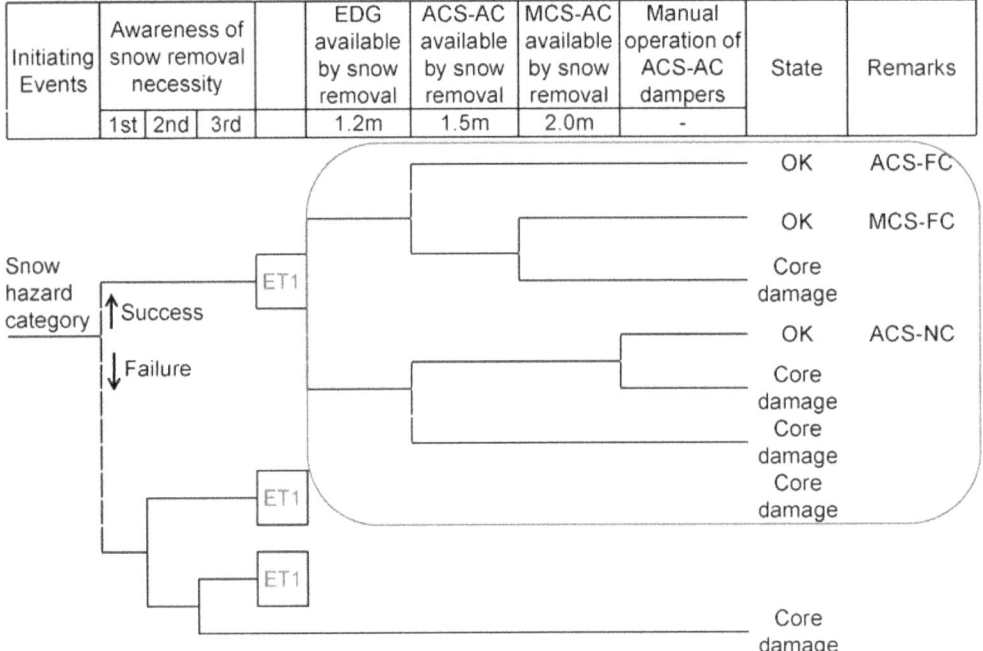

NB) EDG: Emergency Diesel Generator, ACS: Auxiliary Cooling System, MCS: Maintenance Cooling System, AC; Air Cooler, FC: Forced Circulation, NC: Natural Circulation

Figure 7: Main Event Tree for Snow PRA

3.4. Failure Probability for Each Heading

Based on NUREG/CR-1278 [13], the failure probability of the awareness of snow removal necessity was obtained 8.5×10^{-4}/demand using a human error probability with 60 minutes of grace period assuming a multiplier (one share) of an optimum stress level, step-by-step task, and skilled personnel.

Vulnerability against the snow hazard was focused in the present snow PRA, and thus we neglected component random failures involved usually in an internal-event PRA. The filter clogging of the air ventilation channels in EDG, ACS and MCS was assumed to depend only on the failure of snow removal task. Figure 8 shows snow removal failure probability model developed in this study. Its failure probability was modeled assuming a normal distribution with 1σ of 0.5 m/day. The average value of nominal snow removal speed was assumed 3 m/day in this paper. The present PRA assumed that the frequency of snow removal task increased according to the snowfall duration. For example, the snow removal times (days) were assumed five in the case of 5.0–5.5 m depth and 1 m/day (5.0–5.5 days) and once in the case of 5.0–5.5 m depth and 4 m/day (1.3–1.4 days). Figure 8 (b) indicates the increase of failure probability in many chances of snow removal (longer snowfall duration).

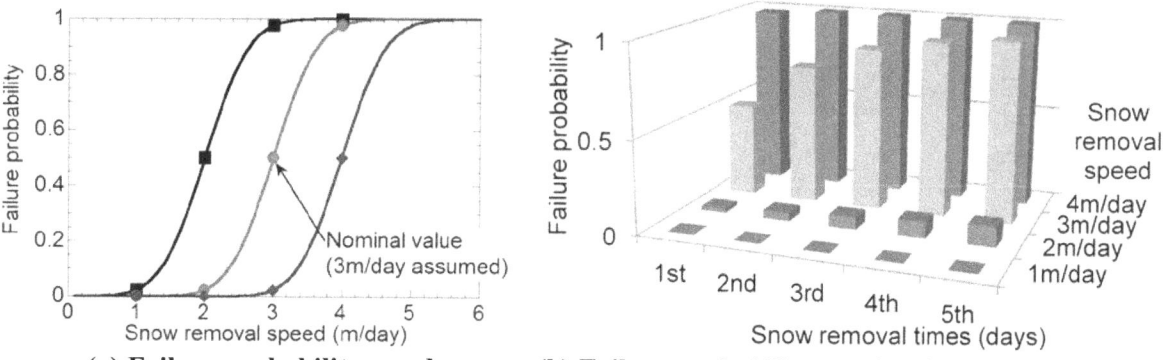

(a) Failure probability per day (b) Failure probability considering snow removal times (days) with 3 m/day of nominal speed of snow removal

Figure 8: Assumed Snow Removal Failure Probability

In manual operation of the AC dampers, sodium temperature measurement is necessary to prevent sodium freezing due to excessive cooling by keeping the damper opening. The reactor coolant temperature usually decreases to about 250°C in three days and then approaches to about 200°C in several days under the natural circulation heat removal condition with three loop ACSs. In the present PRA, the snowfall duration was considered six days at longest. Within six days, sodium freezing in the ACS can be neglected judging from the sodium temperature decrease history mentioned above. In the present PRA, therefore, the sodium temperature measurement was not necessary, and the failure of the natural circulation cooling was assumed to be dependent on the manual operation failure of the AC damper opening. Based on NUREG/CR-1278 [13], the human error probability was specified in regard to the opening operation of two dampers in one loop AC. The manual operation failure probability was estimated 2.4×10^{-4}/demand assuming a high dependence of recovery by a two-personnel implementation task and a low dependence of recovery by plant parameter diagnostics after the task. This estimate was multiplied by 5 assuming very high stress level, step-by-step task, and skilled personnel. Finally, the failure probability of the AC damper manual operation was estimated 6.5×10^{-3}/demand as an average value using 10 of error factors.

3.5. Event Tree Quantification

The decay heat removal failure probability of each event sequence was obtained by introducing the failure probability in Section 3.4 into the event tree in Section 3.3. Figure 9 shows the heat removal failure probability by the snow hazard category. The higher the snowfall speed is, the higher the failure probability is. The failure probability increases when the snowfall duration is long (expressed as the snow depth in this figure).

The CDF by the snow hazard category can be calculated by multiplying each heat removal failure probability described above and each snow hazard occurrence frequency. The CDF brings total to less than 10^{-6}/year. Figure 10 shows the CDF by the snow hazard category. The snow hazard curve allowed the conditional CDFs to appear at relatively low snowfall velocities and short snowfall duration. Although the CDF is highly visible in the snowfall duration longer than 6 days (snow depth higher than 6 m in this figure) in 1 m/day, this value can be distinguished because this was assumed as core damage regardless of event sequences under this snow hazard category, mentioned in Section 3.2.3. This visible CDF could disappear if the snow hazard category is extended. As shown in Fig. 10, the dominant snow hazard category was a combination of 1–2 m/day of snowfall speed and 0.75–1.0 day of snowfall duration. Given that such a snowfall condition is not so rare in some areas in Japan, this PRA result is expected to be useful for future considerations against a lot of snow.

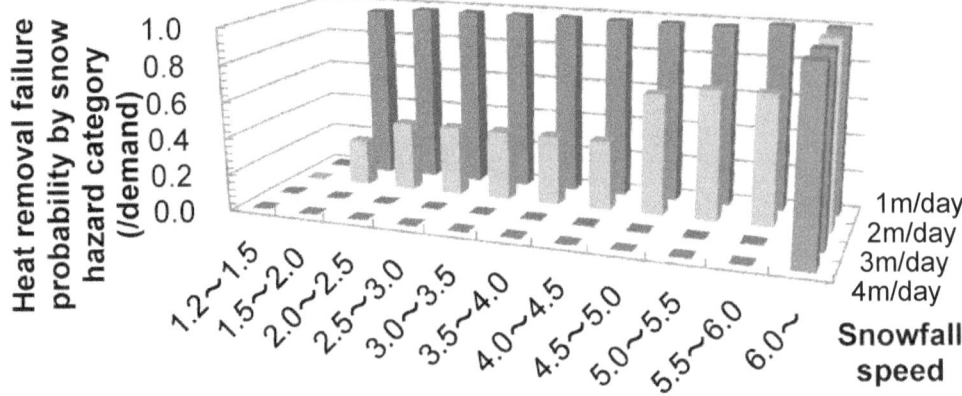

Figure 9: Core Damage Frequency by Snow Hazard Category

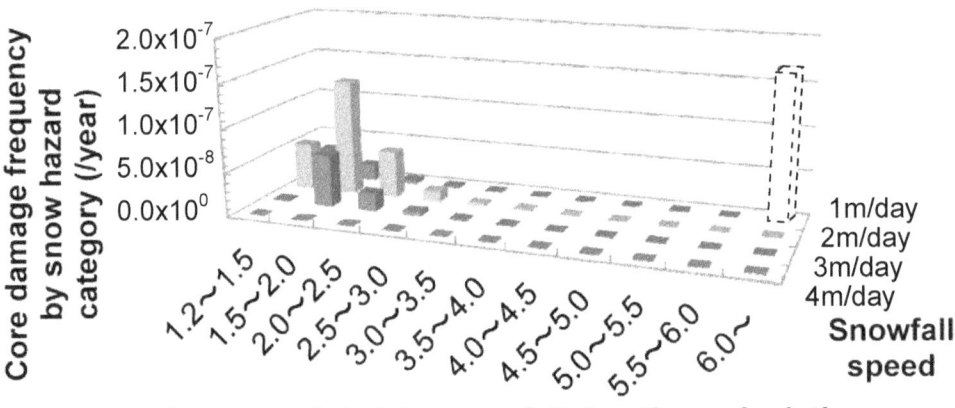

Figure 10: Core Damage Frequency by Snow Hazard Category

3.6. Sensitivity Analysis

A number of various assumptions could bring a large uncertainty in this PRA. Given that the failure probabilities would be changed at the headings in the event tree, their sensitivities should be analyzed for future usefulness and effectiveness of the snow PRA. In this paper, sensitivity analyses were carried out with parameters of the awareness times, the failure probability of awareness, the snow removal speed, and the failure probability of the AC damper manual operation.

The CDFs obtained in the sensitivity analyses are plotted in Fig. 11. When the awareness the snow removal necessity is once, the CDF does not significantly decrease as the snow removal speed. When the frequency of the awareness increased from once to twice, the CDF remarkably decreased by approximately three orders of magnitude. This suggests that the awareness more than once is very important. The CDF significantly decreased as the snow removal speed increased in the awareness both twice and three times. The twice awareness cases were shown similar to the three-time awareness cases even if the snow removal speed was changed. This was because the CDF lower than 10^{-6} cannot be reduced due to the simplification of the present PRA. The effectiveness of many awareness times and high snow removal speed would be investigated for future task. The CDF became three times higher in case of ten times manual operation failure probability and 0.6 times in case of 0.1 times probability. This indicated that the manual operation failure probability was less sensitive.

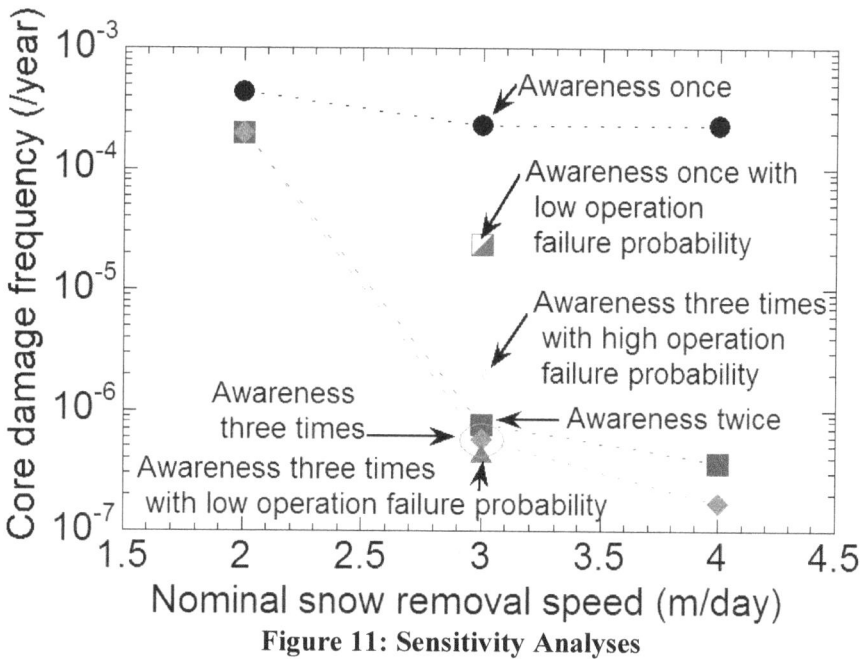

Figure 11: Sensitivity Analyses

4. CONCLUSION

This paper reported the first outcome of the four-year research project, which has started since 2012 to develop a margin assessment methodology of decay heat removal function against external hazards. The scope of external hazards in this study was shown above-ground hazards, which might influence the decay heat removal system of SFR. Through the screening process, this project selected extreme weathers (snow, tornado, wind and rainfall), volcanic eruption and forest fire as representative external hazards. This paper indicated the concept of the assessment methodology developed in this project, which consisted of both the margin assessment and PRA methodologies. On that account, we have initiated to develop the external hazard evaluation, the event sequence and the margin assessment methodologies for each external hazard.

This paper described the preliminary PRA results. The snow hazard evaluation methodology was developed through the assessment using 50-year data of the snow hazard indexes of the annual maximum snow depth and the annual maximum daily snowfall depth at the typical SFR site. Snow hazard categories were defined as the combination of the snowfall speed and snowfall duration. For each snow hazard category, the accident sequence was evaluated by producing event trees that consist of several headings representing the failure of the decay heat removal. In this paper, the snow PRA showed less than 10^{-6}/year of CDF. The dominant snow hazard category was the combination of 1−2 m/day of snowfall speed and 0.75−1.0 day of snowfall duration. Sensitivity analyses indicated important personnel actions, which were the snow removal speed improvement and frequent awareness of snow removal necessity.

In this snow PRA, the simplified evaluation brought conservative CDF by rough categorization of the snow hazard, thus finer hazard categorization than the present PRA is expected to improve the CDF estimation. Furthermore, plant walk-down should be carried out for the improvement of event tree development. The snow removal speed should be also investigated to reduce the uncertainty of sensitive parameter. In addition to the PRA, the margin assessment methodology will be developed against snow. In this project, the PRA methodology will be developed against other hazards: tornado, wind, rainfall, volcanic eruption and forest fire.

Acknowledgements

The present study is the result of "Research and Development of Margin Assessment Methodology of Decay Heat Removal Function against External Hazards" entrusted to Japan Atomic Energy Agency by the Ministry of Education, Sports, Science and Technology (MEXT). The authors wish to thank K. Naruto of NESI, Inc. and H. Sakaba of Mitsubishi Heavy Industries, Ltd. for their assistance of the investigations.

References

[1] Atomic Energy Society of Japan (AESJ), "*Implementation Standard of Seismic Probabilistic Safety Assessment for Nuclear Power Plants: 2007,*" AESJ-SC-P006:2007, AESJ, 2007, Tokyo, (in Japanese).

[2] Atomic Energy Society of Japan (AESJ), "*Implementation Standard Concerning the Tsunami Probabilistic Risk Assessment of Nuclear Power Plants: 2011,*" AESJ-SC-RK004E:2011, AESJ, 2012, Tokyo.

[3] European Nuclear Safety Regulators Group, "*Peer Review Report; Stress Tests Performed on European Nuclear Power Plants,*" 2012-04-25 (online).

[4] M. Mattéi, E. Vial, V. Rebour, H. Liemersdorf, M. Türschmann, "*Generic Results and Conclusions of Re-evaluating the Flooding in French and German Nuclear Power Plants,*" Eurosafe 2001, INIS-FR-1069, Nov. 5-6, 2001, Paris.

[5] International Atomic Energy Agency, "*External Events Excluding Earthquakes in the Design of Nuclear Power Plants,*" Safety Guide, NS-G-1.5, 2003, Vienna.

[6] Committee on the Safety of Nuclear Installations, "*Probabilistic Safety Analysis (PSA) of Other External Events Than Earthquake,*" NEA/CSNI/R(2009)4, OECD Nuclear Energy Agency, 2009, Paris.

[7] International Atomic Energy Agency, "*Site Evaluation for Nuclear Installations, Safety Requirements,*" NS-R-3, 2003, Vienna.

[8] International Atomic Energy Agency, "*Meteorological and Hydrological Hazards in Site Evaluation for Nuclear Installations,*" Specific Safety Guide, SSG-18, 2011, Vienna.

[9] J. A. Lambright. et al., "*Analysis of Core Damage Frequency: Peach Bottom, Unit 2 External Events,*" NUREG/CR-4550, Vol.4, Rev.1, Part3, U.S. Nuclear Regulatory Commission, 1990, Washington D.C.

[10] M. P. Bohn. et al., "*Analysis of Core Damage Frequency: Surry Power Station, Unit 1 External Events,*" NUREG/CR-4550, Vol.3, Rev.1, Part3, U.S. Nuclear Regulatory Commission, 1990, Washington D.C.

[11] Japan Meteorological Agency, http://www.jma.go.jp/jma/index.html.

[12] Japan Meteorological Agency, "*Extreme Weather Risk Map,*" (in Japanese) available at http://www.data.kishou.go.jp/climate/riskmap/

[13] U.S. Nuclear Regulatory Commission, "*Handbook of Human Reliability Probability with Emphasis on Nuclear Power Plant Applications,*" NUREG/CR-1278, 1983, Washington, D.C.

Screening of Seismic-Induced Fires

James C. Lin[a]*, Donald J. Wakefield[a], and John Reddington[b]
[a] ABSG Consulting Inc., Irvine, California, United States
[b] First Energy Nuclear Operating Company, Akron, Ohio, United States

Abstract: Seismic-induced fire has been an issue not addressed quantitatively in both the nuclear plant seismic PRAs and fire PRAs mainly because of the lack of data and a method to estimate the likelihood of a seismic-induced fire. One approach to identify the seismic-induced fire scenarios and evaluate the occurrence frequencies of these scenarios is to perform a screening analysis based on both the likelihood and the impact of such scenarios. Based on frequency of seismically induced fire initiation, there are two aspects to screening fire scenarios: (1) to assess the subset of seismic failure modes that may contribute to fires, the fragility for the structural failure modes including support/anchorage failures conservatively bounds the seismic failure potential; (2) the other factor that can be considered is the conditional probability of potential fire ignition. The seismic screening capacity can be determined by identifying an assumed fragility with which a convolution of the seismic hazard exceedance curves will result in a frequency of SSC failure integrated over the entire seismic hazard acceleration range below an acceptable screening value. For the remaining SSCs that survive the seismic capacity screening, additional screens based on fire consequences can be performed to reduce the number of scenarios to a minimal set for further detailed, quantitative evaluations.

Keywords: Seismic-Induced Fire, Structural Failure Mode, Ignition Probability, Screening, Seismic PRA.

1. INTRODUCTION

Seismic-induced fire has been an issue not addressed quantitatively in both the nuclear plant seismic probabilistic risk assessments (PRA) and fire PRAs mainly because of the lack of data and a method to estimate the likelihood of a seismic-induced fire. Furthermore, the locations of the seismic-induced fires and the possibility of multiple seismic-induced fires are also difficult to identify. However, given a seismic-induced fire at a specific location, the impact of the seismic-induced fire can be characterized in a relatively straightforward manner using information from the seismic PRA and fire PRA. Nevertheless, identification of seismic-induced fire scenarios is still a challenging task.

One approach to identify the seismic-induced fire scenarios and evaluate the occurrence frequencies of these scenarios is to perform a screening analysis based on both the likelihood and the impact of such scenarios. Before we attempt to identify the seismic-induced fire scenarios, let us examine first how a seismic-induced fire may occur.

2. IGNITION SOURCES AND ENERGY

Based on the past experience, many of the seismic-induced fires that occurred initiated in non-seismically qualified equipment, perhaps due to their higher likelihood of being structurally damaged during a large earthquake. As such, in our search for the potential fire sources, the non-seismically qualified equipment should certainly be considered both as ignition and fuel sources.

To cause a fire, the ignition source with sufficient ignition energy must come in contact with the fuel (i.e., combustibles) or its vapor. During an earthquake, sparks may result from both mechanical and electrical effects. Mechanical friction and impact of a metal object during a seismic event can cause a

---

* jlin@absconsulting.com

spark. An electrical spark can be induced by pulling loose an electrical wire connections or contacts between exposed wires/junctions with metal objects. Sparks from a broken/damaged wire during an earthquake are considered more likely than making contacts with exposed wires/junctions. In addition to sparks, ignition energy can also come from hot surfaces.

For sparks generated by mechanical friction or impact with a metal object, the amount of energy contained in the spark is relatively limited because the duration of impact or contact is very short. During a large earthquake with strong vibration and significant displacement, supports for structures, equipment, cabinets, and piping, as well as ceiling/wall materials, etc., may be damaged resulting in falling of their broken pieces and causing impacts with other objects. In addition, vibration and the resultant differential displacements may also cause contacts or friction between two objects that are normally separated spatially. As such, this type of spark may be generated during the time of strong shaking and can occur inside any part of the plant buildings where structural/mechanical failures or differential displacements take place.

The energy contained in the sparks caused by a loose electrical wire or bus bar connection varies depending on the voltage level of the wire or bus bar. Typically, the sparks that may result from a loose electrical wire or connection with the bus bar involving a voltage level below 480V (e.g., 125VDC, 120VAC, or even lower voltage), the energy content is somewhat limited. The arcing that may be generated by a loose electrical wire or connection with a bus bar at voltage 480V and above can contain sufficient energy to ignite most of the combustibles.

A high energy arcing fault has caused explosions in the past. Since the cable trays, conduits, and their support structures are generally built with very strong seismic capacity, they do not fail easily during an earthquake. Most likely, the high energy arcing would only occur in areas where switchgears/buses/motor control centers (MCC) having a voltage level of 480V and above are located. These types of electrical cabinets are typically present in separate rooms (especially for switchgears/buses of 4.16KV or higher) or in areas without substantial combustibles nearby, except for cables. In addition, if the raceways routed into and out of this type of electrical cabinets are sturdily supported or anchored to the wall, the amount of combustibles that could be in contact with any electrical sparks that are generated by overturning of the cabinet would be significantly reduced.

An example electrical arc flash could result from an earthquake-induced severing of the conduits connecting to the top of the electrical cabinets and of the cables routed inside the conduits. Such electrical cabinets as switchgears and MCCs are typically anchored to the concrete pads on the floor. Conduits entering from the top of the cabinets are sometimes braced to the steel frames. Adequate cable flexibility across building joints is thus an important consideration in evaluating the likelihood of a seismic-induced fire due to electrical arcing.

Ignition sources involving equipment with hot surfaces are typically at fixed locations. The energy from hot surfaces can be transferred to the fuel that is in contact with these surfaces. Therefore, to cause fire ignition, the most likely mechanism is for the fluid fuel to spread to the location of the hot surfaces and get in contact with the hot surfaces. These hot surfaces may be the exterior casing of rotating equipment (e.g., pumps, motor-generators, compressors, fans, chillers); motor, generator, or transformer windings; exposed steam pipes (e.g., a small segment of pipes with damaged insulation or with insulation removed for imminent maintenance activities), etc.

3. FUELS/COMBUSTIBLES

There are many different types of combustibles in a nuclear plant. They include cable insulation/jacket materials, plastic casing/materials, hydrocarbon fuels (including such liquid fuels as diesel fuel oil, lubricating oil, and hydraulic oil, as well as gaseous fuels such as hydrogen and propane), etc. In general, solid fuels are more difficult to ignite. Flammable gases and liquid fuels can not only be more easily ignited but also could result in larger fires due to their substantial heat content and the greater potential for fire propagation.

To cause an ignition of the cable insulation/jacket and/or plastic materials, significant ignition energy is needed. It is extremely difficult for mechanical impact/friction sparks or hot surfaces to ignite these types of materials. Therefore, without an external exposure fire, the most likely ignition source for cable insulation/jacket and plastic materials is the high energy arcing generated by a seismically broken/damaged electrical wire or bus bar connection. However, cables and plastic materials can be present in most locations inside the plant buildings.

Hydrocarbon fuels, however, are only present in specific locations inside the plant buildings. In nuclear plants, diesel fuel oil is found in the diesel oil storage tank, diesel oil day tank, and diesel oil piping inside the building that houses the diesel generators or any diesel driven pumps. The largest amount of lubricating oil is the turbine lube oil stored in the Turbine Building. In addition, oil-filled transformers and such rotating equipment as pumps may also contain non-negligible amounts oil (especially larger transformers and pumps). Hydraulic oil is typically contained in the hydraulic equipment in the Turbine Building. These types of hydrocarbon fuel are limited to those locations where the corresponding equipment is located. Due to the relatively low vapor pressure, it is difficult for fuel oil, lubricating oil, and hydraulic oil to be ignited by mechanical sparks because of their limited spark energy available. They can, however, be ignited by hot surfaces and electrical arcing generated from seismically damaged/broken wire (with a voltage of 480V and higher) connections for the oil-filled transformers or motor windings.

For gaseous fuels such as hydrogen and propane, they can be ignited much more easily by any sparks or hot surfaces due to their dispersion characteristics so long as their concentrations in the building atmosphere is within the flammability limits. In general, hydrogen is used in the Turbine Building for generator cooling. It is also used in, for example, the Auxiliary Building for selected Chemical and Volume Control System functions. Typically, hydrogen bottles are stored outside the plant buildings to minimize the impact of its fire and explosion hazards on the plant equipment. Inside the plant buildings, there are primarily small hydrogen tubes (e.g., approximately 1" to 2" diameter lines) which usually are seismically robust. In addition, these tubes are generally well supported (e.g., supported alongside the building walls and not protruding like a cantilever). Unless failure of the building walls occurs, it is unlikely to damage these tubes resulting in a release of the hydrogen gas. Although rarely, it is possible, however, that there may be a very small number of hydrogen bottles inside the Auxiliary or Turbine Building. In nuclear plants, propane is used primarily by the auxiliary boiler in the Turbine Building. Propane tanks are almost always located outside the plant buildings. There may be small propane lines that connect these tanks with the auxiliary boiler inside the Turbine Building. Again, these lines generally have high seismic capacity and are unlikely to fail during an earthquake. However, seismic failure of the structural support/anchorage for the auxiliary boiler could cause a rupture of the connecting propane line.

Based on the preceding considerations, potentially, the most likely seismic-induced fire sources are expected to include:

- High energy arcing of seismically damaged wire or bus bar connections in the areas of switchgears/buses/MCCs (480V and higher) may cause ignition of the nearby combustibles (primarily cables). These are mostly in the switchgear rooms and other plant locations for MCCs. This could lead to damage to the switchgears/buses/MCCs and any other PRA equipment in the area or spread of the fire to additional areas.
- High energy arcing of seismically loosened bus bar connections in the switchgear rooms (480V and higher) may cause explosion. The explosion overpressures could blow open the switchgear room doors and perhaps lead to spreading the ensuing fire to outside the room.
- Seismically damaged piping or containers of hydrogen in the Auxiliary Building may cause a release of hydrogen followed by an ignition of the hydrogen gas by sparks or hot surfaces. The hydrogen fire or explosion could also cause ignition of additional combustibles in the area leading to damage to PRA equipment.

- Seismically damaged piping, seals, blowout panel, or containers of hydrogen or propane in the Turbine Building (e.g., seismically damaged seal to the generator or blowout panel on the covers) may cause a release of hydrogen or propane (e.g., seismic failure of auxiliary boiler) followed by an ignition of the hydrogen or propane gas by sparks or hot surfaces. The hydrogen/propane fire or explosion could also cause ignition of additional combustibles in the area leading to damage to PRA equipment.
- Seismically damaged piping or containers of diesel fuel oil in the Diesel Generator Building may cause a severe leakage of diesel fuel oil followed by an ignition of the diesel fuel oil by hot surfaces or high energy electrical sparks. The resulting oil fire could damage any other PRA equipment in the area.
- Seismically damaged container of turbine lubricating oil in the Turbine Building may cause a major leak followed by an ignition of the lube oil by hot surfaces or high energy electrical arcing. The resulting large oil fire in the Turbine Building could damage PRA equipment in the Turbine Building.
- Seismically induced failure of the pump/compressor (with a significant oil inventory) supports/anchorage in plant buildings may cause a large oil leak followed by an ignition of the oil by hot surfaces or high energy electrical arcing; e.g., from the damaged wire connection for the pump motive power cables.
- Seismically induced failure of the oil-filled transformer supports/anchorage in plant buildings may cause a large oil leak followed by an ignition of the oil by hot surfaces or high energy electrical arcing; e.g., from the damaged wire connection for the transformer winding.
- Seismically induced failure of the supports/anchorage for hydraulic equipment inside plant buildings (e.g., Turbine Building) may cause a large hydraulic oil leak followed by an ignition of the hydraulic oil by hot surfaces or high energy electrical arcing.
- Seismically induced soil failure underneath an oil-filled transformer pad in the yard may cause a structural failure of the transformer and a large oil leak followed by an ignition of the oil by hot surfaces or high energy electrical arcing; e.g., from the damaged wire connection for the transformer winding.

The seismically-induced fire ignition of transient combustibles is expected to be less likely because these solid or liquid fuels must be located nearby a seismically-induced ignition source (e.g., an electrical arc flash, a hot surface) at the time of a large earthquake.

4. FAILURE MODES

To cause an ignition of the hydrocarbon fuel, the containing equipment of the hydrocarbon fuel must first seismically fail causing a release of the hydrocarbon fuel. The released hydrocarbon fuel must then be ignited by an ignition source with sufficient energy. The seismic failure of the containing equipment for the hydrocarbon fuel occurs when the seismic excitation force exceeds its seismic strength capacity. This failure likelihood increases as the seismic acceleration increases. The probability of ignition of the released hydrocarbon fuel is less dependent on the magnitude of the earthquake, although not completely independent because the stronger the seismic excitation, the more structural failures would occur which may result in more impacts and sparks.

For an ignition caused by high energy electrical arcing, sufficient differential displacement must occur causing the electrical wire or bus bar connection to be pulled loose or apart. Only then, arcing may occur due to an electrical discharge across the air gap. Ignition will occur if the spark is in contact with a combustible and the arcing energy is sufficient to cause the combustible to ignite. To pull loose the electrical wire or bus bar connection of a switchgear/bus/MCC, typically, it may involve seismic failure of the cabinet support/anchorage leading to overturning of the electrical cabinet. Therefore, the failure mode is primarily the seismic failure of the switchgear/bus/MCC cabinet anchorage/support. The failure probability is the fragility of the cabinet anchorage/support. Given that there is a seismic-induced differential displacement sufficient to pull the electrical wire or bus bar connection loose, the likelihood of arcing and ignition is dependent on whether the loose connection would be in contact with a combustible and whether sufficient spark energy would be imparted to the combustible. This conditional probability of arcing and ignition is perhaps nearly independent of the magnitude of the earthquake, given that the wire or bus bar connection is already pulled loose.

For a tank containing hydrocarbon fuel to seismically fail causing a release of the hydrocarbon content, the likely failure modes include the support/anchorage failure induced by sufficient vibration energy which leads to buckling failure followed by crack and leakage (for flat bottom tanks), movement of the tank causing failure of the anchor bolts and then failure of the attached piping, support/anchorage failures leading to overturning and ruptures of the tank on impact (less likely than the other two modes), etc. For a pipe to release its hydrocarbon content during a seismic event, the most likely failure modes include seismic failure of the pipe support or excessive differential displacement of the pipe resulting in mechanical impact with an adjacent structure which leads to the rupture of the pipe. This could imply that the evaluation can be focused in areas containing structures, systems, and components (SSC) of different seismic categories, pipe anchorage, and its vicinity.

Similarly, for an oil-filled transformer or a pump to release its oil, there must be sufficient vibration energy to fail its anchorage/support causing the equipment to overturn and rupture its oil casing on impact; i.e., requires structural/anchorage failure to cause a release of the oil contained inside this equipment. For a piece of equipment with a high aspect ratio (e.g., a vertical pump compared to a horizontal pump), there is certainly a higher likelihood to overturn following anchorage/support failures. For an oil-filled transformer with a very low aspect ratio, strong earthquake excitation can cause the anchorage/support to fail and thus move laterally, but it would be more difficult to overturn resulting in a greater impact stress. Note that most of the oil-filled transformers (including Non-Seismic Category I transformers) have similar construction and anchorage. Therefore, most oil-filled transformers have similarly high seismic structural capacity. To also create an arcing, the seismic-induced differential displacements must rip open the electrical wire connections to the transformer or pump/compressor motor. This could also occur when the anchorage fails and the transformer/pump/compressor overturns. Therefore, the most important failure mode is judged to be seismic failure of the anchorage/support causing the transformer/pump/compressor to overturn. Given that both the oil is released and a high energy arcing is generated, the likelihood of ignition is judged to be more of a random event nearly independent of the earthquake magnitude.

5. PROBABILITY OF IGNITION

As discussed in the preceding, the likelihood of a seismic-induced fire given an earthquake is the joint probability of a seismic structural/mechanical failure and ignition. The probability of seismic failure (e.g., seismic failure of the anchorage/supports of an electrical switchgear, pump, transformer, compressor, and tank) can be estimated by the seismic fragility analysis method. The seismic fragility is a function of the seismic acceleration value; e.g., pga. The probability of ignition given a seismic failure leading to a release of hydrocarbon fuel or resulting in the electrical wire or bus bar connections being pull loose/apart is relatively independent of or much less dependent on the earthquake magnitude. Ignition probability is mainly based on the presence and the density/amount of ignition sources in the area, whether the ignition sources can be in contact with the combustibles in the area, how much ignition energy is required and whether there is sufficient energy from the ignition source to ignite the combustibles, etc.

The energy required to ignite the combustible is dependent on the type of the fuel. For gaseous hydrocarbon fuel (e.g., hydrogen and propane), the least amount of energy is required to ignite the vapor cloud of the fuel. Besides, due to the dispersion characteristics of this type of fuel, it can be most easily in contact with ignition sources present in the area. Also, it can be ignited by most ignition sources considered, including the mechanical sparks that may be generated during an earthquake due to falling, collapsing, or movement of objects. As such, the ignition probability of this type of fuel should be the highest. However, the ignition probability may vary depending on the release rate of the hydrocarbon gas. The larger the release rate, the more ignition sources can be encompassed by the flammable gas cloud, prior to dispersion, and thus the greater the chance of ignition. Since the flammable gas cloud released can also be ignited by the mechanical sparks generated during an earthquake event, this portion of the ignition probability contribution may increase as the earthquake magnitude increases. However, it is possible that this portion of the ignition probability may only have a small contribution. Besides, as the extent of seismic failures increases, seismic-induced core damage would be more dominated by seismic failures other than the impacts of seismic-induced fires.

For the liquid hydrocarbon fuel present inside the nuclear plant buildings (e.g., diesel fuel oil, lube oil, hydraulic oil, etc.), the energy required to cause an ignition is significantly higher than that for a flammable gas due to the relatively low vapor pressure. In addition, the ignition sources that can be in contact with this type of fuel are much more limited than those for a flammable gas cloud. The ignition sources must be located where the liquid fuel can spread to. As a result, the probability of ignition given a release of a liquid hydrocarbon fuel should be significantly lower than that for a flammable gas. Furthermore, the probability of ignition for lubricating oil and hydraulic oil should be lower than diesel fuel oil. Nevertheless, the higher the release rate of the liquid fuel, the faster and the farther it can spread and thus the more ignition sources can be encountered; i.e., greater ignition probability.

To ignite the solid combustibles such as cable insulation/jacket and plastic materials, in general, the highest ignition energy may be required. As such, without an external exposure fire, only certain ignition sources (e.g., high energy arcing) can cause the ignition of these materials during an earthquake. Besides, these combustibles are generally fixed in locations. The loose wire or bus bar connections must be close enough to these combustibles for the sparks to cause an ignition which will require a significant differential displacement between the loose wire or bus bar connections and the combustibles. Therefore, it is believed that ignition of solid combustibles during an earthquake should have the lowest frequency of occurrence due to the smallest probability of ignition.

Table A.5 of Reference [1] shows the following ignition probabilities in the process areas inside a petrochemical facility:

Type of Release	Size of Release	Probability of Ignition
Gas	Minor (< 1 kg/s)	0.01
	Major (1 – 50 kg/s)	0.07
	Massive (> 50 kg/s)	0.3
Liquid	Minor (< 1 kg/s)	0.01
	Major (1 – 50 kg/s)	0.03
	Massive (> 50 kg/s)	0.08

The above ignition probabilities can be considered as the probability of immediate ignition; i.e., ignition within a short period from the time of release (e.g., within a few minutes). For flammable gases, a delayed ignition may also occur because the flammable gas can continue to disperse until encountering an ignition source (as long as the concentration of the flammable gas is still within the flammability limits). During an earthquake, immediate ignition appears to be more applicable because mechanical sparks are typically only generated during the period of vibration or during a very short period following the shaking. In addition, due to the compartment design, the ignition sources within the compartment in which the combustibles are located have the most chance of contributing to the seismic-induced fires.

Considering the sizes of the hydrogen/propane lines and the possible break sizes of the oil leakage (from tanks, transformers, pumps, compressors, etc.) in a nuclear plant, it is likely that the release rate resulting from a seismic failure is no greater than 50 kg/s. Even though additional ignition sources may be created during an earthquake (e.g., mechanical sparks due to falling, impact, and friction of objects), it is judged that the probability of ignition should still be less than 0.1 (which is significantly greater than the ignition probability for flammable gases released at less than 50 kg/sec) for all of the ignition source and fuel/combustible combinations considered in this evaluation. Since this is a relatively conservative estimate of the probability of ignition, this conservative value can be considered as bounding for earthquake conditions; i.e., even with the additional ignition sources of mechanical sparks. Because the ignition of combustibles by high energy arcing in the switchgear/bus/MCC area or at the transformers, pumps, or compressors is expected to be significantly less likely than that for the ignition of a flammable gas cloud, these scenarios can also be bounded by this conservative ignition probability of 0.1.

Considering both the seismic structural/mechanical failure and the probability of ignition, the conditional probability of ignition given an earthquake is thus the joint probability of the fragility of the seismic failure mode considered (anchorage/support failure in most cases) and the ignition likelihood; where the seismic fragility is a function of the seismic acceleration and the bounding ignition probability is treated as independent of the seismic acceleration. It is conservatively assumed that the resulting seismic failure mode is an ignitable configuration, with probability 1.0.

6. SCREENING OF SEISMIC-INDUCED FIRE SCENARIOS

Using the preceding considerations, both qualitative and quantitative screening of the seismic-induced fires can be performed. The qualitative screening of the seismic-induced fire scenarios can consider both the likelihood and the impact. The compartment by compartment evaluation may start with a selected set of compartments that contain specific ignition/fuel source combinations that can potentially cause a seismic-induced fire. Additional evaluation criteria that directly influence the likelihood and consequence of a seismic-induced fire may then be used to further screen potential seismic-induced fire scenarios.

The discussion of the ignition source and fuel/combustible combinations in the preceding is largely from the likelihood standpoint. We consider flammable gas and liquid fuel because they are more likely to be ignited. Additionally, the emphasis on the fluid fuel is partly attributed to the potential for a larger fire and thus greater impact. Ignition by high energy arcing is also considered because of its higher likelihood of occurrence. Therefore, the following equipment (which can either be a fuel source or an ignition source) is considered based on this perspective:

- Tanks, Bottles, and Piping (including turbine-generator, auxiliary boiler) That Contain Hydrogen, Propane, and Any Other Flammable Gases
- Above-Ground Tanks and Piping That Contain Diesel Fuel Oil
- Tanks, Equipment, and Piping That Contain Lubricating Oil
  - Turbine-Generator
  - Turbine Lube Oil Storage Tank
  - Oil-Filled Transformers
  - Pumps (especially large pumps)
  - Compressors
  - Piping
- Tanks, Equipment, and Piping That Contain Hydraulic Oil
- Equipment with Electrical Wire or Bus Bar Connections at 480V and Above
  - Switchgears/Buses/MCCs
  - Pumps
  - Oil-Filled Transformers
  - Compressors
  - Others (e.g., other applicable NUREG/CR-6850 fire source bins from Fire PRA that are unique and significant for specific plants)

From the impact standpoint, the screening will examine if the seismic-induced fire (which can be derived from the internal fire PRA analysis of the compartments where the ignition sources/fuel are located) will result in additional impacts than those functions already modeled by seismic failures; i.e., the full impacts of the seismic fragility items, including functional effects. In addition, one needs to determine if additional fire propagation pathways may be created by the seismic failures modeled if the compartment does not screen; e.g., collapse of a block wall separating two fire compartments, structural failure of penetration assemblies. Of course, if the conditional impact of seismically induced fires in a compartment is relatively limited or small (e.g., as reflected by a small conditional core damage probability, CCDP, and a small conditional large early release probability, CLERP), seismic-induced fires in that compartment may not need to be considered. If the combined impact of seismic failures and seismic-induced fire is not significantly more severe than that for the seismic failures alone, the postulated seismic-induced fire scenario may also be screened.

For each fire compartment where the credible ignition sources/fuel considered are located (e.g., an oil-filled transformer, a pump, or a compressor containing significant amount of lubricating oil), one will examine if additional PRA equipment is located in the vicinity that can be damaged by the seismic-induced fire source. It must be noted that this is only considering the additional PRA equipment without counting the seismically failed fuel source (which leads to the release of gas or liquid fuel) or ignition source. Although this additional PRA equipment located in the same fire compartment could also be failed by the seismic force (i.e., modeled as a seismic fragility item), the seismic-induced fire is of interest if it increases the likelihood of failure of this additional PRA equipment. Besides, evaluation could also consider the distance between the ignition sources, combustibles, and fire damage susceptible PRA equipment if the likelihood of fire source igniting is high enough to not allow screening out otherwise. If the fire damage susceptible PRA equipment is located at sufficient distance away from the fire source, seismic-induced fire would not lead to additional impacts and can therefore be screened out.

Furthermore, if relevant seismic structural failure is also modeled for this fire compartment, evaluation needs to be performed to determine if this seismic structural failure would alter the potential of fire spread and the overall fire impacts. Also, the seismic-induced fire may impact the post-initiator operator actions by increasing the diagnosis difficulty and stress for the control room operators and by creating a harsh environment for local operator actions that need to travel through or perform specific action in the fire affected area. Therefore, for each such compartment, the operator actions that may potentially be affected should be identified and evaluated if the compartment is not already screened out by other considerations.

For quantitative screening, we propose to use the joint likelihood of seismic structural failure of anchorage/support and ignition probability (i.e., the product of seismic fragility for this failure mode and the ignition probability as a function of seismic acceleration level in pga) in conjunction with the seismic hazard frequencies and the PRA model (which accounts for the combined impact of both seismic failures and seismic-induced fires) to determine the risk significance of seismic-induced fire scenarios. By varying the seismic fragility value, it is possible to identify a seismic capacity curve versus acceleration (for the seismic structural failure of the anchorage/support considered for seismic-induced fire scenarios) beyond which a bounding estimate or a more realistic analysis of the core damage frequency (CDF)/ large early release frequency (LERF) risk contribution is below an acceptable cutoff value; e.g., 1.0E-7/year for CDF. Because the ignition probability is also considered in this evaluation, it is expected that this seismic capacity screening value obtained for the inclusion of seismic-induced fire scenarios would be noticeably lower than the seismic capacity screening value for the inclusion of direct seismic failures in the seismic PRA. Once this seismic capacity screening value for seismic-induced fires is identified, it can be used during walkdown to screen the possible ignition and fuel sources for the inclusion of the seismic-induced fire scenarios in the seismic PRA.

Therefore, based on the frequency of seismically induced fire initiation, there are two aspects to screening fire scenarios:

- To assess the subset of seismic failure modes that may contribute to fires, the fragility for the structural failure modes including support/anchorage failures (but not the functional failure modes which are often lower) conservatively bounds the seismic failure potential.
- The other factor that can be considered is the conditional probability regarding potential fire ignition. Based on data from other industries (e.g. the oil and gas industry where the ignition

probability is a key component of the analysis), this ignition probability is bounded by 0.1 for all SSC types that may lead to fire.

The seismic screening capacity can be determined by identifying an assumed fragility (e.g., in terms of the high confidence low probability of failure [HCLPF] capacity in pga) with which a convolution of the seismic hazard exceedance curves will result in a frequency of SSC failure integrated over the entire seismic hazard acceleration range below an acceptable screening value. If 5.0E-7 per year is taken to be the single SSC screening fire initiation frequency, the frequency of a single structural failure would be 5.0E-7 divided by the bounding conditional probability of ignition of 0.1; i.e., 5.0E-6 per year. Note that seismically initiated fires do not necessarily, by themselves, cause core damage. Other failures that must also occur could drive the fire-induced core damage frequency lower.

In addition, at high accelerations, there is overlap between the seismic-induced fire impacts and other contributors to seismic core damage. So, the added contribution to core damage frequency from the seismic-induced fire scenarios is not the full 5.0E-07/year due to this overlap. Therefore, because of the additional failures reflected by the CCDP associated with the seismic-induced fire scenario that must occur to lead to core damage and the overlap of seismic-induced fire damages with other seismic failures at high accelerations, an added contribution of lower than 1.0E-7 is thus expected if the SSC has a single SSC seismic failure frequency of 5.0E-7 per year. Further, considering the bounding ignition probability, the frequency of the structural failure mode considered can be taken to be 5.0E-6 per year.

If the seismic screening capacity thus determined is bounded by an acceleration of, for example, 0.35g, it means that SSCs with structural failures modes with a HCLPF greater than 0.35g could be screened by just considering the 0.1 conditional probability of ignition. Such a screen would eliminate many potential fire sources from further investigation. Furthermore, if we exclude the seismic-induced fire contribution from accelerations greater than the level above which the conditional seismic core damage probability is 1.0 (e.g., only consider seismic-induced fires at seismic levels less than 0.5g), the screening seismic capacity for seismic-induced fire scenarios could be as low as 0.25g.

For the remaining SSCs that survive the seismic capacity screening, additional screens based on fire consequences can be performed to reduce the number of scenarios to a minimal set for further detailed, quantitative evaluations.

The procedure that can be used for the identification and screening of seismic-induced fires is to perform the evaluation compartment by compartment. This is mainly because many of the seismic-induced fires may be initiated from the non-seismically qualified equipment which may not be included in the seismic equipment list (SEL) or in the seismic PRA (SPRA) model. However, during the seismic PRA walkdown of the SSCs included in the SEL, the potential effects of Seismic Category II SSCs over Seismic Category I are examined by the fragility analysts. Nevertheless, this walkdown evaluation is performed from the standpoint of seismic failure interactions; i.e., not from the perspective of seismic-induced fires. As such, the SEL developed may not be complete for the analysis of seismic-induced fires. However, a special table can also be compiled for the likely fire ignition sources that are not included in the SEL; e.g., hydrogen lines, fuel oil lines. If no credit is taken for equipment inside a specific building, SSCs in that building can be excluded from this table because the seismic-induced fire impacts resulting from these SSCs cannot add to the core damage frequency. As such, SSCs in fire compartments that are located in buildings that are not credited for seismic events can be screened.

7. CONTAINMENT INTEGRITY

The impact of seismic-induced fires on the containment integrity is primarily the possible effects of fire on the failure of containment isolation or spurious opening of valves leading to interfacing systems loss of coolant accident (LOCA) (ISLOCA). These could result from the fire impacts on the electrical cabinets/MCCs containing the circuitries for the control of the containment isolation function/valves, and for the control of the isolation valves involved in ISLOCA. Therefore, the plant areas containing these electrical cabinets should also be examined for the potential of a seismic-induced fire. For the isolation functions associated with the containment isolation valves, most of the relevant electrical cabinets would be at low voltage level (i.e., 480V and below). Since containment isolation has a significant impact on LERF, fires caused by seismic-induced arcing related to these electrical cabinets may also need to be considered.

However, almost all of the air-operated and perhaps, at some plants, selected motor-operated containment isolation valves are designed to fail in the close position on loss of power or air. Fire damage to their control circuits could result in a hot short preventing the air-operated (or selected motor-operated valves) to fail in the close position. But, based on the results of the fire tests conducted in recent years [2], essentially all of the hot shorts would eventually turn into the open-circuit failure mode. The longest duration of hot short in the previous fire tests was not longer than 12 minutes. For an air-operated (or selected motor-operated) valve with a fail-safe design, the open circuit failure mode will cause the valve to close. Therefore, for containment isolation valves or ISLOCA related isolation valves, only cabinets associated with motor-operated valves need to be examined. At some U.S. nuclear plants, the power supplies to the motor-operated valves at high-low pressure interfaces are removed (e.g., de-energized with the breaker racked out) to prevent inadvertent opening (e.g., fire-induced) of these motor-operated valves, regardless of the cause of the fire.

8. MULTIPLE, CONCURRENT SEISMIC-INDUCED FIRES

Due to the correlation in both the seismic excitation and SSC strengths, seismic PRAs typically model like equipment located in the same building at the same elevation as dependent; i.e., they would be treated as failing concurrently. For seismic-induced fires, this implies that multiple fires may occur concurrently in the plant. However, while the structural failure aspects may be correlated, the conditional probability of ignition at different locations (i.e., in different compartments) is largely independent. Therefore, overall, the occurrence of multiple, concurrent fires are still, to a large extent, a random phenomenon. For the initial screening of seismic-induced fire scenarios, the evaluation can be performed individually for each compartment since seismic-induced fires in different compartment are treated as independent. Once all of the seismic-induced fire scenarios and their corresponding locations have been identified, one can re-evaluate if it is possible that any of these seismic-induced fire scenarios identified are correlated.

9. WALKDOWN IDENTIFICATION AND SCREENING

During seismic-induced fire walkdown, inspections of the ignition and fuel sources discussed previously should be performed to determine:

- Could the seismic structural (e.g., anchorage/support) failure occur with significant likelihood by determining if the seismic capacity for the failure mode considered is above the screening value?
    - Is the anchorage/support sufficiently strong? Well supported/anchored? Co-located SSCs of different seismic categories?
    - Is it possible for the equipment considered to overturn causing a severe impact stress?
    - Is it possible for seismic-induced differential displacement or impact stress due to overturn to cause rupture of the pressure boundary for the fuel source?
    - Is it possible for seismic-induced differential displacement or impact stress due to overturn to pull loose/apart electrical wire or bus bar connections? Adequate cable flexibility across building joints?
- Are there ignition or fuel sources nearby to permit ignition?
    - Given a release of fluid fuel, are there ignition sources nearby with sufficient ignition energy to cause ignition? Are these ignition sources properly secured (i.e., can be free of seismic damage)? Can these ignition sources be in contact with the fuel?

- Are there combustibles nearby? Are these combustibles properly secured (i.e., can be free of seismic damage)? Can these combustibles be in contact with the ignition sources?
 - Are the possible ignition or fuel sources adequately secured to greatly reduce the likelihood of ignition?
 - Can the ignition lead to a significant fire?
 - Are there additional combustibles available to permit fire spread?
- Can the seismic failures create additional fire propagation pathways in the area?
- In addition to the SSCs that fail seismically, is there additional PRA equipment in the area that can be impacted by the seismic-induced fire?
 - Is there additional PRA equipment in the compartment that can be damaged by a seismically induced fire?
 - Are the fire-induced CCDP and CLERP for the fire compartment below the screening values (if the seismic failures do not introduce additional fire propagation pathways and enlarge the fire impacts evaluated in Fire PRA)?
 - Is the fire damage susceptible PRA equipment located with sufficient distance away from the seismic-induced fire sources (i.e., beyond the zone of influence for fire impacts)?
- Are there any post-earthquake operator actions performed in the area or that must pass through the area?
 - Would these post-earthquake operator actions be further affected by the seismic-induced fire effects?

The above considerations and evaluations will help to identify the seismic-induced fire scenarios that can realistically occur based on the actual plant configurations if the potential fire compartments do not all screen.

10. CONCLUSION

Seismic-induced fire scenarios can be evaluated by first performing identification and screening of the potential scenarios. Both qualitative and quantitative screening can be conducted. Qualitative screening can be based on the potential for the types and locations of equipment that may cause a seismic-induced fire as well as the potential impacts that may result. Quantitative screening can be performed using the frequency of seismic-induced fire initiation to determine screening seismic capacity value for a single SSC which should be noticeably lower than the screening seismic capacity for direct seismic failure contributors to CDF/LERF because a conditional ignition probability can also be considered. Quantitative screening can also use the fire consequence reflected by the fire compartment CCDP and CLERP as additional criteria.

References

[1] WS Atkins Consultants Ltd. *"Development of a Method for the Determination of On-Site Ignition Probabilities"*, Health and Safety Executive, RSU 4014/R04.081, (2004).

[2] Electric Power Research Institute, *"Characterization of Fire-Induced Circuit Faults, Results of Cable Fire Testing"*, EPRI Final Report 1003326, December 2002.

Minimization of Vulnerability for a Network under Diverse Attacks

Jose Emmanuel Ramirez-Marquez[a*] and Claudio Rocco[b]

[a] School of Systems and Enterprises, Stevens Institute of Technology, Hoboken, NJ, USA
[b] Facultad de Ingeniería, Universidad Central de Venezuela, Caracas, Venezuela

Abstract: This paper describes an approach to minimize the vulnerability of a network under a defender attacker context. To do so, vulnerability is defined in the context of a resilience-building framework and corresponding mathematical formulations are provided. The solution to network optimization model is based on a three-phased approach consisting on identifying Pareto optimal defense strategies with respect to cost and vulnerability for a known set of network attacks. These solutions are then utilized to identify the network defense strategy that can offer the best protection against any of the attacks. Examples are used to illustrate the approach.

Keywords: Resilience; vulnerability; networks; multi-objective; optimization.

1. INTRODUCTION

During the last decade the concepts of reliability, vulnerability, survivability and resilience as applied to systems have become commonplace and widely discussed. For the last 50 years reliability engineering, theory and methods, have been continuously used to satisfy key stakeholder requirements in a myriad of systems and applications [1]. Among reliability engineers, analysts and researchers there is a standard theory that is understood throughout these communities.

When considering the concepts of vulnerability, survivability and resilience there is neither standard theory nor common language understood among and within these different communities. For example, in the transportation context [2] presents vulnerability as a concept describing "...susceptibility to incidents that can result in considerable reductions in road network serviceability". This definition immediately adds two additional paradigms to be considered: susceptibility and serviceability. Recently, in the same context, [3] describe vulnerability as "...the weakness of a network..." Similarly, survivability has been described as "the capability of a system to fulfill its mission, in a timely manner, in the presence of threats such as attacks or large-scale natural disasters [4]. However, DoD Regulation 5000.2-R states that survivability is "...the capability of a system and crew to avoid or withstand a man-made hostile environment without sustaining an impairment of its ability to accomplish its designated mission. Survivability consists of susceptibility, vulnerability, and recoverability." In fact, Castet and Saleh [5] note that in the engineering context, the concepts of survivability and resilience are sometimes used interchangeably. Finally, for the concept of resilience [6] describes it as related to "...the speed at which an entity or system recovers from a severe shock to achieve a desired state...". However, according to [7], "resilience can be expressed as the post-disruption fraction of demand that can be satisfied by using specific resources while maintaining a prescribed level of service." The reader should note the different measurement in these two definitions: speed in [6] and demand in [7].

From the authors' perspective, the issue at hand is first of definition: a single concept, resilience, is currently used to define one too many ancillary concepts. Thus, due to the conflicting perspectives in the paradigms discussed, this paper has a two-fold contribution: first, to clarify the concepts of vulnerability and survivability as complementary to the resilience framework described in [8] and [9] and second, to provide an optimization based vulnerability reduction approach against diverse number of attacks on a network.

The remaining sections of the paper are organized as follows: Section 2 presents the first contribution of the paper, discussing in detail the resilience framework. Section 3 describes the approach to reduce vulnerability in networks when considering a defender attack contest while section 4 presents examples and results. Finally conclusions are given in section 5.

---

* jmarquez@stevens.edu

2. RESILIENCE FRAMEWORK

With respect to the first contribution, Figures 1.a and 1.b present the resilience building framework as described by [10] and developed based on the model by [9]. In this illustration a system provides a service that is measured or assessed via service function $\varphi()$. The system experiences three different states:

Stable Original State (Reliability Theory) – The normal behavior of the system is considered in the interval t_e-t_0. The theory of Reliability Engineering [1] provides models and techniques to analyze and measure the probability that under normal conditions the failure time is greater than some value t: $R(t)=P(T>t)$, $t \in (t_0, t_e)$. In reliability engineering, the period of time t_e-t_0 corresponds to the system time to failure, where at time t_e, a failure event occurs. In the context of reliability failures occur due to events that are intrinsic to the system.

System Disruption (Vulnerability Theory) – The methods [11,12,13] in this area are used to: i) understand how disruptive events affect the service function –for example by analyzing probability that a disruptive event does not affect the service function below some threshold b: $P(\varphi(t)>b|e_j)$– and ii) identifying the components that are critical to the system (i.e. those components that when "degraded" affect system service function the most). As described in figures 1.a and 1.b, the vulnerable period is contained in the interval t_d-t_e. The difference between these two figures is that Figure 1.a considers service functions for which decreasing values correspond to system degradation: throughput, flow, jobs, number of satisfied costumers, etc.. In contrast, Figure 1.b considers those service functions for which increasing values correspond to system degradation: delay, unsatisfied customers, areas without power, etc… Vulnerability and Survivability are strongly related; from this manuscripts perspective, survivability is the study of methods to minimize the vulnerability of systems, mathematically (for the case of Figure 1.a): Min $\varphi(t_e)$-$\varphi(t_d)$.

System Recovery (Resilience Theory) – Recently, mathematical models and methods have been proposed in different areas to understand the recovery of the system service function from some disruptive event e_j. As described in both figures 1.a and 1.b the recoverability period is contained in the interval of lenght t_f-t_d. At the end of this period the service function enters a new recovered state, which may or may not be identical to the original state. The main research question in this area is to understand how restoration policies affect the system recoverability [4, 9, 14, 15].

It is important to note that the system resilience process is a function of time that can be quantified for different service functions and for different disruptive events. To clarify, one cannot discuss system resilience in the absence of system vulnerabilities (i.e. no system disruption implies no system resilience process) and resilience should be discuss in the context of time (i.e. the resilience of the system at time t.) Based on the framework described in Figure 1.a, and for deterministic cases, system resilience has been defined by [9] as the ratio of restoration at time t_r, $\varphi(t_r|e_j)-\varphi(t_d|e_j)$, to losses up to time t_d, $\varphi(t_0)-\varphi(t_d|e_j)$, mathematically as in equation 1:

$$Я\varphi(t_r|e_j) = \frac{\varphi(t_r|e_j)-\varphi(t_d|e_j)}{\varphi(t_0)-\varphi(t_d|e_j)} \qquad \forall e_j \in D, t_r \in (t_s, t_f) \qquad (1)$$

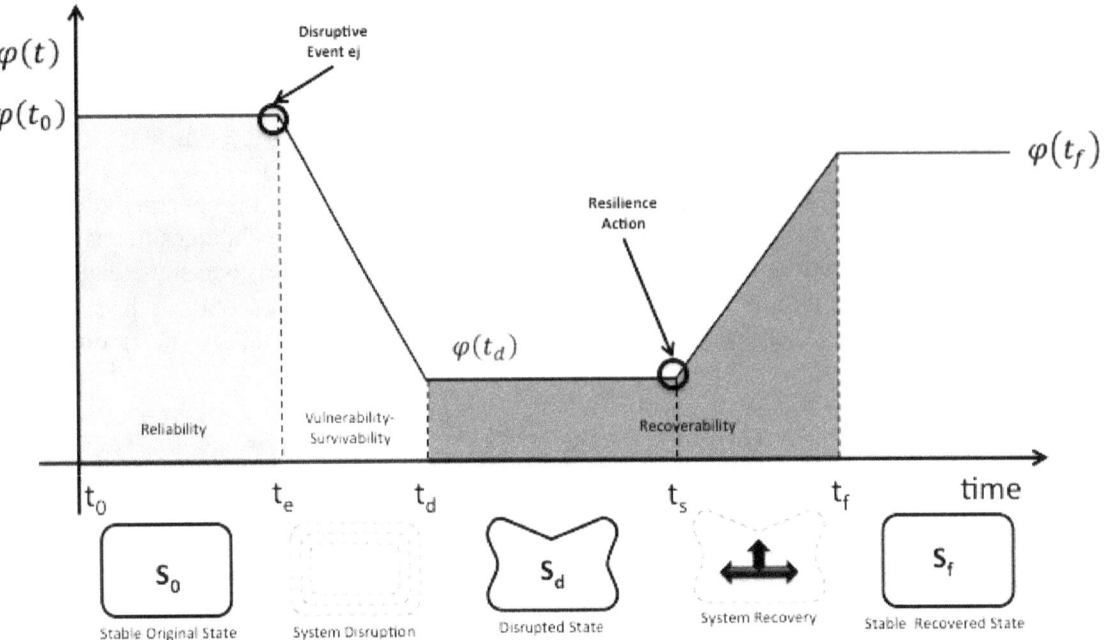

Figure 1a: Decreasing Service Function Resilience Process Illustration

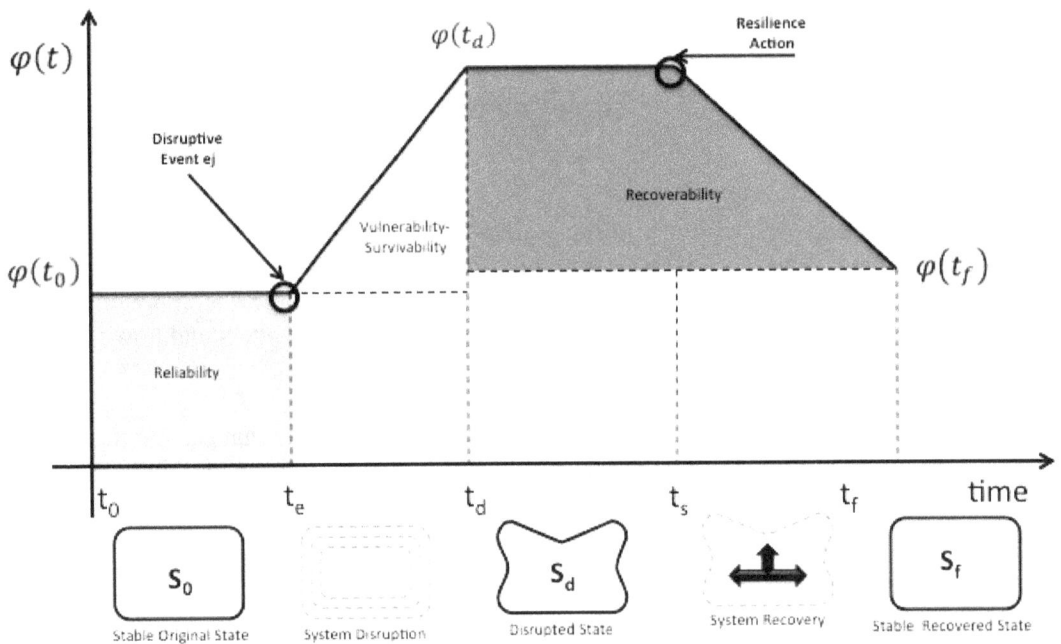

Figure 1b: Increasing Service Function Resilience Process Illustration

Equivalently, based on the framework described in Figure 1.b, and in for deterministic cases, system resilience can be defined as the ratio of restoration at time t_r, $\varphi(t_d|e_j) - \varphi(t_r|e_j)$, to degradation up to time t_d, $\varphi(t_d|e_j) - \varphi(t_0)$, mathematically as in equation 2:

$$Я\varphi(t_r|e_j) = \frac{\varphi(t_d|e_j) - \varphi(t_r|e_j)}{\varphi(t_d|e_j) - \varphi(t_0)} \qquad \forall e_j \in D, t_r \in (t_s, t_f) \qquad (2)$$

A number of studies have described related metrics, for example: [15] provide a temporal description of resilience but no mathematical formulation, [7] provide a demand based perspective, and [Rose 2007] analyzes at the economic impact of resilience. However the time dependent description of these two equations is novel and equation (2) is for the first time proposed.

3. NETWORK VULNERABILITY

To address the second contribution, consider that as described by [10] the denominator of equation 1 represents how vulnerable the system is with respect to event e_j: $V(e_j)=\varphi(t_0)- \varphi(t_d)$. And note that $\Re_F(t_r|e_j) \to \infty$ as $V(e_j) \to 0$. Moreover, $V(e_j) \to 0$ as $\varphi(t_d) \to \varphi(t_0)$. Thus, one can claim that the system is survivable to an event as $V(e_j) \to 0$.

To consider a network context, let G (N, A) represent a capacitated network with known source node s, and sink node t. N represents the set of nodes, and $A=A_1 \cup A_2$ where, $A_1=\{(s,i), (j,t) | 1 < i,j < n\}$ and $A_2 = \{(i,j) | 1 < i, j < n\}$ represent the set of links. For G(N, A), $k_{ij}(a_{ij})$ an element of network state vector $\mathbf{k}$, represent the capacity vector of link (i,j), where $a_{ij}=0$ if link (i,j) has been destroyed and $a_{ij}=1$ if link (i,j) is in its normal state. Under this description, $0=k_{ij}(0)< k_{ij}(1)$ and $\mathbf{k}= (k_{s1}(a_{s1}), k_{s2}(a_{s2}),.., k_{st}(a_{st}), k_{12}(a_{12}),..., k_{ij}(a_{ij}),.., k_{nt}(a_{nt}))$ describes the current capacity of each link in the network.

In the context of this paper, G(N,A) can be disrupted by disruptive event e_k initiated by an adversary, where e_k contains a disruption scenario $\mathbf{e}_k= (e_{s1k}, e_{s2k},.., e_{stk}, e_{12k},..., e_{ijk},..., e_{ntk})$, where $e_{ijk} \in \Re^+$ defines disruption resources e_{ijk} allocated to each link i,j of G(N,A). The assumption in this paper is that a network defender is aware about possible attack scenarios, $\mathbf{e}_k$, in set D, $|D|=K$, but unaware of the specific event that will take place.

To minimize how vulnerable the network is, the defender can implement a defense strategy $\mathbf{h}= (h_{s1}, h_{s2},..., h_{st}, h_{12},..., h_{ij},..., h_{nt})$, where $h_{ij} \in \Re^+$, describes the defense resources invested to protect link i,j. Based on the defender and attacker strategies, the vulnerability $v_{ij}(\mathbf{t},\mathbf{h})$ of network link i,j can be mathematically described using the ratio form of the attacker-defender contest success function as originally presented in [16, 17]:

$$v_{ij}(\mathbf{e}_w,\mathbf{h}_v) = \begin{cases} \dfrac{e_{ij}^m}{e_{ij}^m + h_{ij}^m} & \text{if } e_{ij}^m > 0 \\ 0 & \text{if } t_{ij}^m = 0 \end{cases}$$

(3)

In (3), the attackers' and defenders' resource allocation for attacking/defending the link between nodes i and j, is dictated by the specific attack and defense strategies $\mathbf{e}_w$ and $\mathbf{h}_v$, respectively. In practical terms, as per (3) the vulnerability of the link i, j can be described as the probability that given attack and defense strategies, the flow capacity for link (i,j) is reduced from $k_{ij}(1)$ to $k_{ij}(0)$. It is important to note that that as described in Ramirez-Marquez et al. [18] the contest intensity m is motivated by the history of warfare. While used in this paper, (3) can be substituted for the appropriate contest function relating attack and defense resources to a probability value. Based on the network and vulnerability representation, $\varphi(\mathbf{k}): Z^{|A|} \to Z^+$ maps a network state vector into a maximum network flow between s and t. Note that whenever $\mathbf{e}_w$ and $\mathbf{h}_v$, are known, the capacity of each link in state vector, $\mathbf{k}$, is a random variable taking the following values with corresponding probabilities:

$$k_{ij} = \begin{cases} k_{ij}(1) & 1-v_{ij}(e_w,h_v) \\ 0 = k_{ij}(0) & v_{ij}(e_w,h_v) \end{cases}$$

It is important to realize that as defined, k_{ij} is a random variable that takes values as dictated by v_{ij}. Then, $\varphi(\mathbf{k})$ can be analyzed for any possible realization of $\mathbf{k}$ given $\mathbf{e}_w$ and $\mathbf{h}_v$. In this manuscript the performance function or figure-of-merit is the expected flow of the network between nodes s and t given the defense and attack strategy vectors $\mathbf{e}_w$ and $\mathbf{h}_v$. It can be defined as: $E(\varphi(\mathbf{k})|\mathbf{e}_w, \mathbf{h}_v)$.

3.1. Bi-Objective Optimal Network Protection

The Model BO-Vulnerability illustrates the optimization model considered for identifying the best defender's strategy against all events included in set D and at minimum cost. Note that in the first objective, the expected network s-t flow in $G(N,A)$ is computed for every event w, $\mathbf{e}_w \in D$ to identify a defense strategy that maximizes flow or minimizes the flow reduction. The second objective minimizes total defenders cost. The constraints of the model include the flow balance conservation equation (where $f(k_{ij})$ describes the flow through link (i,j), $f(k_{ij}) \in (0, k_{ij}(1))$ and the non-negativity behavior of the decision variable h_{ij}.

Model BO-Vulnerability

$$\underset{v}{Max}\; E\left[\varphi(\mathbf{k})|\mathbf{e}_w, \mathbf{h}_v\right] \quad \underset{v}{Min}\; C(\mathbf{h}_v) \quad \text{for every } \mathbf{e}_w \in D$$

subject to

$$\sum_{i|h_{ij}} f(k_{ij}) - \sum_{k|h_{jk}} f(k_{jk}) = 0 \quad \forall j \in N - \{s,t\}$$

$$\sum_{j|h_{sj}} f(k_{sj}) - \sum_{k|h_{kt}} f(k_{kt}) = 0 \quad \forall j,k \in N-\{s,t\}$$

$$h_{ij} \geq 0$$

The solution of Model BO-Vulnerability can be obtained via the following heuristic:

Step 1: Determination of Pareto Fronts for every $\mathbf{e}_w \in D$
For each $\mathbf{e}_w \in D$ and based on Ramirez-Marquez et al [19] identify the strategies in set H satisfying Pareto optimality as defined by the following conditions:
Condition 1: Feasible defense strategy $\mathbf{h}'(\mathbf{e}_w)$ dominates a feasible strategy $\mathbf{h}(\mathbf{e}_w)$, iff $C(\mathbf{h}'(\mathbf{e}_w)) \leq C(\mathbf{h}(\mathbf{e}_w))$, $E(\varphi(\mathbf{a})|\mathbf{e}_w,\mathbf{h}'(\mathbf{e}_w)) \geq E(\varphi(\mathbf{a})|\mathbf{e}_w,\mathbf{h}(\mathbf{e}_w))$ and $C(\mathbf{h}'(\mathbf{e}_w)) < C(\mathbf{h}(\mathbf{e}_w))$ or $E(\varphi(\mathbf{a})|\mathbf{e}_w,\mathbf{h}'(\mathbf{e}_w)) > E(\varphi(\mathbf{a})|\mathbf{e}_w, \mathbf{h}(\mathbf{e}_w))$. If no solution dominates $\mathbf{h}(\mathbf{e}_w)$, it is said to be non-dominated.
Condition 2: A defense strategy $\mathbf{h}'(\mathbf{e}_w)$ belongs to the Pareto set H^*, $\mathbf{h}'(\mathbf{e}_w) \in H^*$, iff $\neg \exists\; \mathbf{h}(\mathbf{e}_w) \in H: \mathbf{h}(\mathbf{e}_w)$ dominates $\mathbf{h}'(\mathbf{e}_w)$.

In this manuscript, any defense strategy $\mathbf{h}'(\mathbf{e}_w)$ satisfying conditions 1 and 2 is considered a Pareto optimal solution of the Model BO-Vulnerability with H^* its corresponding true Pareto set. Based on the description of any $\mathbf{h}(\mathbf{e}_w)$, the set H is of infinite cardinality since there are an infinite number of partitions for h_{ij}. For MO problems with infinite solution spaces the true Pareto set can rarely be completely characterized and solution procedures are based on approximating such a set.
The result of this step is a set of Pareto fronts $\mathbf{h}'(\mathbf{e}_w)$, for every $\mathbf{e}_w \in D$. Note that the number of possible defense strategies $\mathbf{h}'(\mathbf{e}_w) \in H^*$ is given by the cardinality of H^*.

Step 2: Behavior of defense strategies for every $\mathbf{e}_w \in D$.
This step analyzes the performance of $\mathbf{h}_v(\mathbf{e}_w) \in H^*$ under every other attack scenario $\mathbf{e}_u \in D$, $u \neq w$.
Montecarlo Simulation is used to generate: $E\left[\varphi(\mathbf{k})|\mathbf{e}_u, \mathbf{h}_v(\mathbf{e}_w)\right]$ for every $\mathbf{h}_v(\mathbf{e}_w) \in H^*$. At the end of

this step, each defense strategy, $h_v(e_w) \in H^*$ is characterized by the expected maximum flow or the expected flow reduction achieved for every attack $e_w \in D$, and its associated cost (i.e., $|D|+1$ values).

Step 3: Determination of the most convenient defense strategy.
As a result of step 2, each defense strategy can be represented as a multi indicator matrix with $|D|+1$ indicators. Multi-indicator matrices represent a set of objects characterized simultaneously by several indicators, criteria or attributes. This structure allows assessing each object, by considering simultaneously different criteria, and defining a ranking to synthesize the global characteristic of each object. Assuming that a defense strategy with lower flow reduction and lower cost is preferred, the strategies could be ranked, for example, from best to worst using a multi-criteria technique.

Multi-criteria ranking techniques are classified as parametric and non-parametric. The first group requires information about decision-maker preferences (e.g., weights assigned to each criterion), while non-parametric techniques do not use such information. In this paper the use of the Copeland Score (CS), a non-parametric technique is used, due to its simplicity.

The approach in this case selects the defense strategy with the largest CS, understood as the number of times a defense strategy is better than other defense strategies and subtracting the number of times that defense strategy is worse than other defense strategies, when they are compared pair-wise for each criterion [20]. Comparisons are made for each criterion and no normalization is required. Copeland Scores assume that each criterion has equal importance. Given a set of n objects, characterized by m criteria $q_j()$, j=1, .., m, the method builds a comparison matrix C. Each position $C(i,l)$ represents the count of comparison between object i and object l, considering each criterion q_j. If $q_j(i) \geq q_j(l)$ then $C(i,l)=C(i,l)+1$. If $q_j(i) \leq q_j(l)$, then $C(i,l)=C(i,l)-1$. Summing up $C(i,l)$ over all objects ($1 \leq l \leq n$), yields the CS(i) of object i. Objects are then ranked using the corresponding CS(i).

For the present case, the Copeland approach is able to identify the best "over all" defense strategy given the set of ALL possible defense strategies derived from the attack scenarios considered, along with the effects derived from Step 2.

4. ILLUSTRATIVE EXAMPLES

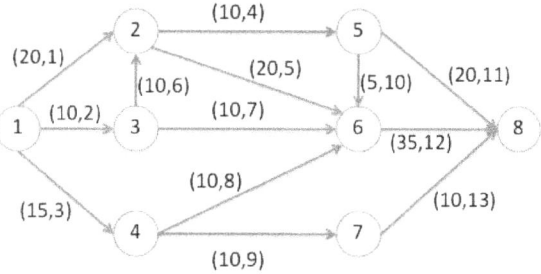

Figure 2: Illustrative Flow Network

To illustrate the proposed approach consider the network presented in Figure 2. Each link in Figure 1 has been assigned two values: capacity and index number, respectively. So for example, the link between nodes 1 and 2 has a capacity of 20 units and is indexed as link 1. In the case of no link failures, the network can handle a maximum flow of 45 units between the source node (node 1) and the sink node (node 8). To illustrate the optimization model and its solution as described in Section 3, an attack budget equal to 520 has been considered for contest intensity $m=1$ and three different attack scenarios e_u. In each scenario e_u the attack budget has been equivalently distributed among the following links: e_1: 2,5,9 and 12; e_2: 2,5,6 and 12; and e_3: 1, 2, 11 and 12.

Step1: Determination of Pareto Fronts - The procedure in [18,19] is used to derive the Pareto front for each scenario. The graphical results of the approximate Pareto front obtained for each scenario is displayed in Figure 3. Each point in the frontier represents a defense strategy with its corresponding maximum flow reduction and associated cost. The number of defense strategies derived for each scenario is: 225, 194 and 201 respectively.

Figure 3 allows for an initial understanding of the vulnerability of the network in Figure 2 for the attacks considered. For example, Attack 2 does not has the lowest effect in flow reduction as a function of cost and when compared against attacks 1 and 3. Table 1 shows 7 out of 225 defense strategies of the Pareto optimal set generated for attack 1. Pareto optimal defense strategies associated with attack 1 and evaluating each against attacks 2 and 3. Clearly from Figure 4 it is becomes evident that the defense strategies obtained for attack 1 do not provide as good defense against attack 3 but do relatively fine against attack 2. Table 2 shows the flow reduction for the selected defenses described in Table 1. As illustrated by Table 1, the point with the highest cost in Figure 1 has a value equal to 3709 with an associated expected flow reduction equal to 5.21. In this case, the defender must allocate resources of 911, 892, 709 and 1197 to links 2,5,9 and 12 respectively.

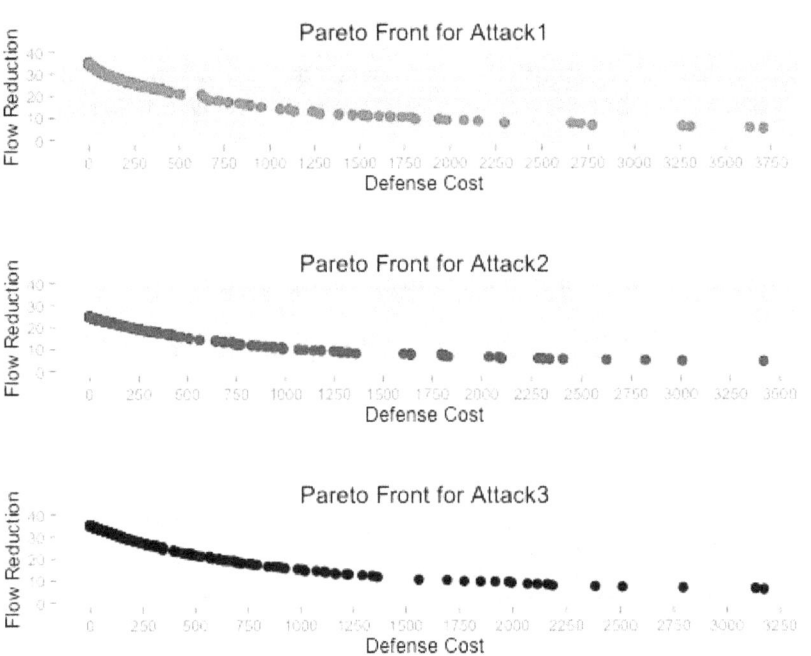

Figure 3: Pareto Fronts for Each Attack

Table 1: Selected Pareto Optimal Defense Strategies under Attack Scenario 1

Def Str	Flow Red.	Cost	Defense Resource Allocated to Links												
			1	2	3	4	5	6	7	8	9	10	11	12	13
a	35.00	0	0	0	0	0	0	0	0	0	0	0	0	0	0
b	29.95	84	0	0	0	0	0	0	0	0	49	0	0	35	0
c	25.08	266	0	15	0	0	33	0	0	0	90	0	0	128	0
d	20.42	626	0	236	0	0	108	0	0	0	157	0	0	124	0
e	14.90	951	0	198	0	0	267	0	0	0	183	0	0	303	0
f	10.24	1783	0	397	0	0	587	0	0	0	191	0	0	608	0
g	5.21	3709	0	911	0	0	892	0	0	0	709	0	0	1197	0

Step 2: Behavior of defense strategies under different attacks - The next step consists on evaluating each of the defense strategies identified in each of the three Pareto fronts described in Figure 3. Figure 4 shows the effect of selecting each of the Defense Strategies Against Attack 1 evaluated on every attack. Figures

7

5 and 6 show the corresponding plots when considering the Pareto optimal defense strategies associated with attack 2 and 3 respectively and evaluating each, against the remaining attacks
The analysis of figures 4, 5 and 6 illustrates that both the optimal defenses against attacks 1 and 3 do relatively good against attack 2. Yet, both the optimal defenses against attacks 1 and 3 do relatively bad in protecting against attack 3 and attack 1 respectively.

Step 3: Selection of the most convenient defense strategy - The determination of the most convenient defense strategy considering all possible attack scenarios is performed using the Copeland approach. Each defense strategy derived form the optimal defense strategies against attacks 1, 2 and 3 (for a total of 620 strategies) is represented by four criteria: the flow reduction under the three attack scenarios and the cost of the strategy. The defense strategy with the highest Copeland Score is the best "over all" strategy to be selected. Figure 7 shows the Copeland score (when considering all Pareto fronts) for each defense strategy in the Pareto fronts described in Figure 4. Table 3 shows the attributes of the best twenty strategies identified using the Copeland Score approach. Note that, no defense strategy is selected from the third Pareto front.

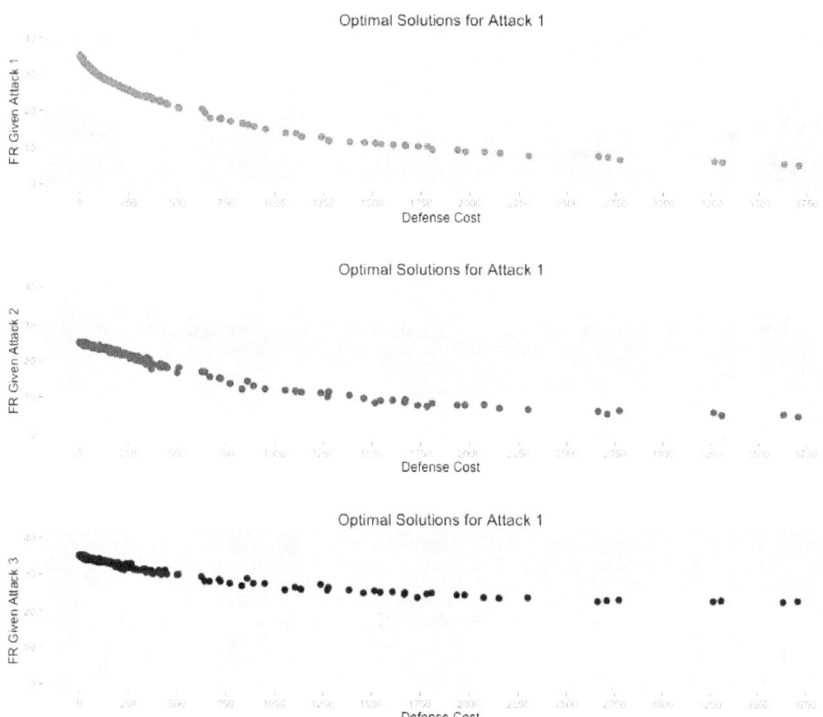

Figure 4: Optimal Defense Strategies Against Attack 1 evaluated on Every Attack
Table 2: Selected Pareto Optimal Defense Strategies under different Attack Scenarios

Defense Strategy	Flow Reduction Given Attack 1	Flow Reduction Given Attack 2	Flow Reduction Given Attack 3	Defense Cost
a	35.00	25.00	35.00	0
b	29.95	24.20	34.11	84
c	25.08	22.18	32.81	266
d	20.42	13.08	27.49	626
e	14.90	16.93	29.33	951
f	10.24	12.29	27.45	1783
g	5.21	7.90	24.35	3709

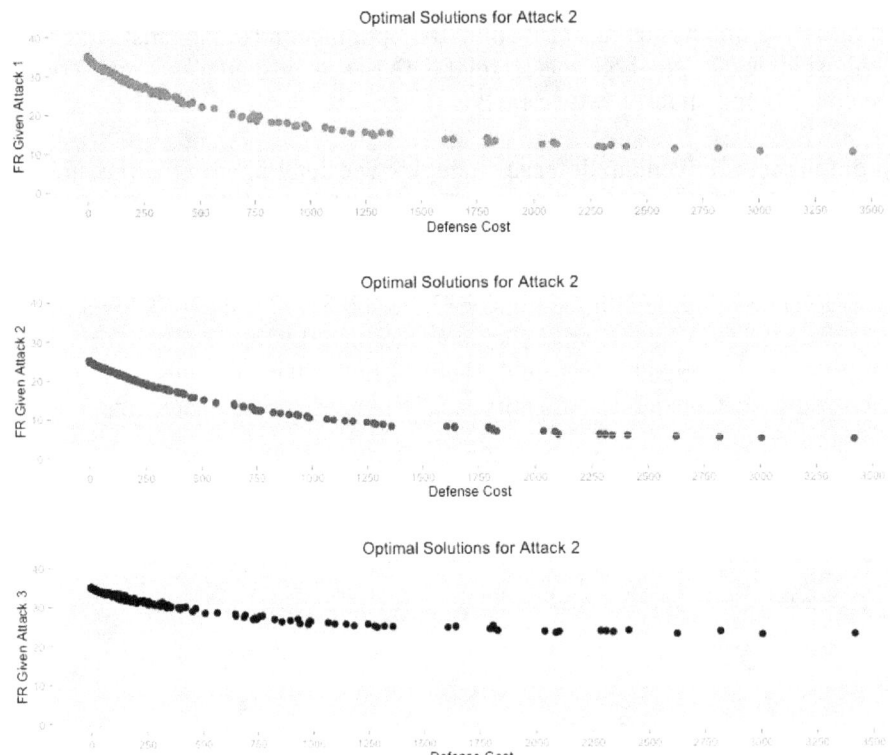

Figure 5: Optimal Defense Strategies Against Attack 2 evaluated on Every Attack

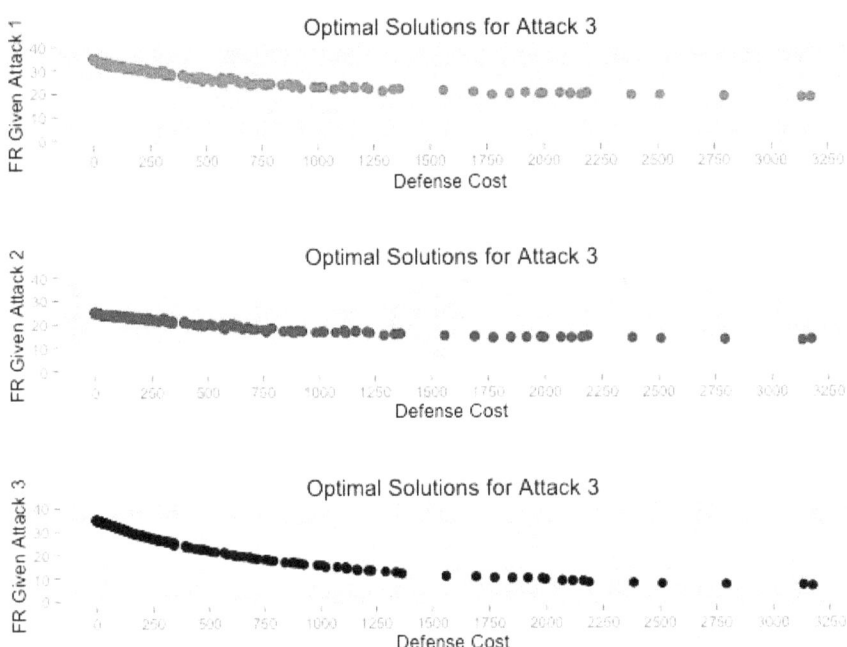

Figure 6: Optimal Defense Strategies Against Attack 3 evaluated on Every Attack

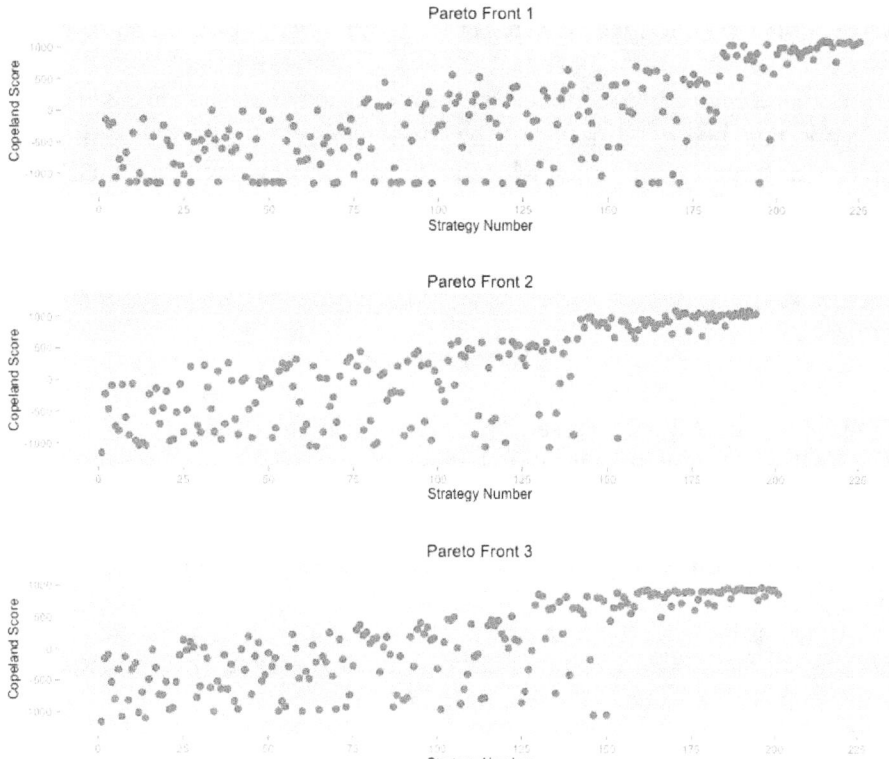

Figure 7: Copeland Score for Each of the Optimal Defense Strategies

Table 3: Attributes for the first best ranked defense strategies.

Rank	Pareto Front, #defense	Flow Reduction Given Attack 1	Flow Reduction Given Attack 2	Flow Reduction Given Attack 3	Defense Cost
1	1,214	5.30	5.35	22.47	3633
2	2,192	10.11	4.15	22.48	3006
3	1,215	5.08	4.83	22.74	3709
4	1,225	7.56	5.58	22.91	2707
5	1,216	6.04	5.18	22.86	3304
6	2,170	9.81	4.67	22.56	3419
7	1,222	8.54	6.29	22.67	2660
8	1,219	6.85	5.99	22.67	3263
9	2,190	10.74	5.65	22.66	2624
10	1,221	7.59	6.46	23.09	2771
11	1,213	8.69	7.02	23.50	2152
12	2,177	11.15	5.87	23.35	2309
13	2,173	10.74	4.99	23.30	2819
14	1,220	11.30	7.81	23.66	1736
15	1,197	9.83	7.53	24.62	1783
16	1,224	8.72	6.71	23.66	2297
17	2,188	12.02	6.35	23.28	2101
18	2,181	12.49	6.39	22.94	2086
19	2,186	11.38	5.92	23.41	2285
20	2,194	12.03	6.48	23.33	2037

5. CONCLUSIONS AND FUTURE RESEARCH

This paper clarifies the concept of resilience as a time based metric in the systems context. The resilience framework has been defined for the first time for both increasing and describing service functions as a function of time. Moreover, the framework presented ties together the engineering concepts of reliability, vulnerability, survivability and recoverability as a continuum in the resilience analysis. The manuscript also provides an optimization approach to reduce the vulnerability of a network for different attack scenarios as a function of flow and cost.

References

[1] Elsayed, E. (1996) " Reliability Engineering", Prentice Hall, New Jersey.

[2] Berdica, K. (2002), "An introduction to road vulnerability: what has been done, is done and should be done", Transport Policy, Vol. 9, No. 2, pp. 117-127

[3] Knoop, V., Hoogendoorn, S. and Van Zuylen, H. (2007) Approach to Critical Link Analysis of Robustness for Dynamical Road Networks, In Traffic & Granular Flow, Springer Verlag, Berlin, pp. 393–402

[4] Sterbenz, J., Hutchinson, D., Çetinkaya, E. Jabbar, A., Rohrer, J, Schöller, M. and Smith P. (2010) "Resilience and survivability in communication networks: Strategies, principles, and survey of disciplines", Computer Networks, Vol. 54., pp1245-1265.

[5] Castet, J. and Saleh, J. (2008) "Survivability and Resiliency of Spacecraft and Space-Based Networks: A Framework for Characterization and Analysis" American Institute of Aeronautics and Astronautics, AIAA Technical Report 2008-7707.

[6] Rose, A. (2007) "Economic resilience to natural and man-made disasters: Multi- disciplinary origins and contextual dimensions," Environmental Hazards, Vol. 7, No. 4, pp. 383–98.

[7] Nair, R., Avetisyan, H. and Miller-Hooks, E. (2010) "Resilience Framework for Ports and Other Intermodal Components" Transportation Research Record: Journal of the Transportation Research Board, No. 2166, Transportation Research Board of the National Academies, Washington,D.C., pp. 54–65

[8] Ramirez-Marquez, J.E. and Rocco, C. (2012) "Vulnerability Based Robust Protection Strategy Selection in Service Networks", Computers and Industrial Engineering, Vol.63, No.1, pp.235-242

[9] Henry, D. and Ramirez-Marquez, J.E., (2012) "Generic Metrics and Quantitative Approaches for System Resilience as a Function of Time" Reliability Engineering & System Safety, Vol. 99, No. 1, pp.114-122

[10] Ramirez-Marquez, J.E. and Rocco, C. (2012) "Towards a Unified Framework for Network Resilience" Proceedings of the Third International Engineering Systems Symposium CESUN 2012, June 18-20, Delft, Netherlands.

[11] Apostolakis, G.E., Lemon, D.M. (2005), "A Screening Methodology for the Identification and Ranking of Infrastructures Vulnerability Due to Terrorism." Risk Analysis, vol. 25, No. 1, pp. 361-376

[12] Bier, V., Haphuriwat, N., Menoyo, J., Zimmerman R. and Culpen, A., (2008) "Optimal resource allocation for defense of targets based on differing measures of attractiveness" Risk Analysis, Vol. 28, No. 3, pp. 763-770.

[13] Hausken, K. (2008) "Strategic defense and attack for series and parallel reliability systems", European Journal of Operational Research, Volume 186, Issue 2, pp. 856-888

[14] Haimes, Y.Y. 2009. On the Definition of Resilience in Systems. Risk Analysis, 29(4): 498-501.

[15] Sterbenz, J. P., Çetinkaya, E. K., Hameed, M. A., Jabbar, A., Qian, S., & Rohrer, J. P. (2011). Evaluation of network resilience, survivability, and disruption tolerance: analysis, topology generation, simulation, and experimentation. Telecommunication Systems, 1-32.

[16] Skaperdas, (1996) "Contest success functions", Economic Theory Vol. 7, pp. 283-290
[17] Levitin, G. and Hausken, K. (2008) "Protection vs. redundancy in homogeneous parallel systems" Reliability Engineering & System Safety, Vol. 93, pp. 1444–1451.
[18] Ramirez-Marquez, J. E., Rocco, C. and Levitin, G. (2011) "Optimal Network Protection Against Diverse Interdictor Strategies" Reliability Engineering & System Safety, Vol. 96, No.3, pp. 374-382.
[19] Ramirez-Marquez, J.E., Rocco, C. and Levitin, G. (2012) "Network Protection Against Diverse Attacks - A Multi-objective Perspective" Proceedings of ESREL 2012, Helsinki, Finland.
[20] Al-Sharrah G. (2010) "Ranking Using the Copeland Score: A Comparison with the Hasse Diagram", Journal of Chemical Information Models, Vol. 50, pp :785–791

Applications of Bayesian Networks for Evaluating Nuclear I&C systems

Jinsoo Shin[a], Rahman Khalil Ur[a], Hanseong Son[b], and Gyunyoung Heo[a*]

[a]Kyung Hee University, 1732 Deogyeong-daero, Giheung-gu, Yongin-si, Gyeonggi-do 446-701, Korea
[b]Joongbu University, 201 Daehak-ro, Chubu-Myeon, Geumsan-gun, Chungnam, 312-702, Korea
*Corresponding Author: gheo@khu.ac.kr

Abstract: The research presented, in this article, has been performed under the Korean research reactor project, an ongoing program to develop an optimized instrumentation & control (I&C) architecture and cyber security assessment of research reactors. The optimization of instrumentation and control systems and cyber security issues have been emphasized due to competitiveness of business (i.e. cost). Furthermore, these issues became more significant with the introduction of digital I&C systems. In this article, we have presented research activities performed for I&C architecture analysis and cyber security assessment for a reactor protection system (RPS). In I&C part, the architecture formulation, reliability feature analysis, cost estimation and cost-availability optimization of I&C architectures has been presented. In cyber security part, the cyber security risk evaluation model has been developed by integrating architecture model and activity-quality evaluation model, and analysis for cyber security evaluation for I&C system is presented. A probabilistic Bayesian network approach has been applied for I&C and cyber security analysis.

Keywords: Instrumentation and Control, Cyber Security, Reactor Protection System, Bayesian Network

1. INTRODUCTION

The various programs related to research reactor study such as instrumentation and control (I&C), cyber security, human factor engineering etc. are ongoing research activities in South Korea under Advanced Research Center for Nuclear Excellence (ARCNEX) project. Research activities and outcome related to I&C and cyber security are described in this article. Rahman and his co-authors [1] explained in one study that I&C architecture of nuclear power plants has been established to certain level, yet these are design dependent and not standardized for all industry. They also highlighted need for research to find suitable architecture for research reactor. The advent of digital technology in I&C has introduced novel kind of problems such as threat to cyber security, highlighted common cause failure (CCF) and software processing unit [2] failure and a comprehensive research is required to verify and get confidence on use of digital technology [3]. The optimization of I&C system architecture with respect to cost and its availability [4] and its resistibility to cyber-attack is also one of the basis of this research. Bayesian network has been selected for analysis in this study because this approach is suitable to model complex dependencies among components and has potential to count for uncertainties in failure data and modeling. Since I&C architecture has complex relation, so BN model has been developed for sensitivity, availability analysis and large uncertainties, due sparse failure data, has been handled effectively in BN model for cyber security evaluation.

In this article, we are presenting reliability and importance analysis of I&C components and modules of reactor protection system (RPS) I&C architecture configurations. Sensitivity study is important to get the insight of risk contribution from each component in a complex system. In this regard, many methodologies such as fault tree analysis, BN etc. have been implemented to find the sensitivity of software and software induced common cause failures to RPS using fault tree technique [5-7]. Four configurations of a single channel of RPS are formulated in the current article and BN models were developed to get the unavailability and I&C component sensitivity analysis. This study is performed for the standardization of an optimized I&C architecture for low & medium power research reactors. In this study, we also suggested the cyber security risk model to analyze the cyber-attack risk. The model utilizing the benefit of BN can analyze the risk that cyber-attack occurs at RPS. It can be utilized for the quantitative analysis by the proposed measure, cyber security risk as well as for various qualitative analyses[8-9]. Cyber security risk model is composed of the activity-quality analysis model and the architecture analysis model. The activity-quality analysis model was proposed to check how people and/or organization comply with the cyber security regulatory guide. It helps to analyze the relationships of the activity-quality checklists and their influences to cyber security. The architecture analysis model was also developed, particularly for the RPS of a research reactor as an illustrative purpose. For the definition of the critical cyber-attack scenarios on research reactors, the vulnerabilities and mitigation measures were analyzed. Then, the two models were integrated to cyber security risk model by using BN. A few kinds of analysis with respect of cyber security were performed by using the cyber security risk model.

The objective of this research, in this article, is to identify a configuration of architecture which gives highest availability with maintaining low cost of manufacturing and low cyber risk. In this regard, four configurations of a single channel of RPS are formulated in the current article and BN models were developed to get the unavailability and I&C component sensitivity analysis. The cyber security risk of RPS has been evaluated by proposing a model based on considerations of vulnerability and the activity-quality checklist. The analysis of the vulnerability and the activity-quality checklist was performed with the assumption that a cyber-attack occurs to a maintenance and test processor in the RPS with BN models. These study are performed for the standardization of an optimized I&C architecture for low & medium power research reactors and the suggestion of cyber security risk evaluation model of I&C architecture with BN.

2. BAYESIAN NETWORK ANALYSIS

The BN is a directed acyclic graph of arc to represent the dependencies between nodes and variables using Bayes' theorem [10]. The Bayes' theorem is represented as the equation (1)

$$P(C|x) = \frac{P(C).P(x|C)}{P(x)} \qquad (1)$$

Where, p(x) is the probability distribution of the variable x at the entire population, p(C) is the prior probability that the some sample belongs to class, p(x/C) is the conditional probability of obtaining the value of the variable x, and p(C/x) is the posterior probability that the value of the variable x belongs to class at given situation. When the learned posterior information on the conditional probability, it can achieve the improvement of the probability by calculating the relationship between the posterior and prior probability. BN is composed of node, arc and node probability table (NPT). The node and arc mean a variable and the cause-and-effect relationship. The nodes have two types like the parent node and the child node. The child node has cause element and the parent node has result element of the child nodes. NPT means the probability table that summarizes the occur probability between the causal relationship nodes. Because NPT value can be used as observable quantities, latent variables, unknown parameters, or hypotheses, it is useful for changing from the qualitative problems to quantitative ones. Although BN has some limitations such as difficulty to defining the NPTs with expert opinions, representing the continuous data, and describing the feedback loops, yet it has strength for application in availability and cyber security of I&C system due to flexibility of input, ease of modelling and less impact of large uncertainties.

The BN has been selected for reliability analysis because this approach works better than Fault Tree Analysis (FTA) for two reasons. For last few decades, BN models have been applied to dependability analyses, such as Boudali and Dugan [11] transformed Dynamic Fault Tree (DFT) to BN for probabilistic analysis and Torres-Toledano and Succar [12] developed BN models for reliability analysis of complex systems based on Reliability Block Diagram (RBD). But these techniques require the development of dynamic fault tree or identification of path sets of system as a pre-request. Identification of path sets becomes difficult in case of complex system and it can produce misleading results because of incorrect or insufficient identification of path sets. The development of BN by mapping Reliability Block Diagram with General Gates (RBDGG) such as 'AND', 'OR' and 'K out of

N (KooN)' has been realized by Kim [13] in 2011. RBDGG is an extended form of RBD. The construction of BN model is easier than developing a fault tree and BN yields exact results because its analysis is based on conditional probabilities. In this article, we are more interested in reliability features of system and importance of components in terms of risk contribution not in detail failure mechanism of system. Here mapping of RBDGG to BN modeling technique has been adopted for desire analysis, in which all the logic and function has been kept preserved for each node. Therefore, it is beneficial to use BN, which will reduce the effort and give the reliable numbers for analysis.

It is also used to develop the cyber security risk model for I&C system for overcoming lack of information when analyzing and modelling about cyber security against cyber-attack by using the benefit of BN. The BN is often used in order to overcome this difficulty by the conversion from the qualitative value to quantitative value [14]. The model with BN can analyze the cyber security risk when cyber-attack occurs to I&C system. It can be utilized for the quantitative analysis by the cyber security evaluation index (CSEI), which means the probability of cyber-attack occurrence or the completeness of mitigation measure and/or the extent of activity-quality, as well as for various qualitative analyses. The CSEI is represented the node of BN model.

3. BN FOR I&C ARCHITECTURE FEATURES

I&C architecture of RPS is selected for analysis in this study and four (4) single channel architecture configurations has been developed. For realization, reliability block diagram (RBD) of architecture configuration-I and BN models as provided in Figure 1. For comparison purpose, a baseline composition of configuration-I, given in (a) part of Figure 1, consists of a single bi-stable processor (BP) BP_A and single coincidence processor (CP) CP_A and circuit breakers to trip with 2/3 logic. This configuration is typical and basic for a channel and has no inter-channel redundancy. Inter-channel redundancy means redundant modules within a channel whereas intra-channel redundancy is based on number of channels. In architecture configuration-II, redundancy is added in BP to evaluate the impact on single channel. In order to observe the sensitivity of CP module on single channel failure, CP is added in the channel for case of configuration-III. This configuration consists of a bi-stable processor BP_A, redundant pair of CP processors CP_A1 & CP_A2 and circuit breakers to trip with 2/3 logic. Configuration IV consists of inter-channel redundancy of BP & CP modules i.e. two modules of each. The differences among configurations are delineated in Table 1.

The RBD of proposed I&C architecture configurations was converted to BN models preserving all the functions and logics of system. BN models, as shown in (b) and (c) of Figure 1, show the propagation of failure from transmitter & Sensor to circuit breaker actuation. Two failure states for each component are considered in this study, which are 0 and 1. State 0 represents the failure state and 1 represents the perfect is representing a node and NPT is prepared for every node based on operational logic and failure data [11-14].

Table 1: I&C architecture configurations composition[†]

Component/Module	Architecture Configuration			
	I	II	III	IV
Bi-stable Processor	1	2	1	2
Coincidence Processor	1	1	2	2
Digital Output	1	1	1	2

[†] All the other components/modules in the architecture are kept the same, as shown in Figure 1 (a).

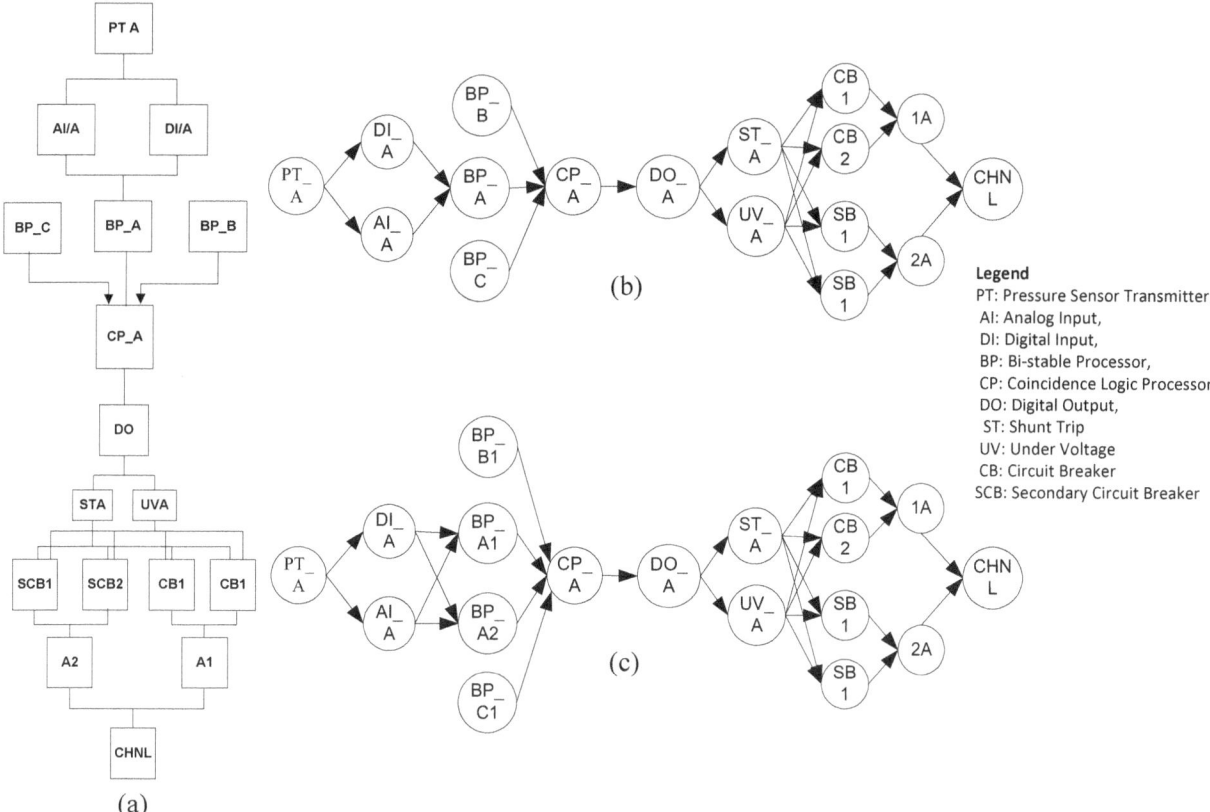

Figure 1: I&C Architecture (a) RBD of configuration-I (b) BN model of configuration-I (c) BN model of configuration-II

3.1. Architecture Availability

Reliability feature analysis of architecture configurations, such as availability, unavailability, has been performed using BN. The channel (CHNL) in BN model gives the output features for states 0 (failure) and 1 (perfect) for single channel. The results for four configurations are presented in Table 2. $P(x=0|\lambda)$ gives probability of failure state whereas $P(x=1|\lambda)$ yields probability of success for single channel. These parameters are also termed as unavailability and availability of I&C architecture.

Table 2: Reliability Analysis of I&C architecture configurations [4]

Configuration	CHNL	Unavailability ($P(x=0|\lambda)$)	Availability ($P(x=1|\lambda)$)
I	(1BP, 1CP)	1.9751E-4	9.998E-01
II	(2BP, 1CP)	3.1525E-4	9.9968E-01
III	(1BP, 2CP)	3.9701E-5	9.9996E-01
IV	(2BP, 2CP, 2DO)	3.1596E-7	0.9999996

3.2. Sensitivity Aspects

The mangers, designers and operators have been keen to recognize importance of equipment, components or system failures on the overall performance of the unit. This is valid for research reactors too. It is very important to know the fact that how much risk will increase/decrease if the failure of component happens frequently or it never fails. The indicator showing the decrease in risk is called Risk Reduction Worth (RRW). The higher the RRW measure, the more sensitive would be the component to risk. It can be calculated by, equation 1 [1, 6], taking the ratio of the failure probability of system with λ for ith component set equal to 0 to channel total failure probability (unavailability). RRW results for four configurations are presented in Table 3.

$$RRW_i = \frac{Q_{CHNL}(\lambda)}{(Q_{CHNL}(\lambda=0))_i} \qquad (2)$$

Whereas index 'i' represents the components/modules in the architecture.
QRPS (λ) is the system unavailability and would be equivalent to P(x=0|λ) in this article. While QRPS (λ=0) shows the system unavailability if ith component never fails (λ equal to 0). In this article, it would be equivalent to P(x=0|λ=0).

Table 3: Sensitivity Results of I&C architecture configurations [6]

Component/Module	RRW			
	Configuration I	Configuration II	Configuration III	Configuration IV
PT	1.000608	1.589843	1	1.000507
DI	1	1.000476	1	1
AI	1	1.000476	1	1
BP	1.002538	1.002385	1.000025	1.000317
CP	4.871858	1.991472	1.00063	1.108632
DO	1.249115	1.14279	131.256	1.025212
ST	1.00076	0.470909	1.003742	1.875022
UV	1.00076	0.470909	1.003742	1.875022
CB	1.000304	0.470853	1.00164	1.257652
SCB	1.000304	0.470853	1.00164	1.257702

3.3. Cost Estimation

It is necessary to mention that information related to cost of safety grade instruments is proprietary and is available for academic researches. Therefore, cost of architecture has been estimated based on certain assumptions. The cost can be discretized into the unit cost for each component and number of components. Cost estimation formula has been proposed in the form of equation (3) [4]. The equation (2) gives cost as the multiple of X and multiple is product of number of components and its unit cost, where X is an arbitrary unit.

$$U_j = \sum_i u_i . n_i . X \qquad (3)$$

Whereas U_j is the cost of jth architecture and j varies from 1 to 4. The parameters u_i and n_i are component unit cost and number of ith component & modules in jth architecture. The components/modules are pressure/level transmitter (PT), analog input (AI), digital input (DI), bi-stable processor (PB), coincidence processor (CP), digital output (DO), shunt circuit (ST), under voltage circuitry (UV). The costs of architecture configurations I, II, III and IV has been estimated 8.5X, 10X, 10X and 12.5X respectively.

In order to observe variation of cast with respect to architecture availability & unavailability, it is plotted in Figure 2. The unavailability of system decreases from 1.9751E-4 to 3.1596E-7 for architecture I to architecture IV and availability increases from 9.998E-01 to 0.9999996 (nearly 1). The physical significance can be realized in terms of cost saving. If we consider an arbitrary unit as 100 US dollar, then cost increases by (4X) or 400USD.

A reliability index (RI) has been proposed for I&C study under this project, based on the equation (4). This index calculates the increase of availability per unit of cost. The architecture availability increases at the rate of 4.99E-05 per X unit of cost.

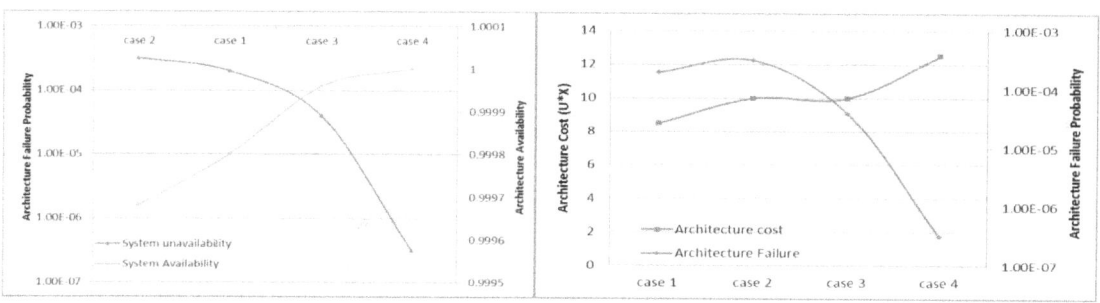

Figure 2: Variation of reliability features and cost for I&C architectures

$$RI = \frac{P_n(X=1\|\lambda) - P_1(X=1\|\lambda)}{U_n - U_1} \quad (4)$$

4. BN FOR CYBER SECURITY RISK MODEL

The cyber security risk model for evaluation of cyber-attack risk was developed based on regulatory guide 5.71 [19] to perform cyber security analysis for I&C systems of nuclear facilities. The model is consist of two parts which are called activity-quality analysis model and architecture analysis model. The activity-quality analysis model is made for qualitative analysis such as whether or not the personal and/or organization carry out the cyber security regulatory guide well. The architecture analysis model has been developed for quantitative analysis such as structural vulnerability of I&C system for cyber-attack. Since the activity-quality analysis model affects the architecture analysis model, two model is integrated into the cyber security risk model by using BN for transformation from qualitative value of the activity-quality model to quantitative value. The cyber security risk model has performed the analysis about case studies with assumption that a cyber-attack occurs to RPS.

4.1. Application of BN for Cyber Security Risk Model

The cyber security risk model was developed with BN to utilize the benefit of it such as converting from the qualitative value to the quantitative value and calculation for back propagation by using Bayes theorem. The model is consist of activity-quality analysis model and architecture analysis model because both the management aspects and the system architecture aspect are important in terms of cyber security.

The activity-quality analysis model was developed based on regulatory guide 5.71 and using cyber lifecycle to check how personal and/or organization comply with the cyber security regulatory guide. We make 27 checklists (ex, one-way data flow, security assurance for safety degree), which is specified by cyber security regulatory guide, and represent as nodes with BN. The model helps to analyze the relationships of each nodes and their influences to cyber security and affect mitigation measure on architecture analysis model.

The architecture analysis model was constructed for RPS with two assuming situations that one is fail to trip timely due to cyber-attack and the other is reactor trip due to maliciously insertion of control rod. It offers a general perspective for the construction of the architecture analysis model for any I&C system. In order to develop the architecture analysis model for RPS, we study the network and structure about each subsystems of RPS such as BP, CP, Interface and Test Processor (ITP), Maintenance and Test Processor (MTP), and Intra-Channel. The model is composed with vulnerability and mitigation measure parts for reflection of extent of vulnerability of architecture and mitigation against penetration [??]. The vulnerabilities and mitigation measures are analyzed for RPS architecture by using this model. The lists of vulnerability are 1) Denial of service (DoS) attacks and malware execution on systems network during maintenance works (V1), 2) system shut-down by contagion of malware from maintenance works (V2), 3) data alteration by contagion of malware from maintenance works (V3), 4) Dos occurrences and malware carrying out on other systems by vulnerabilities existing in the system (V4), and 5) data alteration by using recognized vulnerabilities of standard communication protocols (V5). The lists of mitigation measure are 1) Establishment of managing infection detection systems for external storage media like USB or PC used for PLC maintenance works (M1), 2) Establishment of security system such as firewalls / Intrusion detection system / intrusion prevention system (M2), 3) Check for running services (M3), 4) Network monitoring (M4), 5) Establishment of device validation policies (M5), and 6) Vulnerability patches (M6). The architecture analysis model with BN is developed by using these analysis results which are system network for RPS and vulnerability and mitigation measure against cyber-attack.

The activity-quality analysis model for administrative aspects evaluation is linked to architecture analysis model for evaluation of architectural system aspects for development integrated cyber security risk model. The integrated model as cyber security risk model make it possible to evaluate and analyze the final risk in view of cyber security for I&C system. Figure 3 shows the cyber security risk model for RPS with BN.

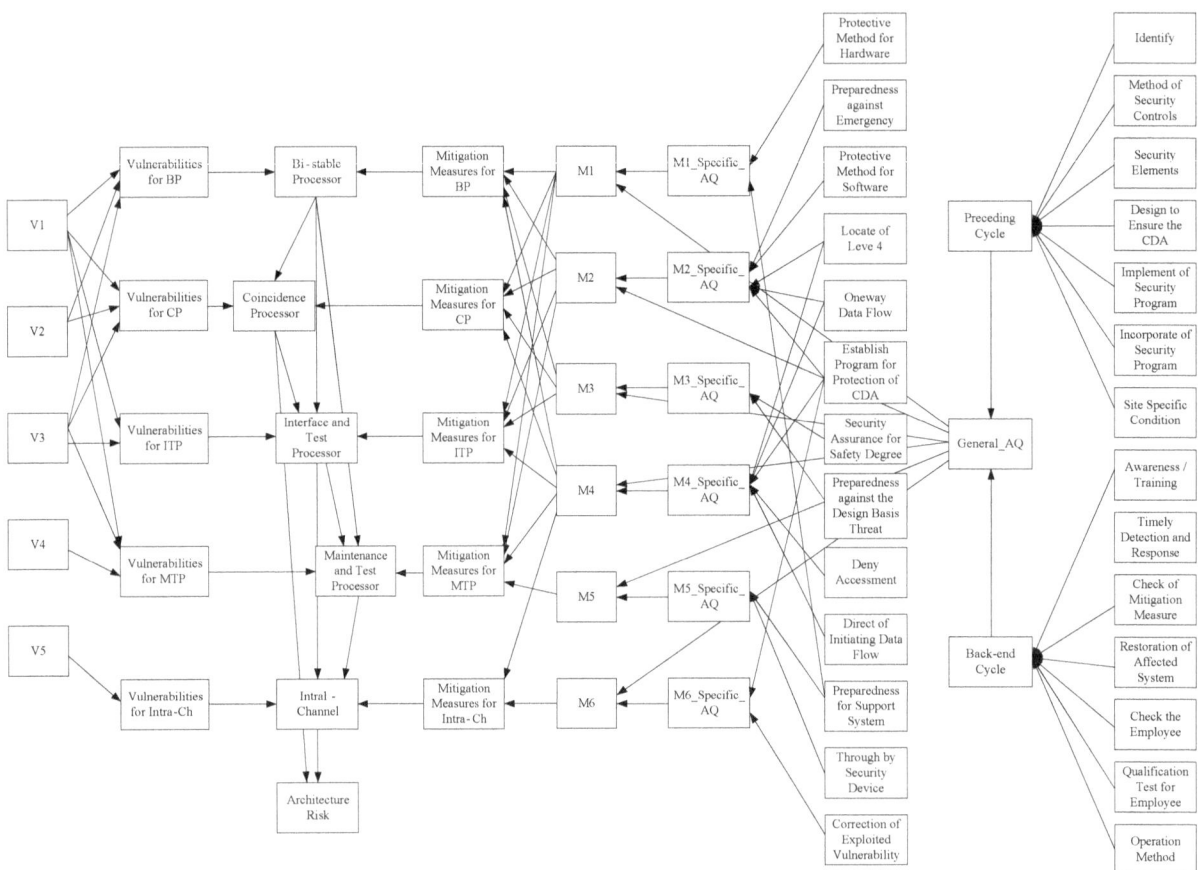

Figure 3: Cyber Security Risk Model with BN

4.2. Cyber Security Evaluation Index (CSEI)

The model used the cyber security evaluation index, called CSEI, to change from qualitative data to quantitative data for quantitative analysis. The CSEI means the extent of compliance with regulation guide or the probability of cyber-attack occurrence or the completeness of mitigation measure. It is calculated by multiplying the numeric values of each stage of node and the percentage of each stage evaluation of node as the equation. The CSEI is performed according to equation (5).

$$CSEI = \sum_{s=1}^{5} 10(2s-1) \times EP \qquad (5)$$

S is a numeric value of each stage and EP is a percentage ratio of evaluation for each stage. The numeric value for activity-quality model and architecture model are represented in Table 4.

Table 4: The numeric value of CSEI

At activity-quality checklist node		At architecture vulnerability	
Numeric value of stage	Meaning	Numeric value of Stage	Meaning
1	Very Well	1	Very Low Occur Probability
2	Well	2	Low Occur Probability
3	So so	3	Medium Occur Probability
4	Bad	4	High Occur Probability
5	Very Bad	5	Very High Occur Probability

The CSEI of each node varies from 10 to 90 points. The CSEI score of the architecture vulnerability node has the minimum 10 points when the cyber-attack probability is the highest.

4.3. Case Studies with Cyber Security Risk Model

We have performed some case studies for by using the model with BN. In this article, we introduce a results representatively. The vulnerability is analyzed by using the model when cyber-attack occurs to the MTP. The purpose of this analysis is to get information on which vulnerabilities and activity-quality checklists should be prioritized considering cyber security by using the back propagation of BN function. Assuming that any preliminary evaluation is not performed, the point of 20% is assigned to each stage of a node resulting in the average 50 points given to the node. Then the BP, CP, and ITP nodes are assigned to 70 points as hard evidences. This means that these subsystems of RPS have low possibilities to be attacked. After this, 10 points are assigned to the MTP node as hard evidence, which means the MTP was attacked. The simulation results of the model are analyzed. The result for vulnerability is shown in Table 5.

Table 5: Analysis of vulnerabilities for RPS when cyber-attack occur to MTP

Points of MTP Node \ Vulnerability	V1	V2	V3	V4	V5
Points before Cyber-attack	64.98	60.16	64.98	52.50	62.39
Points after Cyber-attack	25.49	77.67	25.49	14.15	81.99
Gap between Points	-39.49	17.51	-39.49	-38.35	19.60

The second row means CSEI score for MTP before cyber-attack happening and the third row means CSEI score for MTP after cyber-attack occur to the MTP. The last low means CSEI gap between before and after cyber-attack occurrence. The negative values at last low represent the vulnerability related with MTP and the positive values represent the non-related vulnerabilities. In addition, the results can show that completeness of the mitigation measures affecting MTP have become lower in the following order: M1, M2, M4, and M5. The activity-quality checklist evaluating each mitigation measure is affected since the points of mitigation measures node are changed. The checklists that have influence of significant with MTP decrease considerably than the checklists that have influence of relatively a little.

5. SUMMARY AND CONCLUSION

BN has is some advantages such as modeling ease, suitability for modeling complex dependencies among components and potential to count for uncertainties in failure data and modeling. Since I&C architecture has complex relation, so BN model has been developed for sensitivity, availability analysis and large uncertainties, due sparse failure data, has been handled effectively in BN model for cyber security evaluation.

The study has been performed to get the cost optimized results in terms of architecture availability and analyze the cyber-attack risk for I&C system during cyber-attack with these strength of BN. Four configurations of I&C architecture of RPS has been proposed and their BN models have been developed to get the sensitivity and availability analysis. Cost estimation model for I&C architecture has been proposed and cost-availability relation has been found out and it is defined as RI. RI provides increment in architecture availability with respect to cost, in this study this index has value of 4.99E-05 per X unit of cost. The risk due to cyber security has been evaluated for RPS in terms of administrative aspects and architectural aspects and has been measured in terms of index CSEI.

The selection of architecture has many aspects such as safety concern, designer & operator desire, availability criteria, cost etc. Based on the reliability analysis results exclusively, architecture configuration IV can be designed for the research reactor because it has a very high availability of 0.9999996. If we suppose a criteria that single channel availability of the order of 1.0E-05 would be sufficient then cost can lead towards decision of architecture. Then keep the current scenario in perspective, architecture configuration IV cane be suggested for research reactor I&C systems, because its cost varies from 10-11 X units while it has availability 0.99996 (unavailability 3.97E-05 per demand).

The cyber security risk model with BN is developed for whole RPS architecture by integration the cyber security activity-quality model and cyber security architecture model to evaluate the cyber security risk for RPS. A few analyses for RPS were performed by using the model. When cyber-attack occur to RPS, the model provides information such as the prioritized vulnerability, mitigation measure, and checklist orders with the CSEI. Theses analysis proved that the developed model could provide this kind of information through the back propagation feature of the BN. This analysis inferred that the use of cyber security risk model makes it possible to create simulated penetration test scenarios.

Acknowledgements

This work has been supported by Advanced Research Center for Nuclear Excellence (ARCNEX) project funded by the Ministry of Education, Science and Technology of Republic of Korea (Grant Number: 2013-075450).

References

[1] R. Khalil Ur, J. Shin, M. Zubair, G. Heo, and H. Son, "Sensitivity Study on Availability of I&C Components Using Bayesian Network," Sci. Technol. Nucl. Install., vol. 2013, pp. 1–10, 2013.

[2] S. Authen and J.-E. Holmberg, "Reliability Analysis of Digital Systems in a Probabilistic Risk Analysis for Nuclear Power Plants," Nucl. Eng. Technol., vol. 44, no. 5, pp. 471–482, Jun. 2012.

[3] R. Khalil Ur, G. Heo, and H. Son, "Architecture dependent availability analysis of RPS for Research Reactor Applications," in Transaction of Korean Nuclear Society, 2013.

[4] R. Khalil ur, J. Shin, and G. Heo, "Study on Optimization of I & C Architecture for Research Reactors using Bayesian Networks," in Joint IGORR 2013 and IAEA Technical Meeting, 2013.

[5] S. Kamyab, M. Nematollahi, and G. Shafiee, "Sensitivity analysis on the effect of software-induced common cause failure probability in the computer-based reactor trip system unavailability," Ann. Nucl. Energy, vol. 57, pp. 294–303, Jul. 2013.

[6] R. Khalil Ur, M. Zubair, and G. Heo, "Reliability Analysis of Nuclear I & C Architecture using Bayesian Networks," in 11th International Bhurban Conference on Applied Sciences & Technology (IBCAST), IEEE, 2014.

[7] R. Khalil Ur, M. Zubair, and G. Heo, "Sensitivity Analysis of Digital I&C Modules in Protection and Safety Systems," IOP Conf. Ser. Mater. Sci. Eng., vol. 51, Dec. 2013.

[8] J. Shin, H. Son, and G. Heo, "Cyber Security Risk Analysis Model Composed with Activity-quality and Architecture Model," Proc. Int. Conf. Comput. Networks Commun. Eng. (ICCNCE 2013), pp. 2–5, 2013.

[9] "Development of Cyber Security Evaluation Model Using Bayesian Networks."

[10] D. Heckerman, A Tutorial on Learning With Bayesian Networks, vol. 1995, no. November. 1996.

[11] H. Boudali and J. B. Dugan, "A discrete-time Bayesian network reliability modeling and analysis framework," Reliab. Eng. Syst. Saf., vol. 87, no. 3, pp. 337–349, Mar. 2005.

[12] J. G. Toledano Torres and L. E. Sucar Succar, "Bayesian Networks for Reliability Analysis of Complex Systems," Prog. Artif. Intell. — IBERAMIA 98, vol. 1484, pp. 195–206, 1998.

[13] M. C. Kim, "Reliability block diagram with general gates and its application to system reliability analysis," Ann. Nucl. Energy, vol. 38, no. 11, pp. 2456–2461, Nov. 2011.

[14] T. L. Chu, M. Yue, and A. Varuttamaseni, "APPLYING BAYESIAN BELIEF NETWORK METHOD TO QUANTIFYING SOFTWARE FAILURE PROBABILITY OF A PROTECTION SYSTEM 1," in NPIC&HMIT 2012, pp. 296–307.

[15] U.S National regulatory Commission, "Reliability Study: Westinghouse Reactor Protection System, 1984-1995." NUREG/CR-5500, Vol 2, Washington, 1999.

[16] U.S National regulatory Commission, "Industry-Average Performance for Components and Initiating Events at U . S . Commercial Nuclear Power Plants." NUREG/CR-6928, Washington, 2007.

[17] International Atomic Energy Agency (IAEA), "Generic Component Reliability Data for Research Reactor." IAEA-TECDOC-0930, Vienna, 1997.

[18] International Atomic Energy Agency (IAEA), "Component reliability data for use in probabilistic safety assessment." IAEA-TECDOC-478, Vienna, 1988.

[19] U.S National regulatory Commission, "Cyber Security Programs for Nuclear Facilities," Reculatory Guide 5.71, 2010.

Portfolio Analysis of Layered Security Measures

Samrat Chatterjee[a], Stephen C. Hora[a*], Heather Rosoff[a]

[a]CREATE, University of Southern California

Abstract: Layered defenses are necessary for protecting the public from terrorist attacks. Designing a system of such defensive measures requires consideration of the interaction of these countermeasures. In this article, we present an analysis of a layered security system within the lower Manhattan area. It shows how portfolios of security measures can be evaluated through portfolio decision analysis. Consideration is given to the total benefits and costs of the system. Portfolio diagrams are created that help communicate alternatives among stakeholders who have differing views on the trade-offs between security and economic activity.

Keywords: Portfolio decision analysis, homeland security, terrorism risk, systems analysis, expert elicitation, probabilistic risk assessment

INTRODUCTION

To defend against terrorism, governments adopt a strategy of multiple layers of defense to thwart and respond to attacks of various kinds [1,2]. Some security measures, such as shoe inspection at airports, are effective against a relatively narrow set of attacks while others, such as those entailing wireless surveillance, have a potential for stopping a wide variety of attacks. The Transportation Security Administration (TSA) [3] claims to have twenty layers in its airline security system. Building a layered defense system that is effective in preventing successful attacks and efficient with respect to the consumption of resources is complex and requires one to think about the synergy of various security measures.

In this article, an approach to building a system of layered defenses is discussed in terms of portfolios of security measures. The portfolio approach is proposed as the performance of a system that cannot be inferred from the individual performance of security measures without consideration of how the protective systems work together and how effective they are against various threats[4]. An effective system of security measures work together as a team much as one needs football players to have different capabilities. Moreover, when faced with multiple threats, the layered defense system needs to have sufficiently diverse components such that no gaps are left for adversaries to exploit.

This research was part of the Urban Commerce and Security Study (UCASS)[5] funded by the Science and Technology Directorate of the Department of Homeland Security through its National Centers of Excellence at Rutgers University and the University of Southern California and with the cooperation of the Mineta Transportation Institute at San Jose State University. The study demonstrates how to analyze the trade-offs between security and economic activity in an urban environment. Although the study focused on New York's lower Manhattan Security Area including Wall Street and the World Trade Center, the general methodology is independent of location.

The product of this analysis is an evaluation methodology for various systems of security measures. The analysis separates the benefits and costs in such a way that one can examine alternative layered defenses that vary with both risk and cost. Risk is measured in terms of reduction in expected consequences while costs entail the direct capital and operating costs of security measures as well as the unintended costs they may impose on the economy or, in some cases, collateral benefits.

*Hora@USC.Edu

SECURITY PROBLEM

The problem of terrorism is very complex in that the variations of attacks are great in number as are the types of defenses. To bring the problem to a manageable size, we have selected five scenarios and seven security measures that capture elements common to a larger set of possible scenarios and security measures. For the portfolio analysis, the selected scenarios and security measures provides a concrete set of examples to work from as the modeling methodologies were discussed, developed, and integrated.

Scenarios

A scenario serves to provide a realistic description of the attack under consideration for analysis. Each scenario was developed as a brief narrative describing the attack type, location, and frequency. A concern with reducing the number of scenarios included in the analysis is that it might lead to an underestimation of the total threat likelihood, and thus an undervaluing of the portfolios of security measures as the less the threat, the less the possible risk reduction which is the primary value of security measures. We call this the "completeness" problem and deal with it in much the same way as one scales up from a sample to a population. A total threat is assessed and that total threat probability is then allocated to the reduced set of scenarios so that the sum of the threats from the reduced set of scenarios is the same value as if the entire set of potential scenarios were included.

A set of five scenarios were created, each of which were borrowed from historical and well-known terrorist events. The scenarios were modeled after past major terrorist attacks in Mumbai, India; Tokyo, Japan; Madrid, Spain; London, England; and Israel [5]. Each scenario was accompanied with a brief narrative describing the individual event. The five stylized scenarios are as follows:

S1 (Mumbai): *Several small teams of attackers shoot their way into a number of large office buildings and hotels surrounding the World Trade Center construction site and begin a killing rampage.*

S2 (Tokyo): *In five coordinated attacks, perpetrators release a chemical agent on several lines of the New York City Metro and PATH (The Port Authority of New York and New Jersey) trains passing through Lower Manhattan and the World Train Center Station.*

S3 (Madrid): *During the peak of New York rush hour, multiple explosions occur aboard New York Metro subway trains heading into Lower Manhattan. These include the 7th Avenue express and local, Lexington Avenue express and local, 8th Avenue express and local, Queens/ Broadway/Brooklyn express and local, and Nassau Street express and local.*

S4 (London): *Terrorists detonate a large bomb aboard a Manhattan Express bus heading into Lower Manhattan, targeting civilians using New York's public transportation system during the morning rush hour.*

S5 (Israel): *A terrorist with explosives strapped to his chest detonates a bomb at a checkpoint at an entrance outside of the New York Stock Exchange.*

Security Measures

A security measure, or countermeasure, works to reduce the risk of a threat by deterring the attack, by thwarting an attack, or by reducing the consequences of a successful attack. The effectiveness of a security measure may be different for one type of attack scenario compared to another. Some security measures are designed for a specific attack mode, such as a concealed bomb, while others work more generically against many types of threats. For example, a metal detector (magnetometer) is useful in both deterring and thwarting an attack that employs a metallic weapon such as a gun. Closed circuit television cameras (CCTVs), on the other hand, can detect suspicious activities of many kinds and have forensic value that may translate into deterrence. Neither of these security measures, however, would reduce the severity of a successful attack as might a stockpile of medical supplies in the case of a biological attack or emergency escape ladders and stairwell lighting in the case of the bombing of a building.

While the number of possible security measures, policies and initiatives are infinite, a set of seven security measures were defined for this study. These security measures were selected because they not only represent a wide range of measures that are available, but also because they are heavily employed within Lower Manhattan (and other urban environments). The selected security measures range from technological to human, visible to invisible, permanent to temporary, and focused on screening people to screening vehicles [5]. The security measure descriptions are as follows:

> *C1 (Random vehicle inspections)*: A perimeter of checkpoints at the entry/exits points around Lower Manhattan is established. Security inspections that entail the search of persons and their vehicles are conducted on a random basis.
>
> *C2 (Permanent street closures to vehicular traffic)*: A portion of Broad Street in front of the New York Stock Exchange and Federal Hall is closed to vehicular traffic. Pedestrian and bicycle traffic, however, would still be permitted on the sidewalk. Traffic in this area is redirected as appropriate to create minimal disruption for vehicles.
>
> *C3 (Temporary perimeters and access control)*: Street restrictions and security checkpoints affecting pedestrian traffic are put in place. The checkpoints include temporary barricades at various intersections in the area. Anyone traveling into or within the area is subject to a "stop and search" by a uniformed New York police officer.
>
> *C4 (Random bag and parcel inspection)*: Security personnel conduct random inspections of bags and parcels at rail and subway stations heading to or leaving Lower Manhattan. The random searches are carried out 24 hours a day, 7 days a week. Police use visual checks, bomb-sniffing dogs, and explosive detection technology to check the bags for hazardous materials.
>
> *C5 (X-rays & magnetometers in building lobbies)*: Building security is upgraded by hiring additional protection officers and installing magnetometers to detect metal items, such as guns and knives, and x-ray scanners to inspect bags and packages at the entrances of major/large buildings in Lower Manhattan.
>
> *C6 (Increased visible presence of police)*: Police presence is increased throughout Lower Manhattan. Police officers do not target specific individuals, but are instructed to be more vigilant in pursuing tips and leads and analyzing patterns of unusual behavior.
>
> *C7 (CCTV cameras)*: An additional 1,700 close-circuit television (CCTV) cameras (resulting in 3,000 cameras total) are located throughout Lower Manhattan. The CCTV cameras help police assess suspicious activity or actual events, reduce incident response time, and create a common technological infrastructure for security surveillance.

PORTFOLIO MODELING AND ANALYSIS FRAMEWORK

In this section, we describe our modeling and analysis framework and the underlying assumptions. With seven security measures, one can construct $2^7 = 128$ different security portfolios for consideration. Denote the kth security portfolio by set of indicators $\{x_{1k}, x_{2k}, ..., x_{7k}\}$ where $x_{jk} = 1$ if the jth security measure is included in the portfolio and $x_{jk} = 0$ otherwise. The expected annual risk reduction, R_k, and net cost, C_k, are computed as below (see equations 1 and 2). The expected annual risk reduction formulation is based on fundamental concepts of probability theory [6] where we assume that the security measures operate independently of one another in terms of deterrence and the ability to thwart an attack. Expected annual risk reduction not only depends on the efficacy of the portfolio of security measures but also on the likelihood of an attack and the consequences of a successful attack. The net cost is an aggregation of the direct (capital and operating) and indirect (spillover) costs associated with a security portfolio.

$$R_k = \sum_{i=1}^{M} t_i c_i \left[1 - \prod_{j=1}^{N} (1 - d_{ij} x_{jk})(1 - e_{ij} x_{jk}) \right] \quad (1)$$

$$C_k = \sum_{j=1}^{N} x_{jk} \left(k_j + o_j + s_j \right) \quad (2)$$

where:

- t_i is the annual frequency of scenario i without the security measures in place;
- c_i is the expected consequence of scenario i given a successful attack;
- d_{ij} is the deterrence effect against scenario i of security measure j;
- e_{ij} is the interdiction effect against scenario i of security measure j;
- k_j is the amortized annual capital cost of security measure j;
- o_j is the annual operating cost of security measure j;
- s_j is the indirect benefit/cost or spillover effect of security measure j (positive or negative);
- M is the number of scenarios; and
- N is the number of security measures.

Equation (1) is developed by considering deterrence and interdiction as distinct benefits of a security measure. Now suppose a threat has an annual frequency t, and two security measures are available that reduce the threat through deterrence by a fraction d_1 and d_2 respectively. The frequency of the attack with the first security measure would be $t(1 - d_1)$. Adding the second security measure further reduces the frequency by $(1 - d_2)$ and thus, the threat frequency is $t(1 - d_1)(1 - d_2)$ with both security measures in place. Similarly, the security measures provide reductions in the frequency of successful attacks through interdiction of fractions e_1 and e_2, respectively. The overall frequency of successful attacks with both security measures, following a similar line of reasoning, is $t(1 - d_1)(1 - d_2)(1 - e_1)(1 - e_2)$ so that the reduction in the frequency of successful attacks is $t[1 - (1 - d_1)(1 - d_2)(1 - e_1)(1 - e_2)]$. Multiplying this last term by the consequence of a successful attack gives the risk reduction for that threat. Finally, summing across all threats gives the total risk reduction as in equation (1).

Figure 1 presents a notional risk reduction versus cost plot and notional pie charts to indicate scenario and portfolio related computation elements in equations (1) and (2). In this figure, we also present the computational approaches we adopted as part of the risk, economic, and portfolio analyses.

The notional risk reduction versus cost plot in Figure 1 leads to a multi-objective decision making problem of portfolio selection. The decision making objectives include minimizing net cost while maximizing risk reduction. The tradeoff between net cost and corresponding risk reduction may generate an efficient frontier of non-dominated solutions as shown in a notional risk reduction versus cost plot below (see Figure 2). In this figure, for a given level of net cost or risk reduction, the optimal security portfolio falls on the efficient frontier. The efficient frontier consists of those portfolios that cannot be bested simultaneously in both a lower cost and greater annual risk reduction.

Figure 1 Information for Analyzing Security Portfolios

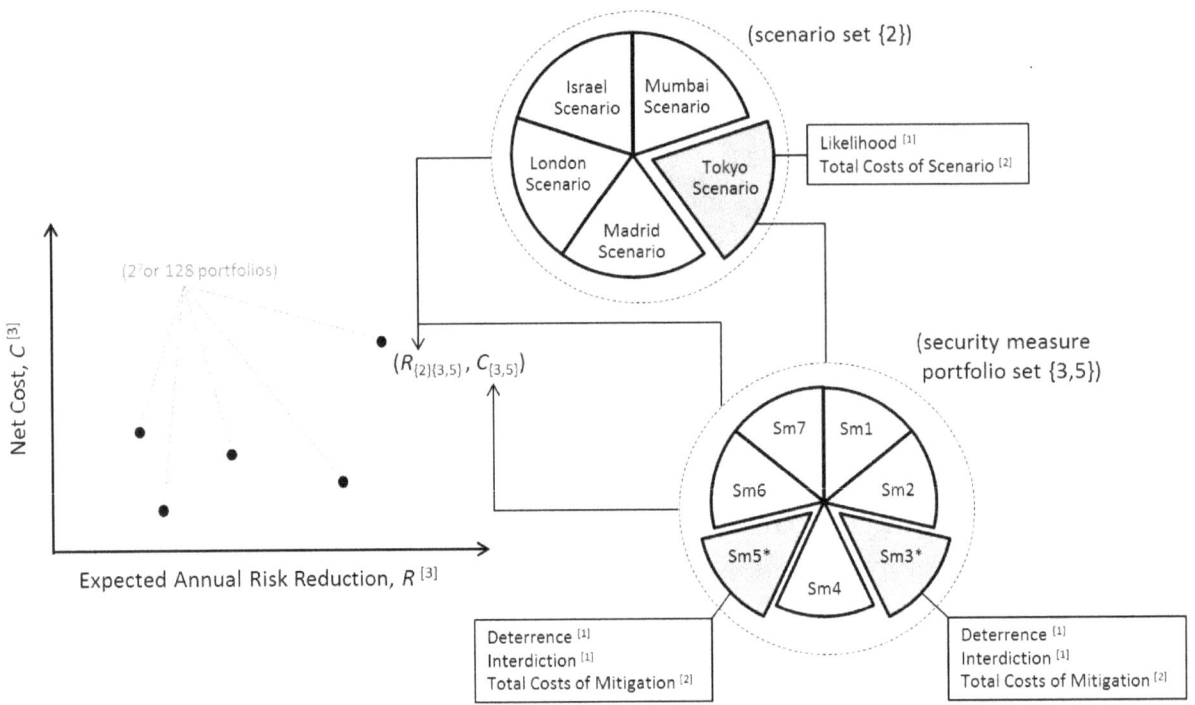

Notes:
[1] Risk Assessment : *Expert Elicitation, Statistical Analysis, Excel VBA Macro*
[2] Economic Analysis : *CGE, Survey, Simulation, Statistical Data Analysis, Literature Synthesis*
[3] Portfolio Analysis : *Excel VBA Macro, Tableau Public Viz*
*Sm3: Temporary perimeters and access control to certain restricted areas of the city
*Sm5: X-rays and magnetometers in building lobbies

Figure 2. Portfolio Analysis Framework

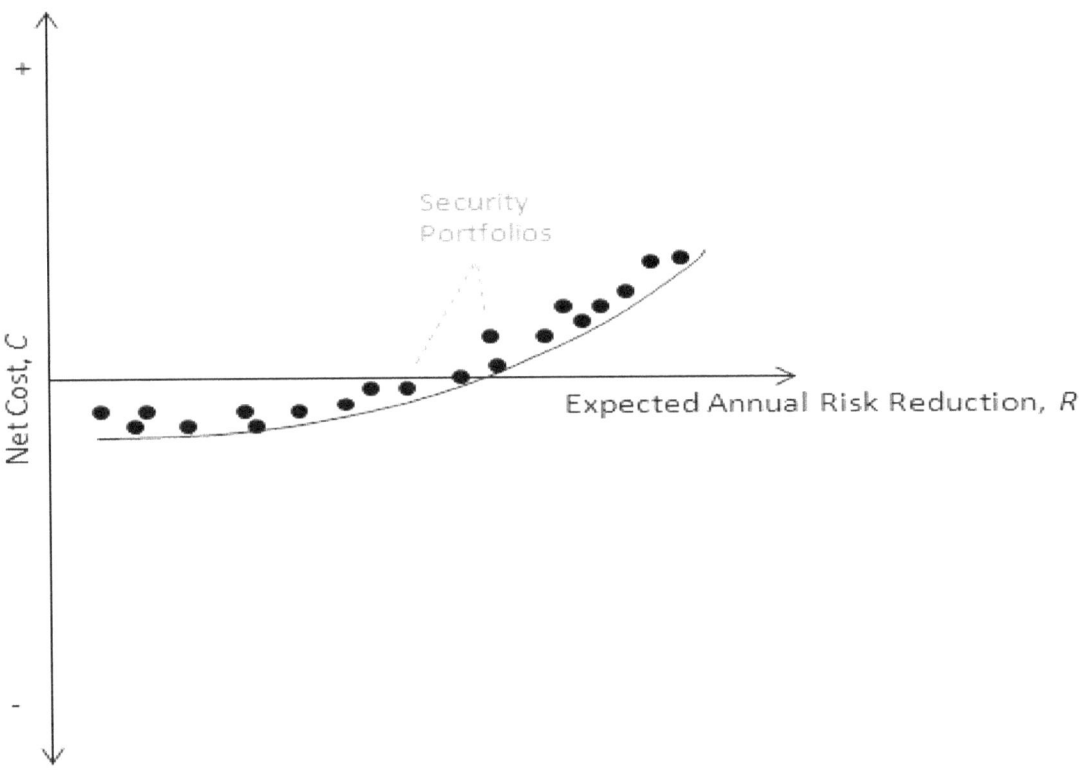

Model Assumptions

Inherent in the portfolio model are assumptions about how the security measures operate, the definition of scenario frequency, and the uncertainties associated with expected annual risk reduction and net costs. More specifically, with regards to security measure operations, it is assumed that they operate independently of one another. This assumption of probabilistic independence is a first order approximation to the interaction of various defensive measures in the portfolio and provides a useful operating assumption unless there are reasons to believe that the measures are redundant or act synergistically. Defensive measures may be redundant if they are different methods of performing the same function. For example, if one has wands to detect metal objects, such as weapons, and has walk through detectors for the same purpose, there is little benefit to adding both measures to the portfolio as long as the capacity of each detector is adequate for the purpose. In this case, one should restrict the set of portfolios considered by excluding redundant security measures.

Sometimes the security measures are partially redundant, so that incremental improvement is obtained. In this case, equation (1) would be modified to include a constructed security measure that represents both security measures in the portfolio and restricts portfolios to contain no more than one of the original two security measures and the constructed security measure. The deterrence and interdiction effects of the constructed security measure would require a separate assessment. The extension to three or more redundant security measures is similar.

Synergistic security measures also may result from one method supporting another. For example, a detection or alarm system provides little interdiction benefit unless forces are available to respond in a timely manner. If the synergism is very high, then a constructed security measure should be used in place of the original security measures, eliminating the indicator variables for both of the original security measures. If synergism is weaker, it may be necessary to employ the same method as when security measures are partially redundant.

The scenario frequencies, t_i, are a parsing of the total estimated annual threat frequency to the set of included scenarios. This is done so that the sum of the likelihoods of the five threat scenarios would be equal to the total threat annual frequency across all possible scenarios whether explicitly included in the analysis or not explicitly included. This allows for a comparison of total security measure costs, which are independent of the specific scenarios, to the estimated total risk reduction. Thus, the five included scenarios are representative of the set of all possible threat scenarios which, while not possible to enumerate, can be assigned a total threat frequency.

In this analysis, the expected annual risk reduction and net cost in equations (1) and (2) are evaluated as point estimates. In reality, these estimates have uncertainty and may be represented as random variables. A thorough explanation of the direct and indirect cost estimates and the consequence estimates are provided in [7].

RESULTS AND DISCUSSION

Once all possible combinations of security measures have been evaluated in terms of their expected risk reduction and total costs, they can be plotted as points on a portfolio diagram as shown in Figure 5. The benefits/costs that are independent of an attack are shown on the vertical axis and are sensed so that lower costs are preferred to higher costs. When costs are negative, the portfolio has a net benefit independent of any attacks. This occurs for some portfolios because the indirect benefits of having a particular security measure in place outweigh the amortized capital and operating costs. In Figure 3, the horizontal axis shows reduction in expected risk and is positively sensed in that greater reductions are preferred. Risk is the probability of a successful attack times the consequences of that attack.

Given the definition of the axes in the portfolio diagram, a policymaker would prefer portfolios that are found down and to the right in the diagram. Some portfolios will be found to dominate other portfolios in the sense that for the same or lesser cost, the dominating portfolio delivers greater risk reduction. Or, conversely, for the same or higher level of risk reduction the portfolio has lower costs. If one portfolio dominates another it will be found below and to the right of the dominated portfolio, or possibly directly to the right or directly below the dominated portfolio.

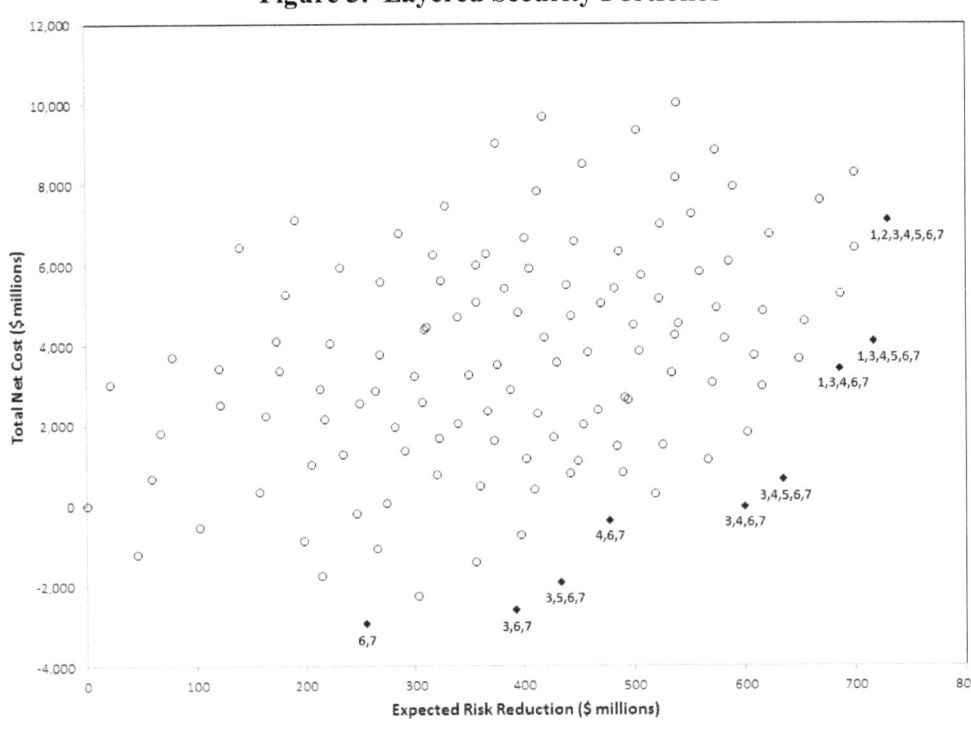

Figure 3. Layered Security Portfolios

The portfolios that are not dominated by any other portfolio are members of the efficient set and are those that, all else equal, should be considered as candidates for implementation as they cannot be improved upon simultaneously in both cost and risk reduction. These efficient portfolios are depicted by diamonds rather than disks in Figure 3. Moreover, they are labeled with the security measures included in that particular portfolio.

As the policymaker considers her decision making priorities among the portfolios, she must consider the tradeoff among security and economic activity. Two lines, representing the security and economic activity tradeoff, have been inserted in Figure 4. The upper line passes through all portfolios that have expected risk reduction equal to expected cost. If one is willing to accept that $1 of expected risk reduction is as valuable as $1 in cost reduction, then portfolios above this line have costs that exceed benefits (risk reduction) while those below the line have benefits that exceed costs. The lower line is a translation of the upper line such that all portfolios on the lower line have equal net benefit (risk reduction minus cost) but the net benefits are greater than those of any portfolio on a parallel line above this line. Thus, to identify the optimal portfolio(s) in Figure 4, one should find the parallel line that passes through a portfolio as far down and to the right as possible.

A more risk averse policymaker might consider Figure 5, where the slope of the straight line is increased to 10 implying a willingness to spend $10 up front to avoid $1 in expected loss from an attack. Many would argue that, to date, the $300 billion spent by the Federal Government [8] on preventing terrorism has greatly exceeded this 10:1 ratio. Changing the slope of this line changes the portfolio that has the best risk-reduction and cost profile. The optimal portfolio now shifts up and to the right.

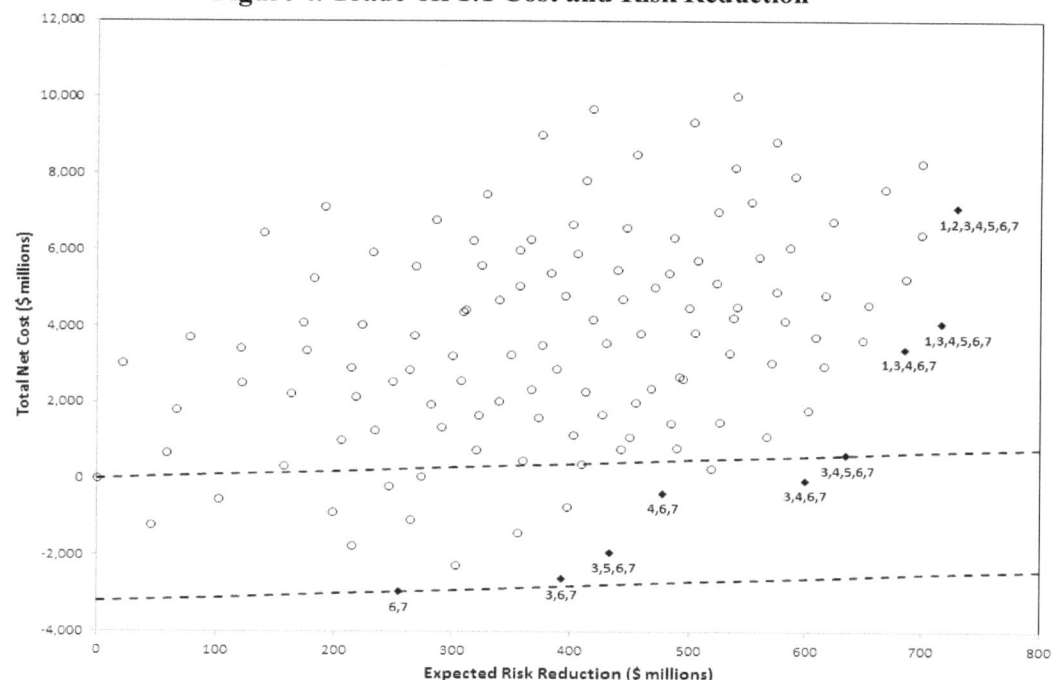

Figure 4. Trade-off 1:1 Cost and Risk Reduction

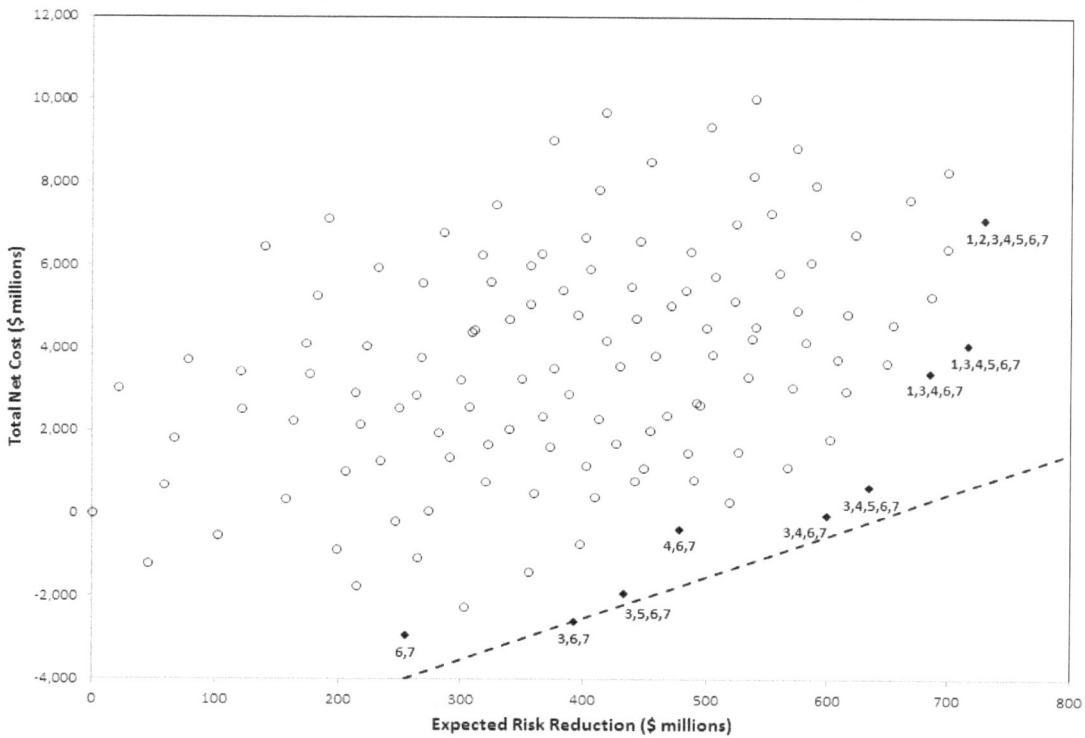

Figure 5. Trade-off 10:1 Cost and Risk Reduction

The portfolio diagram cannot answer the question of which portfolio is best, however, unless one is willing to state the appropriate trade-off between pre-event costs and post-event expected losses. Views on these trade-offs will differ among stakeholders. Law enforcement tends to be conservative and invests heavily in prevention; part of a "not on my watch" attitude. Others, such as real estate investors or tour companies, may put greater emphasis on reducing the dampening effects of nonpassive security measures, such as check points or bag inspections. Their trade-off lines would be less steep than those of their law enforcement counterparts.

One might infer that the U.S. Government views the trade-off line to be rather steep as expenditures on preventing terrorist attacks appear to be high relative to the expected risk reduction. This may be due in part to how a human life is valued. Here, the U.S. Department of Transportation value $6.2 million dollars for the loss of a life [9] has been used. It may be that in allocating funding to counter terrorism activities, the Government implicitly uses a higher figure for the loss of life due to terrorism vs. loss of life due to traffic accidents or other more common causes. See the work by Viscusi [10] for interesting evidence related to this issue.

The set of portfolios that should be considered can be reduced further if one assumes that the trade-off ratio between pre-event costs and risk reduction is constant and thus the trade-off line is linear. Those portfolios that form a convex hull are those that are candidates to be on such a linear segment. This assumption results in the exclusion of some portfolios that, although not dominated by any single portfolio, are dominated by a positive linear combination two other portfolios. This is demonstrated in Figure 6 where there are just 6 portfolios that should be under consideration given the linearity assumption.

Figure 6. Best Portfolios under Linear Trade-offs

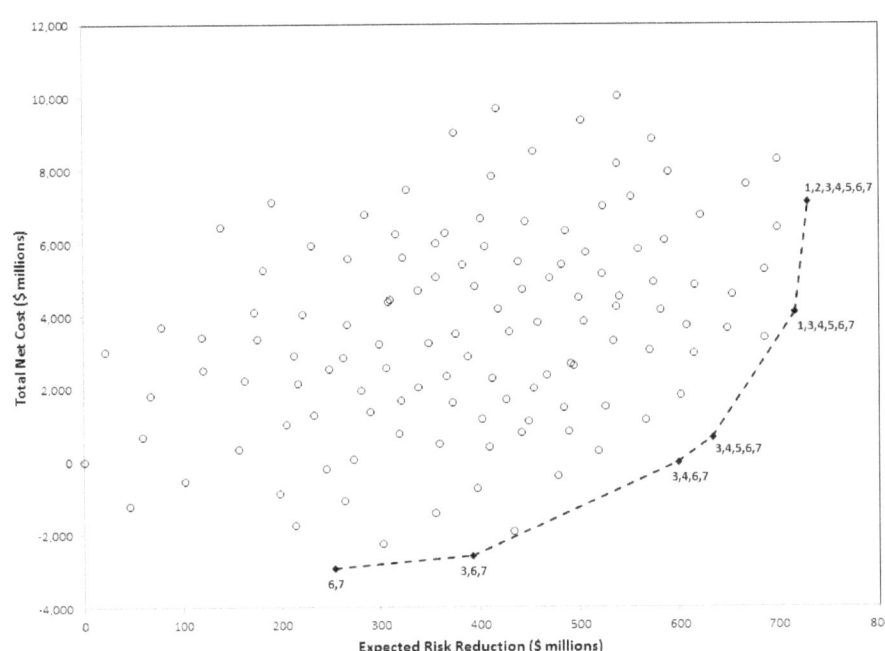

Lastly, as seen in Figure 6, benefits and costs may accrue disproportionately for various stakeholders. For example, law enforcement agencies may be responsible for purchasing and operating CCTVs. For this reason, they may not wish to consider the benefits of CCTVs as the law enforcement mandate is security and not economic development. Law enforcement may consider it inappropriate to use their budget for any purpose other than security and, therefore, may choose to ignore any benefits that are not the direct result of risk reduction. Similarly, the indirect economic burdens brought on by delays, congestion, and inconvenience may not be taken into account because they are not borne by those choosing the security measures. Figure 9 shows the portfolio diagram with spillover costs and benefits removed. This might be the diagram that law enforcement would choose to use in selecting a portfolio rather than the diagram in Figure 6. Note that the scales on the vertical axes in Figures 4 and 7 are very different. The dark lines slightly slanted from vertical in Figure 7 are, in fact, the one-to-one tradeoff lines for risk reduction and cost and serve the same purpose as the similar lines in Figure 4.

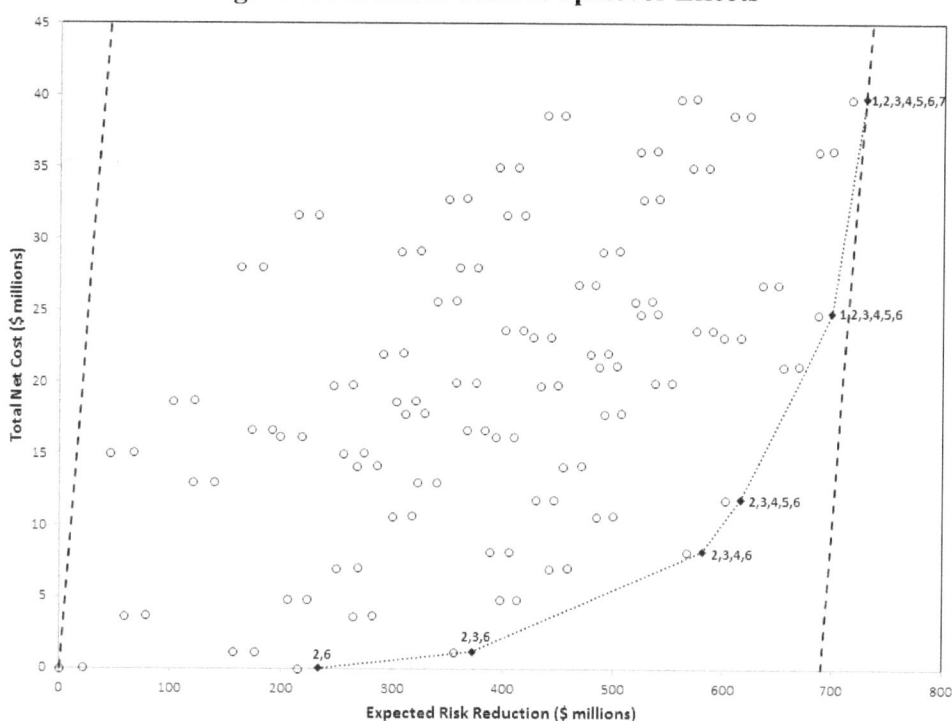

Figure 7. Portfolios without Spillover Effects

One observation derived from the comparison of Figures 4 and 7 is that the entry of CCTV cameras into efficient portfolios is delayed when spillover effects are ignored. This is not surprising, as one might expect, since CCTV systems provide security without active interference and, thus, may enhance perceived wellbeing and economic activity. A more global observation is that with one-to-one trade-offs between costs and risk reduction, ignoring spillover effects makes all portfolios have greater benefits than costs and the best of these becomes the portfolio with all security measures included.

CONCLUSIONS

Decisions about security measures are often made in isolation, considering one measure at a time. A better approach, however, is to consider the security measures together as a system of systems that can be optimized only by considering the interactions among the component systems. The capabilities of the component systems may vary with the type of attack as will the interactions among the capabilities. A portfolio of component systems should be selected to minimize "holes" or weak points which will simply attract attacks.

The benefits and costs of a security system will accrue to different parties unequally. By considering spillover effects, both positive and negative, one attains a more accurate picture of the total benefits and costs to our nation. Myopic decision making where total benefits and costs are ignored can lead to suboptimization. In the example presented here for the Lower Manhattan security area, exclusion of spillover effects can result in significant overinvestment in security and a loss of societal net benefits because too much security can have a dampening effect on economic and social activities.

The portfolio approach to layered security does not prescribe a single solution, but instead provides a number of potential solutions while excluding some configurations that have been identified as dominated. This analysis is not only useful as a pruning device to focus attention on a smaller set of candidate portfolios, but it also provides a basis for communication and negotiation among stakeholder groups who have differing views on the benefits and burdens of enhanced security.

ACKNOWLEGEMENT

This research was supported by the United States Department of Homeland Security through the National Center for Risk and Economic Analysis of Terrorism Events (CREATE) under Cooperative Agreement No. 2010-ST-061-RE0001. However, any opinions, findings, and conclusions or recommendations in this document are those of the authors and do not necessarily reflect views of the United States Department of Homeland Security or the University of Southern California.

REFERENCES

[1] Lehrman TD. Building a Layered Defense to Combat Weapons of Mass Destruction. Remarks to the NPT Conference, Washington College of Law, American University, Washington, D.C., February 9, 2006.
[2] Department of Defense. Strategy for Homeland Defense and Civil Support, Washington D.C., 2005.
[3] Transportation Security Administration. Layers of Security, 2013. Available at: http://www.tsa.gov/about-tsa/layers-security, Accessed on September 5, 2013.
[4] Buede DM. The Engineering Design of Systems: Models and Methods. New York: John Wiley & Sons, 2011.
[5] Roberts F, Hora S, Jenkins B. Urban Commerce and Security Study Technical Report. Division of Science and Technology, Department of Homeland Security, Washington D.C., June 2013.
[6] Ross S. A First Course in Probability. Delhi, India: Pearson Education, 6th Edition, 2004.
[7] Rose, A., M. Avetisyan and S. Chatterjee. A Framework for Analyzing the Economic Tradeoffs between Urban Commerce and Security, forthcoming in Risk Analysis.
[8] Stewart MG. Risk-Informed Decision Support for Assessing the Costs and Benefits of Counter-terrorism Protective Measures for Infrastructure. International Journal of Critical Infrastructure Protection, 2010; 3: 29-40.
[9] Trottenberg P, Rivkin R. Memorandum to Secretarial Offices and Modal Administrators: Treatment of the Economic Value of a Statistical Life in Departmental Analysis- 2011 Interim Adjustment, U.S. Department of Transportation, 2011. Available at: http://www.dot.gov/sites/dot.dev/files/docs/Value_of_Life_Guidance_2011_Update_07-29-2011.pdf, Accessed on September 5, 2013.
[10] Viscusi WK. Valuing Risks of Death from Terrorism and Natural Disasters. Journal of Risk and Uncertainty, 2009; 38: 191-213.

Cyber security: the Risk of Supply Chain Vulnerabilities in an Enterprise Firewall

Marshall A. Kuypers[*a], Greg Heon[a], Philip Martin[a], Jack Smith[a], Katie Ward[a], and Elisabeth Paté-Cornell[a]

Stanford University, Stanford, CA

Abstract: Cyber security is a critical concern for many organizations. One defense approach is to install firewalls, but their effectiveness is uncertain and the cheapest model may not be the best. One may try to inspect them for vulnerabilities that may have been introduced in the product's supply chain. Most existing models that quantify cyber risk do not address that issue, and the risk that corrupted components could be successfully inserted into a secure network is not directly considered other than by characterizing the supplier. We present a probabilistic risk analysis model for a firewall linking its parts to the different stages of production. We then evaluate the tradeoff between cost (system and inspection) and security by comparing two firewalls. We base our analysis on expert opinions, which we aggregate using the Delphi method. The model shows that in the illustrative case presented here, the value of information about the effectiveness of a firewall is actually worth little to a risk neutral decision maker. Therefore, inspecting firewalls for vulnerabilities may not be the most effective way to address the system's security. Gathering information by monitoring for warning signals of a cyber attack could be a beneficial alternative or complement.

Keywords: Cyber security, Supply chain, risk analysis

1. INTRODUCTION

The increased reliance on chip-based electronics has resulted in a greater risk to organizations of data destruction, corruption, or loss of confidentiality. Many organizations are constantly adding new hardware and software to information technology (IT) networks, but the security of new products is often taken for granted. Over the past five years, however, several companies have been compromised by an intentional vulnerability introduced in a product's supply chain. At this time, decision makers have a limited toolset to analyze risk from new products caused by non-secure supply chains. A common heuristic might be to use the national origin of the equipment as a discrimination factor, which is suboptimal. In this paper, we model an organization's decision to purchase an enterprise firewall using decision analysis and risk analysis to assess the likelihood of supply chain vulnerabilities. Quantifying the risk to an organization that a vulnerability has been introduced in its cyber defenses enables a clear breakdown of the costs and benefits of different firewalls. In this paper, we consider the option of inspecting these defenses for undetected vulnerabilities and we assess the value of these inspections.

2. Background

Supply chain attacks have increased in frequency and scope. One of the earliest examples occurred in Chicago in 1982, when several residents died after taking store bought Tylenol that had been laced with cyanide[†] [1]. In response to public alarm across the nation, the pharmaceutical industry developed tamper-proof seals and began analyzing supply chain security. Soon after, new supply chain threats emerged. The liquid in eye drop products was replaced with Hydrochloric acid and other medications were poisoned. Criminals also began targeting manufacturing processes. In 2006, McDonalds recalled promotional mp3 players that included spyware designed to steal users' passwords [3]. The GPS

---

[*] mkuypers@stanford.edu
[†] Authorities eliminated contamination during the manufacturing process as a cause since contaminated bottles came from two manufacturing plants in different states and only surfaced in Illinois. Authorities believe that someone tampered with the medicine shortly before the point of sale [2].

manufacturer Tom-Tom shipped malware infected devices in 2006, and Apple ironically shipped iPods infected with a PC virus [4]. Vulnerabilities were also introduced in the supply chains of flash drives, digital picture frames, external hard drives, and laptops, and customers purchased products that had been compromised before ever being used.

A particularly sophisticated supply chain attack occurred in 2008, when extra computer chips were covertly added to Mastercard credit card readers. These chips then transmitted credit card data to criminals overseas. According to Joel Brenner, a former US National Counterintelligence Executive, the vulnerability was introduced either during the manufacturing process in China or shortly thereafter, since the new card readers appeared to be in their original packaging when they arrived at stores for installation [5]. A small, wireless chip had been inserted behind the card reader's motherboard, which allowed the copied data to be sent to a server in Pakistan [5,6]. Mastercard resorted to sending teams to weigh each card reader, since virtually the only sign of tampered card readers was an extra 3 to 4 ounces in their weight [7]. An estimated $50M to $100M were lost as a result of the attack.

The possibility of the introduction of vulnerabilities by foreign adversaries through a supply chain gained new national attention in 2010, when Sprint was considering bids to upgrade its US telecommunications network [8]. Huawei, the world's second largest telecommunications and internet manufacturer, placed a bid that was estimated to save Sprint $800M in the first year alone [9]. Security concerns arose and a group of senators wrote a letter to national security officials pointing out that Huawei had repeatedly violated intellectual property rights and had ties to the Chinese government[‡] [10]. Huawei's bid was blocked and since then, Huawei has struggled to dispel fears about its products[§].

2.1. Related Work

As supply chain attacks have become more frequent and widespread, governments and organizations have been increasingly interested in the study of risk controls and mitigations. Supply Chain Risk Management (SCRM) is a well-studied area primarily focused on optimizing a supply chain against disruptions [13, 14]. The aftermath of events such as the September 11th attacks on the US and the 2011 earthquake and tsunami in Japan have shown the importance of robust supply chains to avoid production line shutdowns that can cost up to $10,000 per minute of downtime. However, the risk of an intentional introduction of vulnerabilities in specific products is considerably less studied.

Many researchers and organizations have addressed cyber security quantitatively with tools ranging from probabilistic risk analysis (PRA) to game theory [15]. The unique damage caused by cyber attacks includes the loss of physical equipment, network downtime, reputation damage, and other costs. They require new modeling approaches to fully capture the range of impacts. Initially, the complexity and uncertainty of losses led to simple measures of risk, such as Annual Loss Expectancy [16]. As the limitations of simple expectation methods were exposed, new probabilistic tools were developed [17, 18]. The Gordon-Loeb model demonstrated how economic analysis could be applied to the cyber domain for a variety of applications [19]. Buckshaw outlined a decision analysis framework to quantitatively evaluate information system designs [20]. Recently, there has been growing interest in attack trees, kill chains, and the optimization of IT resources [21, 22, 23]. Yet, very few models address security concerns about the supply chain, assuming instead that IT products and controls increase security without introducing additional risks. This problem is the focus of this paper.

At this time, interest in supply chain attacks is growing. In 2012, the White House released the "National Strategy for Global Supply Chain Security", which responded to earlier government studies calling for better management of supply chain risks [24]. The United States Government

---

[‡] The founder and CEO of Huawei, Ren Zhengfei, was a Major in the People's Liberation Army Engineering Corps.
[§] In 2013, Huawei announced that it was not interested in the US market, although a month later, it announced plans to launch a smartphone in the US [11, 12].

Accountability Office has released several reports about risks in IT supply chains, and the National Institute of Standards and Technology has published best practices for businesses regarding supply chain risk management [25, 26, 27, 28]. Many organizations and researchers have also begun to call attention to supply chain attacks [29, 30]. Yet, little quantitative, risk-analytic work has been done in this area. The objective of this paper is to present a PRA for a firewall considering the possibility that vulnerabilities have been introduced in the different parts in the production process.

3. Model Formulation

To study the tradeoffs between price and security in IT products, we constructed an organization comparing two enterprise firewalls (see Figure 1). The client organization is located in nation state Y and owns valuable intellectual property (IP). This organization considers buying firewalls from two companies, A and B. Company A is publically owned, also based in nation state Y, and has a reputation for delivering high quality, secure products. Company B is privately owned and based in nation state X, which is an adversary of nation state Y. It is less well known and may have a less secure supply chain but cheaper products. The client organization must also purchase a maintenance plan from either the firewall's manufacturer, or from company C in nation Y. In spite of their reputations, companies A or C as well as company B, could be collaborating with nation state X, which may introduce vulnerabilities into their supply chains. The decision maker would like to analyze which firewall should be purchased, the value of additional information (e.g., through inspection) about the probability of a vulnerability, and what strategies could be used to secure the organization's IP.

Figure 1: Decision Model for the purchase of a firewall

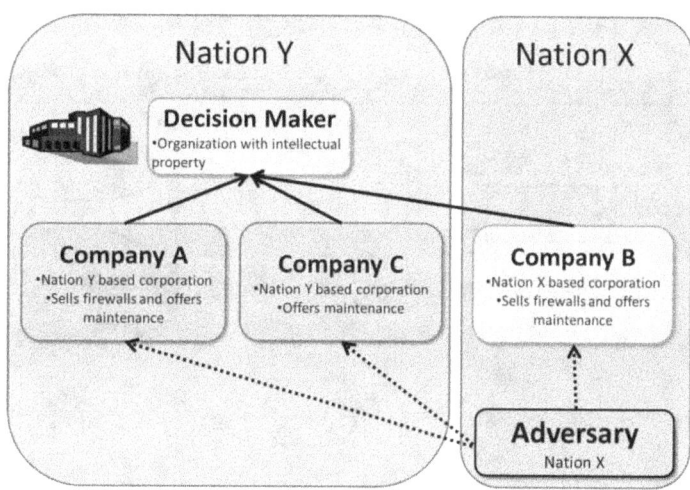

3.1. Vulnerability Analysis

In order to narrow the scope of the analysis, we only model one specific vulnerability for this illustrative study. Firewalls contain a run configuration file, which lists the protocols, the open ports of the network, and the security controls of the machine. The run configuration file is thus valuable to attackers because it provides information about the network defenses and allows the adversary to launch more targeted, effective attacks. Accessing the run configuration file does not guarantee that the adversary will immediately steal or destroy data, but it increases the likelihood that it will compromise the client organization in the future. Therefore, the decision maker is concerned that a vulnerability that copies and sends the run configuration file to an adversary could be introduced in the firewall's supply chain. The nature of the run configuration file also provides a high degree of plausible deniability[**]. Also, firewalls typically maintain a certain configuration for months or years.

[**] The run configuration file is typically on the order of kilobytes, meaning a file transmission could blend into network traffic easily.

Therefore, the information obtained by the adversary about the network security would remain useful for a substantial length of time. This is the vulnerability described in the model presented here.

3.2. Decision Diagram

A decision diagram for the decision maker is shown in Figure 2. He/she will first choose a firewall and maintenance plan. If either of the producing companies is fully cooperating with the adversary nation X, the decision maker assumes that a vulnerability will be introduced with probability one, since it is technically feasible. If this collaboration does not occur, the adversary can attempt to introduce the vulnerability through company's vendors who supply materials such as chips, software, or other components of the firewall. The vulnerability could be introduced in the supply chain several tiers below the main supplier unless effective security practices are followed. The adversary may also introduce an agent into the company. An agent is defined here as a person or group of people directed by an adversary (forcefully or willfully) to implement a vulnerability in the client's system. The definition encompasses blackmail, bribery, and agent infiltration. Agents who attempt to plant the vulnerability may be unsuccessful due to security restrictions, or because it is not technically feasible in the particular production phase that they have infiltrated. Finally, the adversary may introduce the vulnerability through maintenance since code updates may not be as thoroughly tested as the original product.

Figure 2: Decision Diagram for the purchase of a fire wall

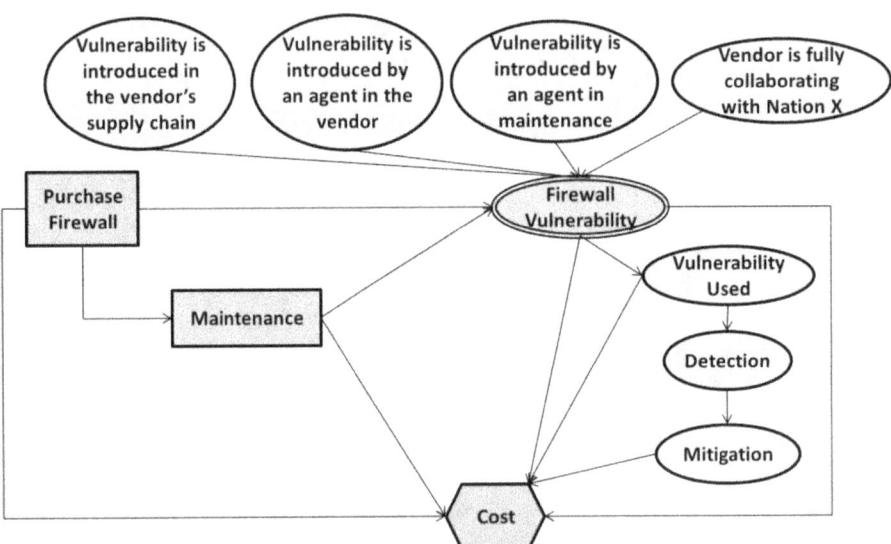

The decision maker, even if he/she is aware of a breach, may not know with certainty whether the adversary will actually use the vulnerability that was introduced. In this study, we assume that the run configuration file is always exploited if a vulnerability is introduced in it, but we include an uncertainty about its actual use, which would increase the probability its detection. An option for the decision maker is to inspect the system to try to detect intrusions.

3.3. Supply Chain Model

The probability that a vulnerability is successfully inserted in a product's supply chain depends on the stage at which it was introduced, the difficulty of insertion at different stages, and the likelihood that it is detected later in the manufacturing process. Our model of the supply chain for an enterprise firewall was developed to study how the likelihood of a successful insertion changes through the supply chain lifecycle, and the chances of detection through inspection of the product.

The structure and secrecy of IT supply chains makes this analysis difficult in practice. Most companies consider supply chain information to be proprietary, leading to a lack of publically available data on the number of suppliers, their locations and the nature and frequency of attacks. Tiered suppliers make tracing components back to original sources difficult as well. Subcontractors are often nested several (often three) layers deep below the main contractor, making it difficult to track products' security. Multiple vendors may be involved in the manufacturing of the components[††]. Supply chains often overlap, but their lack of transparency makes the understanding of common risks nearly impossible[‡‡]. Villasenor presents a comprehensive overview of other challenges in assessing the overall security of a supply chain [30].

A firewall production lifecycle consists of several basic steps: research and development, design, manufacturing, and delivery/ installation[§§]. During this process, components from external vendors may be integrated into the product. Each of these phases presents an opportunity to introduce or detect a vulnerability. Figure 3 shows the possible attack vectors.

Figure 3: Supply Chain vulnerabilities

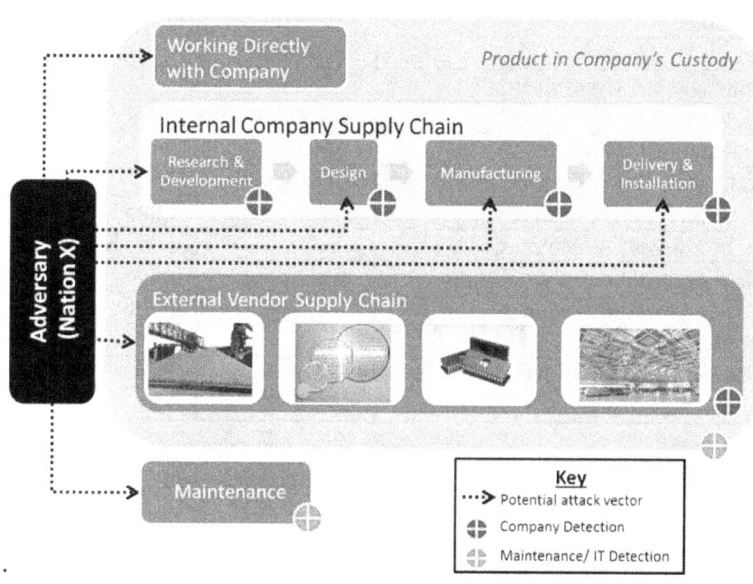

For a vulnerability to exist, it must be introduced in a phase *i* and not be discovered in any subsequent phase *p*. The probability $P(V)$ that the vulnerability remains in the system into its operation is thus simple if one assumes independence of attempts to insert and detect vulnerabilities at different stages of the supply chain:

$$P(V) = \sum_{i=1}^{n} P(S_i) \left(\prod_{p=i}^{n} (1 - P(F_p)) \right)$$

in which:

$P(S_i)$ is the probability that a vulnerability is introduced in Stage *i*
$P(F_p)$ is the probability that a vulnerability is found in stage *p*

Decomposing the probability of a vulnerability introduction in different phases of the supply chain simplifies the data needed to populate the model by conditioning each event on the phase in the supply chain where it takes place.

3.4. Expert Probability Elicitation

[††] For example, a manufacturer may run out of a RAM chip made from vendor A and substitute a chip from vendor B.
[‡‡] For example, a CPU chip vendor may supply company A and company B with the same chip.
[§§] Note that the usage and disposal are not modeled, due to the limit scope of the model.

To populate the model with probability inputs, interviews were conducted with four experts with backgrounds ranging from network security to firewall manufacturing[***]. The Delphi method was used to obtain and integrate expert probability elicitations [32]. Each interview was prefaced with the list of definitions for each supply chain phase and two rounds of interviews were conducted, after which the responses were averaged and used as inputs to the model. Despite considerable efforts to make the questions unambiguous, experts often made implicit assumptions that led to probability estimations that differed by orders of magnitude. Eliminating ambiguities in the elicitation questions will thus be essential in future studies. A summary of the data obtained from the expert probability elicitation and other input data is contained in Table 1, for the comparison of two different firewalls (A and B) and two maintenance options for each (maintenance from supplier or from another firm C).

Table 1: Input data and firewall comparison: expert opinions on introduction of vulnerabilities

Options	Cost	Probability
Firewall A	$250,000	
Firewall B	$150,000	
Maintenance A (lifetime)	$150,000	
Maintenance B (lifetime)	$50,000	
Maintenance C (lifetime)	$100,000	
Run Configuration File Stolen	-$1,000,000	
Vul. in Firewall A, Maint. A		0.00513
Vul. in Firewall A, Maint. C		0.00518
Vul. in Firewall B, Maint. B		0.608
Vul. in Firewall B, Maint. C		0.558
Prob. Detection	Vul. AA		0.25
Prob. Detection	Vul. AC,BB, BC		0.20
% of Damages Mitigated	Detection		0.75

4. Decision Analysis for the Choice of a Vendor and Maintenance Operator

After the expert probability elicitation, the data were entered into a decision tree along with the costs defined in the original problem statement to determine the optimal choice of a supplier and of a maintenance company.

Figure 4: Decision Tree for the choice of supplier and maintenance organization

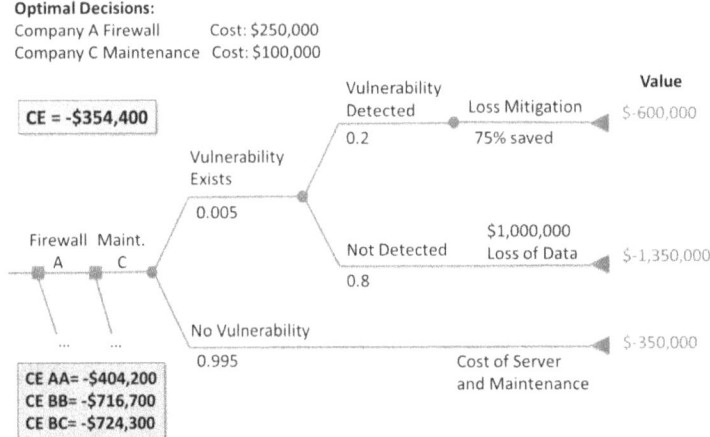

We found that in this illustrative case, the optimal decision for the client organization (assumed first, to be an expected value maximizer) is to buy company A's firewall and company C's maintenance

---

[***] More experts are needed to ensure accurate results, but four experts allowed an effective demonstration of the model for this illustrative scenario. For an example of the Delphi method applied to IT risk, see Herrmann [31].

(see Figure 4). A sensitivity analysis was performed for all input probabilities and different risk tolerance levels. It showed that the decision is fairly robust. The value of perfect information is found to be only $5,800 for a risk neutral decision maker.

4.1. Analysis

The decision between company A and company B is often simple if the client organization places a high value on losing data or if the probability of a vulnerability is significantly different between the two companies. These two metrics are compared in Figure 5. The analysis assumes that the price difference between the two firewalls is $100,000. Therefore, for the risk neutral decision maker, the expected loss due to the increased risk must be less than $100,000 for company B to be chosen. Figure 5 shows the tradeoff, which offers a simple way for decision makers to decide if more analysis is needed.

Figure 5: Sensitivity Graph of the value of data loss for different risks of a vulnerability

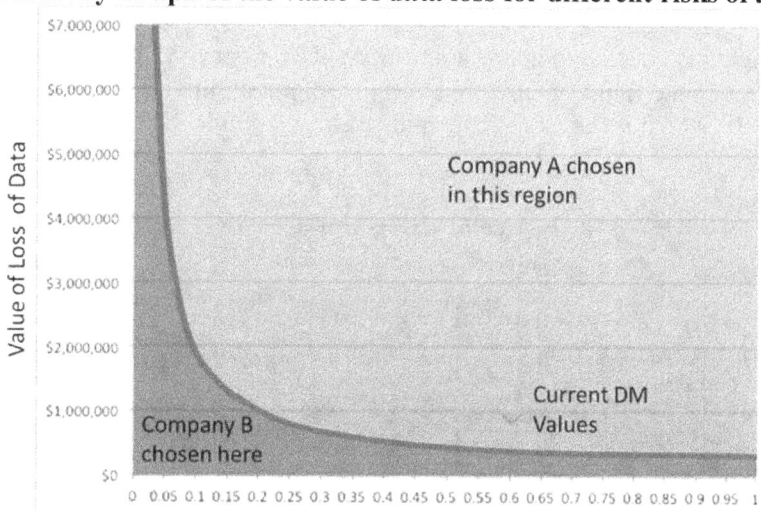

Probability that Vulnerability Exists (For firewall B and Maintenance C)

4.2. Inspection

Various controls have been proposed to reduce the probability that a vulnerability is successfully introduced in a supply chain. Most software comes with a checksum[†††], which detects code differences between the original source and the installed application. However, vulnerabilities introduced during the firewall software development would not be detected using checksums. Large organizations often rely on "many eyes" to detect vulnerabilities, claiming that it is infeasible for an adversary to maliciously tamper with their product. However, empirical evidence suggests that "many eyes" do not prevent vulnerabilities if they look for the same things, even in open source applications. For instance, enough incidents of malicious code insertion by disgruntled employees have occurred to raise doubt about an organization's ability to detect unauthorized insider activity[‡‡‡] [33, 34].

Given the uncertainty surrounding vulnerability insertions and the effectiveness of many proposed controls, a quantitative analysis of security measures is useful. One option is to inspect incoming products. Firewalls could be dismantled to look for suspicious components and executable files could

---

[†††] A checksum function works by computing a string of bits corresponding to the code. Standard algorithms of that type are very difficult to defeat by reverse engineering because any change in the original code will change the checksum value and alert a user that the code has been compromised. In a sense, it is almost a "tamper-evident" seal.
[‡‡‡] The frequent occurrence of "Easter Eggs", or secret features inserted into software or hardware such as a flight simulator in Excel 1997, also suggest that code development is not secure.

be reviewed for strange activity[§§§]. However, dismantling a firewall to look for hardware or software vulnerabilities is labor- and cost-intensive with uncertain outcomes. Our model can help a decision maker determine if products should be inspected and what investment in inspection is cost-effective.

If firewall B is purchased, a perfect inspection that finds all vulnerabilities is worth $142k to a risk neutral decision maker, assuming that a refund is given when vulnerabilities are found in a firewall. If refunds are not given, perfect inspection is worth only $12k. In reality, perfect inspection cannot reasonably be obtained. Instead, a decision maker should evaluate by how much the prior probability of a vulnerability in the product can be reduced by a given level of inspection. For example, if after spending $100k on inspection, the decision maker believes that the probability of a vulnerability in firewall B can be reduced from the prior of 0.608 to a posterior of 0.1, then the investment is cost-effective. However, given the complexity of the firewall hardware and software, if the decision maker thinks that a $100k inspection would only reduce the posterior probability of vulnerability detection to 0.4, then the inspection is not cost-effective. Figure 6 shows the cost-effective investment ranges for firewall B.

A decision maker might also decide to invest some amount of money to inspect firewall A. In this illustrative case and given the figures that were chosen, the probability of a vulnerability can only be reduced from the 0.005, which is the prior. The value of inspection is much lower: a perfect inspection value is only $2.2k.

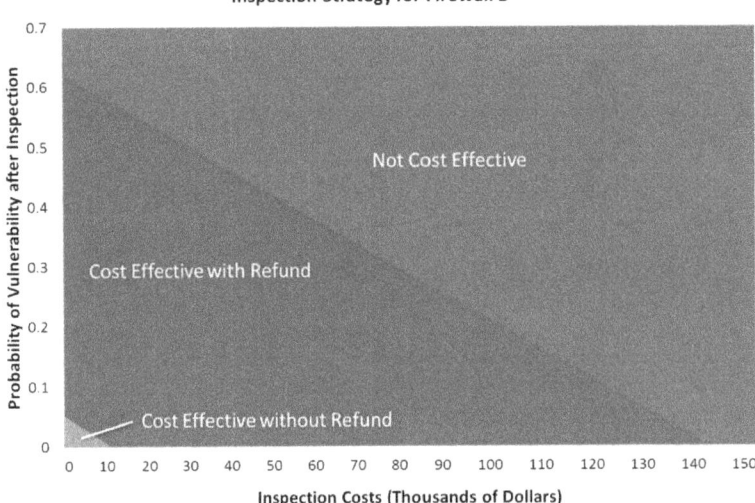

Figure 6: Inspection Costs

5. CONCLUSION

Regarding the cyber security of supply chains and the benefits of inspection to detect vulnerabilities, there are many qualitative reports and best practice statements, but to the best of our knowledge, no quantitative tools have been proposed to help organizations mitigate that risk. Supply chain attacks are complex problems, involving dozens of manufacturers, hundreds of vulnerabilities, and thousands of components. The model presented in here involves only one attacker and one vulnerability. It can be expanded to include multiple adversaries and multiple attacks.

We start here from the observation that the frequency of cyber attacks related to information systems supply chains has increased over the past 30 years. One can reasonably assume that it is likely to continue to increase. Using data from expert probability elicitation, we modeled such a supply chain to calculate the probability that a successful vulnerability is introduced and undetected in a firewall at

---

[§§§] The source code, which contains comments and is designed to aid engineers in the analysis and verification of a code, is not included in most applications. The executable code is considerably more difficult to analyze.

each stage of its production lifecycle. We find that product inspection is rarely cost-effective, and we show how one can quantify and communicate the tradeoff between price and security using simple graphs. In comparing supply options, one should thus not count solely on the chances of detecting vulnerabilities through product inspection, but consider alternatives (or complements) such as signal monitoring, signature- or anomaly-based detection, and other warning systems to improve the security of organizations.

Acknowledgements

The authors would like to thank Steve Hurd, Alex Keller, and Matthew Daniels for their expertise and guidance for this paper. The authors also thank 4 anonymous experts who participated in the expert probability elicitation.

References

[1] K. A. Wolnik, F. L. Fricke, E. Bonnin, C. M. Gaston, and R. D. Satzger. *"The Tylenol tampering incident-tracing the source"*, Analytical chemistry 56.3, 466A-474A, (1984).

[2] M. Beck, M. Hagar, R. LaBreque, S. Monroe and L. Prout. *"The Tylenol Scare"*, Newsweek. 11 October 1982.

[3] C. Arthur. *"Get yer malware with fries, or on your new video iPod,"* The Guardian. 18 October 2006.

[4] R. McMillan. *"Virus located in TomTom GPS systems,"* InfoWorld. 29 January 2007.

[5] A. Modine. *"Organized crime tampers with European card swipe devices"*, The Register. 10 October 2008.

[6] S. Gorman. *"Fraud Ring Funnels Data From Cards to Pakistan"*, Wall Street Journal. 11 October 2008.

[7] H. Samuel. *"Chip and pin scam 'has netted millions from British shoppers',"* The Telegraph. 10 October 2008.

[8] J.S. Lublin and S. Raice. *"Security Fears Kill Chinese Bid in U.S,"* Wall Street Journal. 5 November 2010.

[9] S. Prasso. *"What makes China telecom Huawei so scary?"* Fortune. 28 July 2011.

[10] J. Kyle, C. Bond, R. Shelby, J. Inhofe, J. Bunning, J. Sessions, R. Burr, S. Collins. Letter to Secretary Geithner, Secretary Locke, Administrator Johnson, and Director Clapper. 18 August 2010.

[11] K. Hille and P. Taylor. *"Huawei 'not interested in the US any more'"*, The Financial Times. April 23, 2013.

[12] C. Thompson. *"Huawei plans to tackle US market with huge new smartphone,"* CNBC. 6 January 2013.

[13] S. Chopra and M. S. Sodhi. *"Managing risk to avoid supply-chain breakdown,"* MIT Sloan Management Review. (Fall 2004)

[14] C. S. Tang. *"Perspectives in supply chain risk management,"* International Journal of Production Economics, 103.2, 451-488, (2006).

[15] H. Cavusoglu. S. Raghunathan, W.T. Yue. *"Decision-Theoretic and Game-Theoretic Approaches to IT Security Investment"*, Journal of Management Information Systems, Vol. 25, No. 2, pp. 281–304, 2008.

[16] K. J. Soo Hoo. *"How Much Is Enough? A Risk-Management Approach to Computer Security"*, 2000.

[17] R. Bojanc, B. Jerman-Blazic, M. Tekavcic. *"Managing the investment in information security technology by use of a quantitative modelling"*, Information Processing and Management, 48, 1031–1052, (2012).

[18] N. Xie. *"SQUARE Project: Cost/Benefit Analysis Framework for Information Security Improvement Projects in Small Companies"*, Software Engineering Institute. 2004.

[19] L.A. Gordon and M.P. Loeb. *"The Economics of Information Security Investment"*, ACM Transactions on Information and System Security, Vol. 5, No. 4, Pages 438–457, (2002).

[20] D.L. Buckshaw, G.S. Parnell, W.L. Unkenholz, D.L. Parks, J.M. Wallner, O.S. Saydjari. "*Mission Oriented Risk and Design Analysis of Critical Information Systems*". Military Operations Research, 10.2, 19-38, (2005).

[21] A. Roy, D.S. Kim, K.S. Trivedi. "*Cyber security analysis using attack countermeasure trees*", In Proceedings of the Sixth Annual Workshop on Cyber Security and Information Intelligence Research (p. 28). ACM. (2010)

[22] E.M. Hutchins, M.J. Clopperty, R.M. Amin. "*Intelligence-Driven Computer Network Defense Informed by Analysis of Adversary Campaigns and Intrusion Kill Chains*", Leading Issues in Information Warfare & Security Research, 1, 80, (2011).

[23] R.A. Miura-Ko. "*Modeling and Mitigation of Information Technology Risks*," 2010.

[24] White House. "National Strategy for Global Supply Chain Security", 2012.

[25] G.C. Wilshusen et al. "*IT Supply Chain: National Security-Related Agencies Need to Better Address Risks*", United States Government Accountability Office. March 2012.

[26] M.L. Goldstein et al. "*Telecommunications Networks: Addressing Potential Security Risks of Foreign-Manufactured Equipment*", United States Government Accountability Office. 21 May 2013.

[27] J. Boyens, C. Paulsen, N. Bartol, S.A. Shankles, R. Moorthy. "*Notional Supply Chain Risk Management Practices for Federal Information Systems*", National Institute of Standards and Technology. 2012.

[28] J. Boyens, C. Paulsen, N. Bartol, S.A. Shankles, R. Moorthy. "*Supply Chain Risk Management Practices for Federal Information Systems and Organizations*", National Institute of Standards and Technology. 2013.

[29] J. Villasenor. "*Compromised By Design? Securing the Defense Electronics Supply Chain*", Brookings. 2013.

[30] R.J. Ellison, J.B. Goodenough, C.B. Weinstock, C. Woody. "*Evaluating and Mitigating Software Supply Chain Security Risks*", Software Engineering Institute. May 2010

[31] A. Herrmann. "The Quantitative Estimation of IT-Related Risk Probabilities." Risk Analysis, Vol. 33, 8, 1510-1538. (2012).

[32] H.A. Linstone and M. Turoff. "*The Delphi method: Techniques and applications*", 2002.

[33] G. Schryen. "Security of open source and closed source software: An empirical comparison of published vulnerabilities", (2009).

[34] J. Vijayan. "*Unix Admin Pleads Guilty to Planting Logic Bomb*", PC World. 21 September 2007.

Security Informed Safety Assessment of Industrial FPGA-Based Systems

Vyacheslav Kharchenko*[a,b], Oleg Illiashenko[a], Eugene Brezhnev[a,b], Artem Boyarchuk[a], Vladimir Golovanevskiy[c]

[a] National Aerospace University KhAI, Kharkiv, Ukraine
[b] Centre for Safety Infrastructure Oriented Research and Analysis, Kharkiv, Ukraine
[c] Western Australian School of Mines, Curtin University, Australia

Abstract: The strong interconnection and interrelation of safety and security properties of industrial system which are based on programmable logic (field programmable gate arrays, FPGA) is reviewed. Information security, i.e. system's ability to protect the information and data from unauthorized access and modification, is a subordinate property with respect to safety of many instrumentation and control systems (I&Cs), primarily to the NPP reactor trip systems. Such subordination may be taken into account by implementation of security informed safety (SIS) approach. The methodology for safety assessment of FPGA-based systems which are widely used in industrial critical systems is described. It is based on joint using of security analysis techniques (GAP-analysis and intrusion modes, effects and criticality IMECA analysis) and also their reflection on the final safety assessment picture of the system with two channels. This methodology forms so called security informed safety approach. Additional aspects of safety assessment of diverse instrumentation and control FPGA-based systems for safety-critical application are described.

Keywords: Safety, Security, Security Informed Safety, FPGA, Assessment

1. INTRODUCTION

The program logic devices and Field Programmable Gate Arrays (FPGA) particularly are widely used for development and implementation of safety-critical industrial I&Cs. FPGA-based systems have irrefutable advantages relatively microprocessor (software)-based ones confirmed by their application in critical domains such as NPP I&Cs, aerospace equipment, etc.

However, the use of FPGA in industrial I&Cs causes specific risks for ensuring of safety. I&C projects on FPGA are complex solutions which include both software and hardware components. Overall and precise safety assessment of modern FPGA-based industrial I&Cs is impossible without taking into account its security properties.

The goal of the paper is the presentation of the technique and tool for of SIS-based assessment of the industrial FI&Cs. The structure of the article is as follows: section 2 describes integration of safety and security into overall safety assessment of the FPGA-based I&Cs, the influence of security on system's safety is given. Main stages of GAP-IMECA analysis and concepts of security informed safety approach are given in section 3. Section 4 contains methodology and case study of proposed security informed safety approach for assessment of diverse FPGA-based industrial I&Cs. The description of the tool for automation of GAP and IMECA analysis stages is presented. Finally paper contains conclusions and directions of future research.

2. SAFETY AND SECURITY INTERRELATION

2.1. The principle of unity of safety and security assessment

Presently there are no integrated approaches for assessment of functional safety (further "safety") and information security (further "security") of complex industrial I&Cs. The overall methodological apparatus in the area of safety and security would allow to assess and ensure the safety if I&Cs. It should be based not only on traditional approaches and experience of experts (separated analysis of safety and security) but, primarily considering the allocation of general and private features both for safety and security.

According to well-known international standard ISO/IEC 15408 [1] security is connected with defense of assets from threats, where threats are classified based on abuse potential securable assets. All kinds of threats should take into account, especially those associated with human actions, malicious or otherwise. Figure 1 shows high-level security concepts and their relationships, the security model taking from ISO/IEC 15408 standard series. The area of notions in which influence of I&Cs is occurred is dotted with red color.

Figure 1: Security concepts and relationships according to ISO/IEC 15480 series

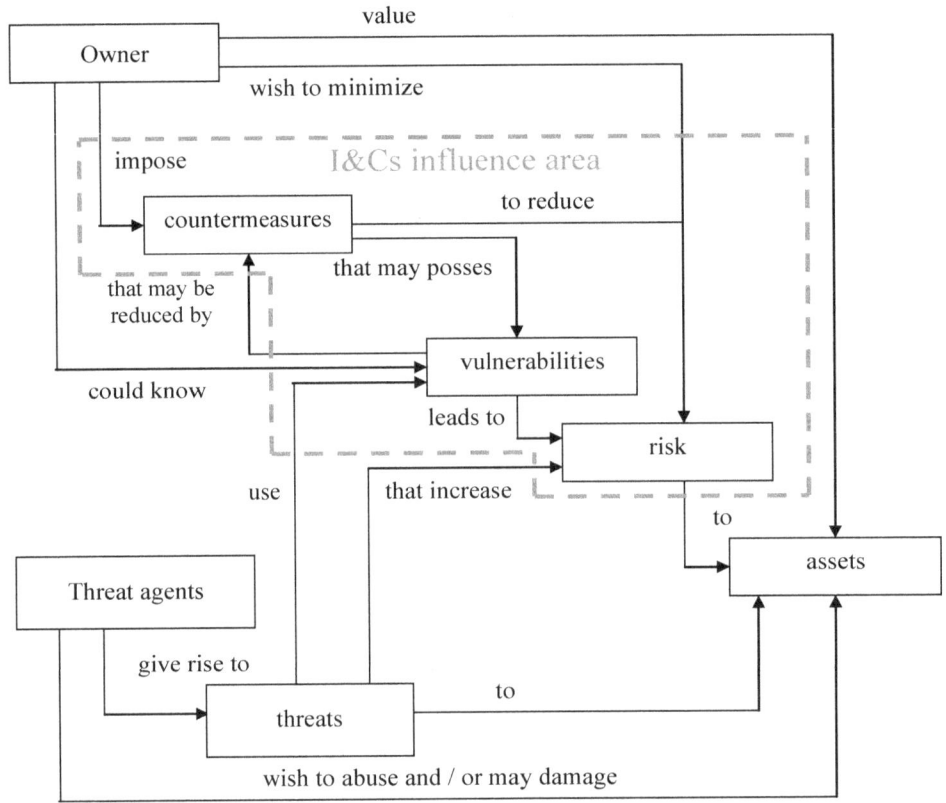

This area is specific for industrial I&Cs in frame of assessment and assurance of its security. It encompasses the following entities [2]:
- countermeasures for risk reduction (because some I&Cs could be one of the such countermeasures, e.g. I&Cs important to safety),
- vulnerabilities (because from the one side I&Cs aimed at vulnerabilities elimination and from the other they could have vulnerabilities itself),
- risks (from the one side I&Cs, as countermeasures itself, aimed to decreasing the risks, and from another they could produce additional risks to the system).

The difference between security analysis and safety analysis is lying in the assets for which the analysis is performed (safety analysis – for critical objects of control and management (OCM), security analysis – for information assets).The appropriate representation of this interrelation is shown on figure 2. Depending on I&Cs assets and safety functions that are performed the "flow" of functional safety into information security and vice versa is taken place. For information assets security is taken into account and in this case safety aimed at ensuring of safety integrity. It should be noted that for information assets safety functions and processes which ensure security integrity are determined in series of ISO/IEC 15408 standards. For OCMs safety functions are determined by specific features of object and they should be regulated by industry standards.

Figure 2: The structure of objects which are used during safety analysis: integration of level of assets and I&Cs

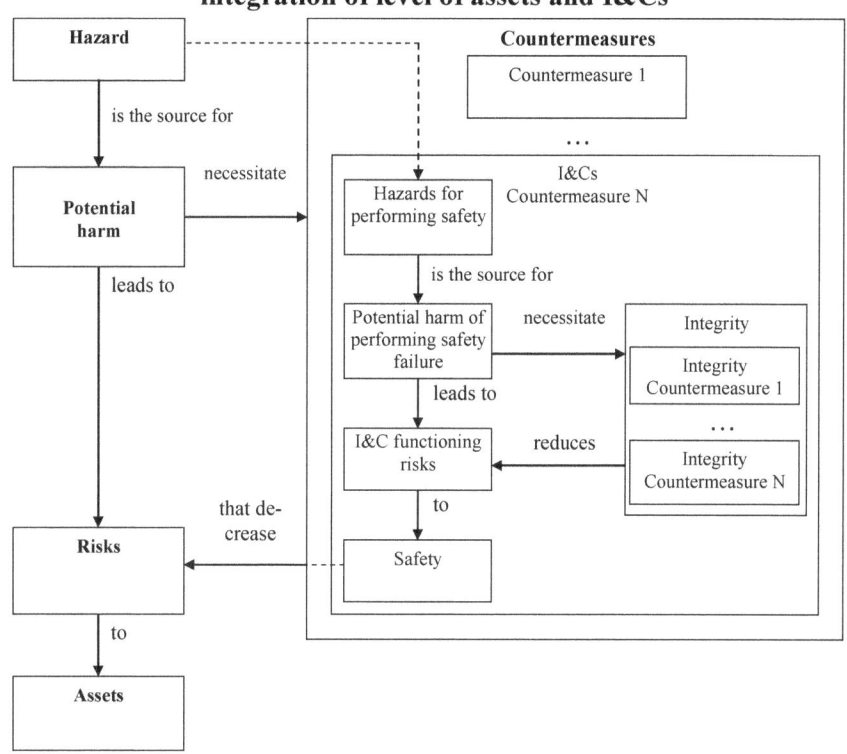

The interrelation of safety and security is presented on figure 3.

Figure 3: Safety and security cross influence

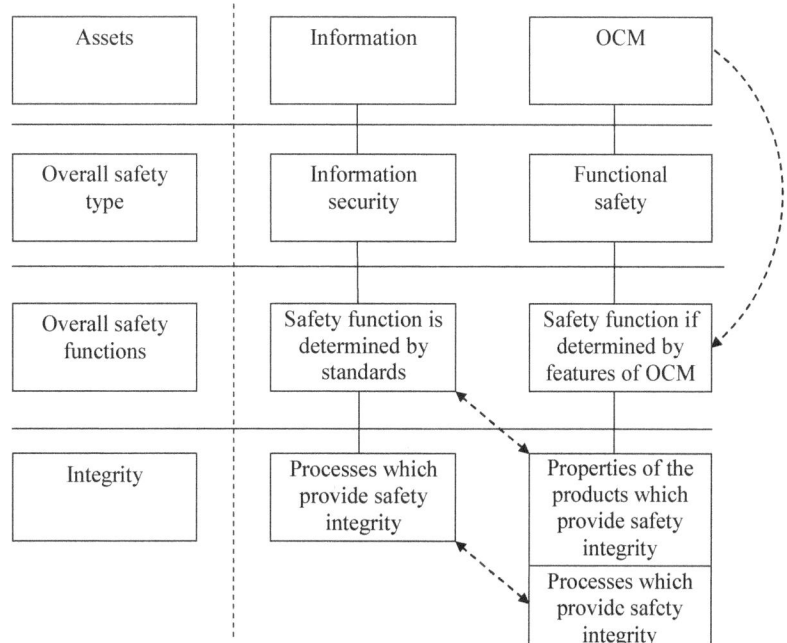

2.2. Safety and security lifecycle model of FPGA-based I&Cs

To assess security of critical FPGA-based I&Cs (FI&Cs) it is needed to refine the life cycle (LC) model and the strategy for reporting about development process, including control of environment and development tools. LC model is a structured and systematic model covering the development and operation phase of a system. Ideally, it shall be possible to verify the output of each stage of development LC, which should be the successful implementation of the considered input stages. Taking into account I&Cs safety and security needs, general safety life cycle model based on the model of application-specific integrated circuit development lifecycle (standard IEC 61508 [3]) and security LC model (based on standards IEC 15408 [1], IEC 62645 [4]), the joint security and safety LC model for FI&Cs, taking into account features of FPGA technology, is proposed.

Figure 4: Safety and security lifecycle model of FPGA-based I&Cs

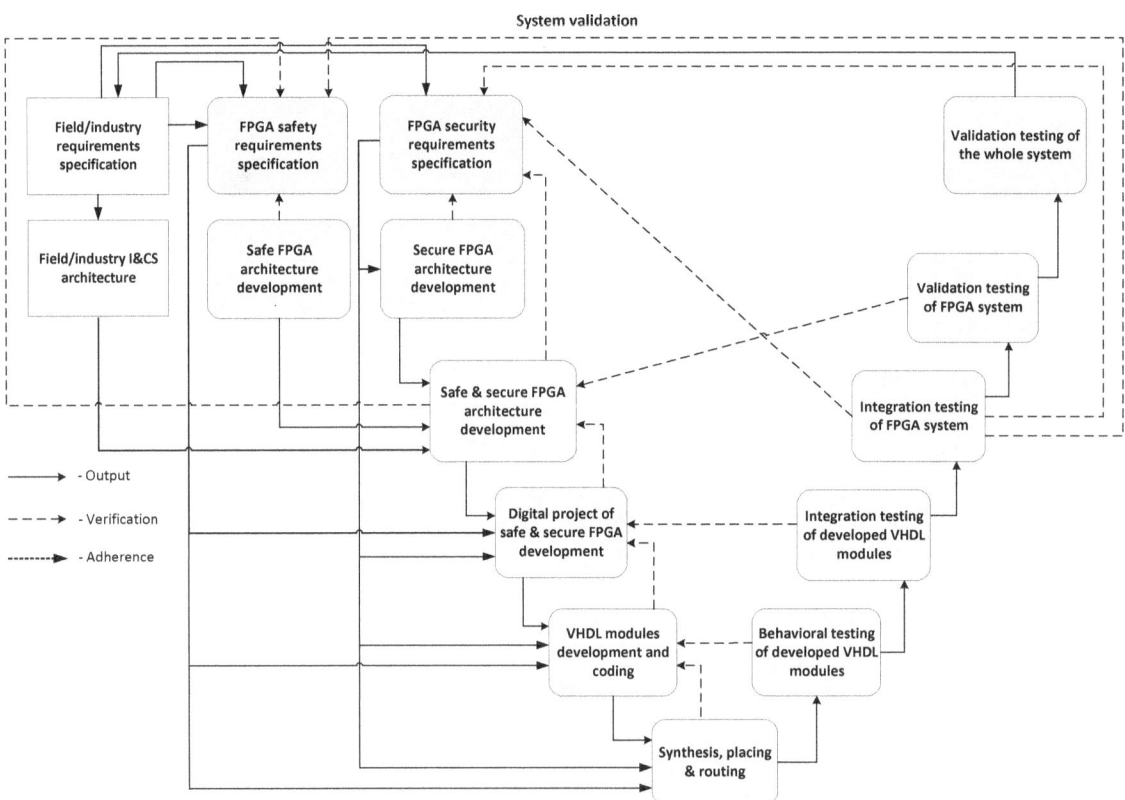

In figure 4 two rectangles depict activities that are not directly related to the project based on FPGA development LC, but they play regulatory role during the development of such systems. Rectangles with rounded corners depict the activities related exactly to the stages of development LC. The arrows of different types are shown the relationships between the above LC activities. Adherence of safety and security depicts with a dotted line.

One of the important challenges is balance between safety and security requirements in critical systems and cost for providing safe and/or secure architecture of critical I&Cs. The weak spot hides in specification of clear and precise safety and security requirements for I&CSs under development to avoid the situations when these requirements will contradict with each other and also to examine the system in accident regimes when safety and security requirements could conflict with each other in order to guarantee that the system will operate in an appropriate way in such situations.

3. GAP-IMECA SECURITY ASSESSMENT TECHNIQUE

3.1. GAP-analysis technique

Key principle in the security assessment is the use of process-product approach, which consists in determination of the possible problems and discrepancies in the final product and product development process. One of the fundamental concepts behind the idea of the approach is the concept of GAP, which is determined as a set of discrepancies of any single process within the lifecycle of I&C system that can introduce some anomalies (e.g. vulnerabilities) in a product and/or cannot reveal (and eliminate) existing anomalies in a product.

Depending on FI&Cs under consideration, each GAP is represented in a form which determines all discrepancies. The formal description should be made for a set of discrepancies identified within the GAP. GAP-analysis technique and used notions is described in detail in [7,8].

3.2. IMECA-analysis technique.

The IMECA analysis is actual refinement of FMECA-analysis (Failure modes, effects and criticality analysis) applied to security (analysis of modes and effects of intrusion to the system). Each identified GAP could be represented by a single local IMECA table and each discrepancy inside the GAP can be represented by a single row in that local IMECA table taking into consideration process-product features of the FPGA and FI&Cs itself. For each GAP, a separate table that contains all the vulnerabilities identified in the GAP analysis is created. All separated tables are combined into general IMECA table. IMECA-analysis technique with all supporting theoretical material is presented in detail in [7,8].

3.3 Criticality matrix

Each row (vulnerability and effect of intrusion) of the general IMECA table is represented by the cell of criticality matrix according with its probability and severity in context of FI&C safety. The integrated metric is calculated using criticality matrix. If any of the vulnerability parameters is not included in the allowed range, a corresponding countermeasure should be implemented.

3.4 Security informed safety approach

Safety systems operate in an open environment and they need to be secure in order to be safe. Both security and safety are sophisticated engineering cultures that emphasize the need for good process, the importance of risk analysis and the need for assurance and justification. Besides, security informed safety (SIS) approach was described in [5,6] in wide context for critical infrastructure safety assessment. This approach is based on the use of structured safety cases.

The problems of assessment and assurance of FPGA-based I&Cs safety and security were earlier researched: [2, 6-9]
- consideration of possible vulnerabilities that may occur in the components due to any anomalies in the earlier phases of the life cycle;
- development of the product security threat models;
- ranging of identified vulnerabilities in accordance with their criticality and severity;

- determination of both sufficient and cost-effective countermeasures either to eliminate identified (or even possible) attacks, vulnerabilities and threats or make them difficult (or even impossible) to exploit by an attacker.

4. SIS ASSESSMENT OF DIVERSE FPGA-BASED I&Cs: CASE STUDY

4.1 Version redundancy principle

To provide reliability and safety for I&Cs the principle of version redundancy or diversity is widely used ("is there an appropriate element of redundancy against each failure condition for which one is required?"). Using version redundancy it is possible to increase the reliability and ensure functional safety of I&Cs [10]. It assumes performance of the same problem with two or more methods (versions), processing the data obtained for the control, selection or formation of a final or intermediate results [11,12]. Need to use of the version redundancy principle is due to a fact that only when it is applied it is possible to confront the most dangerous in terms of their consequences for redundant structures (or redundant processes) mean failures, so-called common cause failures (CCF) [13]. Version redundancy principle is widely used for industrial I&Cs protection [14]. Due to this failure there exists a defect (error), which causes losses of operability of all channels of the system (or causes negative results of processes performed) independently from the number of reserved channels (redundant processes).

4.2 Diverse FI&Cs SIS assessment technique

The example that follows illustrates the problem in general. It describes simplified architecture of 2-channel platform RadICS™, which composed of multiple type of modules, based on the use of FPGA-chips as computational, processing and system-internal control engine for each of the modules [15]. RadICS™ is used for installation and implementation of the biggest ESFAS systems on-line for VVER-1000 type reactors with full "hot" redundancy and double diversion (figure 5). Safety controller of RadICS is based on FPGA.

Figure 5: Diverse system with two channels

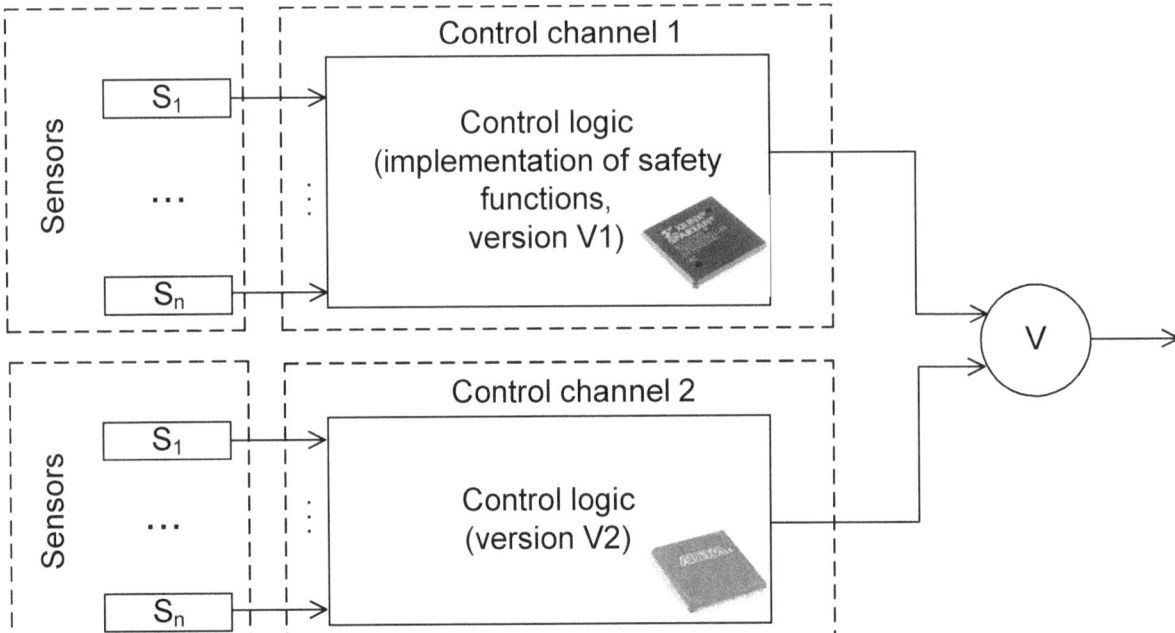

Let us also suppose the architecture is such that for an accident to happen it is necessary that both channels exhibit failures (erroneous behaviour) that would cause an accident if that channel were the whole FPGA-based controller (critical channel failures). That is, the system is safe as long as at least one channel's behaviour satisfies a safety condition. If one out of two channels fails then the remaining channel is able to detect the failure and trigger a transition to a failsafe state. In the mentioned example for implementation of safety functions of safety controller in control channels 1 and 2 different types of FPGA are used (e.g. produced by Xilinx and Altera vendors). The type and number of sensors are the same, so they didn't take into account. It is assumed that voting unit is absolutely safe and reliable (it will not fail and works properly).

Based on the analysis of possible FPGA vulnerabilities for both vendors it is possible to choose several types of intrusions (attacks), among which the brute force will reviewed. Good brute force attack is time and resource consuming and hackers are likely to pass this attack type, but it's is good applicable for the example of analysis provided.

Table 1 shows application of IMECA technique for analysis of abovementioned intrusion (brute force attack) with possible countermeasures which are based on the experience of FPGA use and some recommendations derived from [16]. It describes a regulatory position that promotes a defensive strategy consisting of a defensive architecture and a set of cyber security controls based on standards provided in documents that are based on well-understood threats, risks, and vulnerabilities, coupled with equally well-understood countermeasures and protective techniques [17, 18]. As an example of gap that could lead to successful implementation of brute force attack could be in violation of (C.3.7, Appendix C to RG 5.71, Page C-7: "...Employing hardware access controls (e.g., hardwired switches), where technically feasible, to prevent unauthorized software changes...") requirement from [16]. Going back to the safety and security lifecycle model of FPGA-based I&Cs it is possible to state that this attacks (intrusion) could be mitigated if the appropriate preventive actions would take place during the stage "FPGA safe and secure architecture design".

Table 1: Result of IMECA analysis for brute force attack

Gap №	Attack mode	Attack nature	Attack cause	Occurrence probability	Effect severity	Type of effects	Countermeasures
1	Brute force	Active	• Search for a valid output attempting all possible key values • Exhaustion of all possible logic inputs to a device in order • Gradual variation of the voltage input and other environmental conditions	Low	Moderate	Leak of undesirable information	• Detecting and documenting unauthorized changes to software and information

This type of attack could be applicable to both channels of FI&Cs and both FPGAs. And as soon as both channels performs the same logic, criticality matrixes will be the same (figure 6). The number inside of the matrix represents an appropriate row number of IMECA table.

Figure 6: Criticality matrix of brute force attack

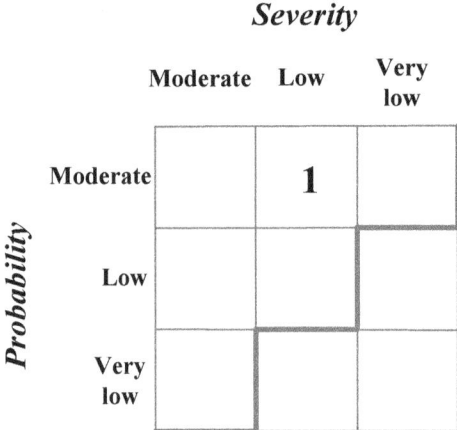

From security assurance point of view, the possible way of risk reduction is in decreasing of attacks' occurrence probability, since related damage is constant. Such decreasing of the probability can be achieved, for example, by implementation of certain process countermeasures.

The next important step is transformation of security-oriented criticality matrix that was received during GAP-IMECA-analysis into safety-oriented criticality matrix, which means the reevaluation of potential risk for the system. New, safety-oriented criticality matrix is shown on figure 7.

Figure 7: Criticality matrix of brute force attack

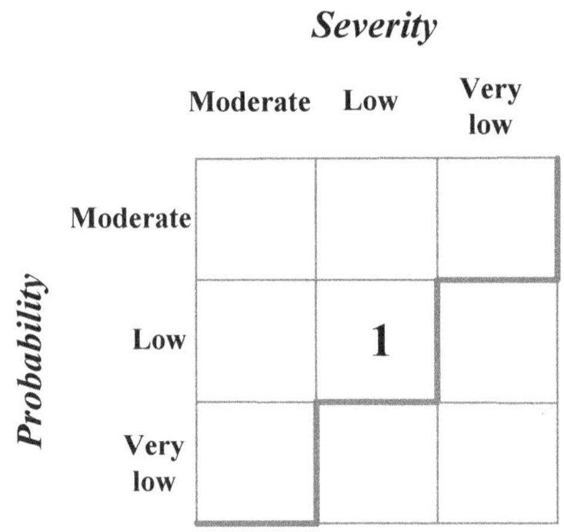

The probability (likelihood) of successful attack of the same type which is implemented into one of the channels was considered in this section.
Taking into account diversity of the control logic functions the following statements are true:
- the probability of successful attack of the same type decreased if resources of attacker is fixed,
- the probability of successful attack of the same type which is applied for both channels decreased if resources of attacker is fixed,
- during analysis of system both vulnerabilities of channels separately and of system (as combination of diverse channels) should be taken into account.

4.3 SIS-GAP-IMECA diversity analyzer tool

To decrease the risk of manual errors, the tool for the SIS-oriented assessment automation is described. The tool is based on joint use of abovementioned models and techniques, is proposed. The tool allows conducting the joint use of the following analysis techniques: GAP and IMECA. The block-scheme of main stages of analysis is shown on figure 8.

Figure 8: The general scheme of analysis process

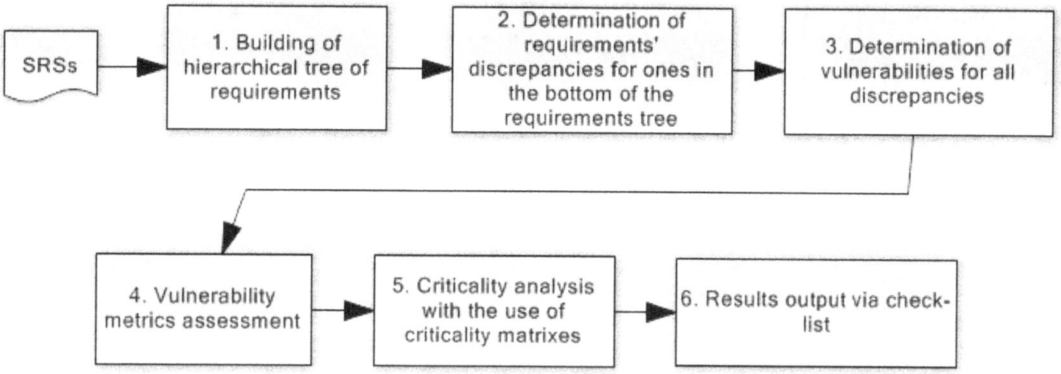

SRSs → 1. Building of hierarchical tree of requirements → 2. Determination of requirements' discrepancies for ones in the bottom of the requirements tree → 3. Determination of vulnerabilities for all discrepancies → 4. Vulnerability metrics assessment → 5. Criticality analysis with the use of criticality matrixes → 6. Results output via check-list

The ideal system is represented by requirements profile (SRS), which contains all elements of the system of process on the different levels of decomposition.

Input data is requirements profile. Requirements could be divided into different levels hierarchically. After determination of quantity of requirements levels the list of requirements for each level is composed. Levels of requirements are filled alternately from top to bottom. When filling one level, for each requirement of the current level the requirements on the lower level, which expand, clarify or detail it, are created. As a result, the requirement at one level can meet one or more requirements of the level below (Step 1 and Step 2).

After input of requirements their analysis at the lowest level is conducted. It is assumed that requirement could be violated, i.e. GAP is introduced artificially and detailed further. During the analysis of the requirement, the specific violations that may possibly occur depending on the nature of requirement are pointed out. In such way GAP is represented as a set of violations of a certain requirement, which could take place in the critical FPGA-based I&Cs under consideration. At this stage the IMECA-tables are formed for each discrepancy (Step 3, Step 4). It could also be defined more options, which could be determined by expert assessment or additional methods of analysis. One of the required parameters is the likelihood and critical impact on the system. The additional parameters also could be defined with the help of expert assessment or with the use of additional methods of analysis. Above the parameters under assessment are the likelihood and impact on the criticality of the system. Quantitative parameters can be determined by peer review or other auxiliary tools and techniques.

For each GAP, a separate table that contains all the vulnerabilities identified in the GAP analysis is created. Each vulnerability is determined by the criticality matrix. With the help of criticality matrix on the basis of vulnerability parameters the metric should be calculated and resulting conclusion for vulnerability shall be made. For the criticality matrix the set of valid parameters is defined.

If any of the parameters of the vulnerability are not included in the allowed range, a decision that the vulnerability is present in the system and requires fixing is made (Step 5).

The presence of discrepancy is determined on the basis of criticality matrix. Check-list is formed from the requirements and a conclusion about their implementation (Step 6). Example of check-list is shown on figure 5.

5. CONCLUSION

Ensuring security and safety of industrial FI&Cs must be done with a special care, because their development is under the strict constraints related to resources and cost. It should be done iteratively, rather than the disposable decision.

Features of project development with the use of FPGA technology are represented in safety and security LC model. Based on FPGA technology a set of safety and security assurance processes is formed. This set of processes allows further implementation of process-product approach to assessment and to optimize choice of countermeasures

Thus it was described the proposed security informed safety approach for safety assessment of diverse industrial FPGA-based instrumentation and control systems. It may be used to provide security analysis and safety related risks. Future research will be dedicated to formalization of assessment procedure and description of different types of attacks effects (effect for 1 channel, effect for the whole system) depending on the attacker's resources, quantity of vulnerabilities, ways of successful attacks on them and quantity of attackers, and extension of tool for the tasks of countermeasures choice. Developed tool could be expanded for GAP-xMECA analysis.

References

[1] ISO/IEC 15408:2009, *"Information technology – Security techniques – Evaluation criteria for IT security"*, 2009.
[2] V. Kharchenko, V. Sklyar, E. Brezhniev, *"Safety of information and control systems and infrastructures"*, Palmarium Academic Publishing, 2013.
[3] IEC 61508:2010, *"Functional Safety of Electrical /Electronic/Programmable Electronic Safety-related Systems"*, 2010.
[4] IEC 62645:2013, *"Nuclear power plants - Instrumentation and control systems - Requirements for security programmes for computer-based systems"*, 2013.
[5] R. Bloomfield, K. Netkachova, R. Stroud, *"Security-Informed Safety: If It's Not Secure, It's Not Safe"*, Software engineering for resilient systems, Springer, pp. 17-32, 2013.
[6] O. Illiashenko, V. Kharchenko, G. Jervan, *"Security of industrial FPGA-based I&C systems: normative base and SIS approach"*, Radioelectronic and computer systems Scientific and technical magazine №3(62), National Aerospace University KhAI, pp.86-91, 2013.
[7] V. Kharchenko, A. Kovalenko, A. Andrashov, A. Siora *"Gap-and-IMECA-based Assessment of I&C Systems Cyber Security"* Complex Systems and Dependability, Advances in Intelligent and Soft Computing, pp. 149-164, 2012.
[8] O. Illiashenko, V. Kharchenko, A. Kovalenko, *"Cyber Security Lifecycle and Assessment Technique for FPGA-based I&C Systems"*, EWDTS-2012 (East-West Design and Test Symposium 2012), Proceedings of IEEE East-West Design & Test Symposium (EWDTS 2012), pp. 432-436, 2012.
[9] V. Sklyar, *"Cyber Security of Safety-Critical Infrastructures: a Case Study for Nuclear Facilities, Information & Security"* An international Journal, Vol. 28, No.1, pp. 98-117, 2012.
[10] V. Akimov, V. Lapin, V. Popov, V. Puchkov, V. Tomakov, M. Faleev, "Reliability of technical systems and technogenic risk", Business Express, 2002.

[11] V. Kharchenko, V. Sklyar, A. Siora, *"Multi-Version FPGA-Based Nuclear Power Plant I&C Systems: Evolution of Safety Ensuring"* Nuclear Power: Control, Reliability and Human Factors, INTECH, 2011.

[12] V. Kharchenko, V. Duzhyi, V. Sklyar, A. Volkoviy, *"Diversity assessment of multi-version NPP I&C Systems: NUREG7007 and CLB-BASED techniques"*, East-West Design and Test Symposium proceedings, pp. 1-5, (2013).

[13] R. Wood, R. Belles, M. Cetiner, D. Holcomb, K. Korsah, *"Diversity Strategies for Nuclear Power Plant Instrumentation and Control Systems"*: NUREG/CR–7007 ORNL/TM–2009/302. – U.S. Nuclear Regulatory Commission, Oak Ridge National Laboratory, pp. 251, 2010.

[14] B. Littlewood, P. Popov, L. Strigini, *"DISPO project: A Summary of CSR Work on Modelling of Diversity"*, Centre for Software Reliability, City University, London, UK, 2006.

[15] http://www.radiy.com/eng/products/fpga_based_systems/fpga_based_platform/

[16] Regulatory Guide 5.71, *"Cyber security programs for nuclear facilities"*, U.S. Nuclear regulatory commission, 2010, 105 pp.

[17] NIST SP 800-52, *"Guidelines for the Selection and Use of Transport Layer Security (TLS) Implementations Computer Security"*, National Institute of Standards and Technology Special Publication 800-52, 2005, 33 pp.

[18] NIST SP 800-53, *"Information Security, Security and Privacy Controls for Federal Information Systems and Organizations"*, National Institute of Standards and Technology Special Publication 800-53 Revision 4, 2012, 375 pp.

Uncertainty of the Thermal-Hydraulic Model Analysis

Yu YU*, Yingqiu HU, Junchi CAI, Shengfei WANG, Fenglei NIU

School of Nuclear Science and Engineering
Beijing Key Laboratory of Passive Nuclear Safety Technology
North China Electric Power University, Beijing, China, 102206

Abstract: Passive containment cooling system is innovatively used in AP1000 to improve the safety of nuclear power plant. By this system the steam produced in the containment can be condensed through natural circulation and independent of outside power. However, since the system is a new design, the uncertainty exists in the thermal-hydraulic (T-H) model especially in the correlations of heat and mass transfer. In this paper, the effect of uncertainties of such correlations on the output of T-H model is analyzed. Since the uncertainty of the correlations is within 20% based on the experiments, at different operation conditions such as different air temperature, we run the T-H model with the heat and mass transfer coefficients 5%, 10%, 15% and 20% higher and lower than their calculation value respectively and compare the results with the result of exact value by correlations. Then the amplitude of variation of the T-H model output and the safety margin of the system can be gained for different operation conditions, from the results, it is illustrated that the uncertainties of heat and mass transfer correlations may have important effect on the system reliability in some operation conditions and should be considered in the system reliability model.

Keywords: Passive containment cooling system, natural circulation, thermal-hydraulic model uncertainty

1. INTRODUCTION

Passive system [1] is widely used in new generation reactor design to improve the safety especially under disaster such as earthquake, flooding and so on. However since the system operates depending on natural circulation but not on outside power, uncertainties in physical model and input parameters [2] have important effect on system performance. In past years, more attention is paid to evaluate the influence of parameter's uncertainty, and some methods are developed to analyze the physical process failure [3-8].

Passive cooling containment system (PCCS) [1,9] is innovatively used in AP1000 nuclear power plant, since it's a new design, the thermal-hydraulic process such as heat and mass transfer way is firstly used, and the heat transfer and mass transfer process are described by empirical correlations fitted based on experiment data [10], uncertainty in thermal- hydraulic model should be considered in the system reliability analysis. For this system the model uncertainty is mainly induced by the heat and mass transfer correlations at inside and outside surfaces of the steel vessel, in this paper, we calculate the pressure and temperature in the containment at 5%, 10%, 15% and 20% lower and higher of the heat and mass transfer coefficients than the exact calculation values, then we evaluate the influence of thermal- hydraulic model uncertainty on the system performance based on the results by calculating the safety margin of the system under different uncertainties of heat and mass transfer correlations.

2. THERMAL – HYDRAULIC MODEL UNCERTAINTY

2.1. System description

The passive containment cooling system in AP1000 [10] is an important safety-related system that functions to reduce containment temperature and pressure following a loss-of-coolant accident (LOCA)

---

* E-mail: yuyu2011@ncepu.edu.cn

accident, a main steam line break (MSLB) accident inside containment, or other events that cause a significant increase in containment pressure and temperature. The system achieves this by removing thermal energy from the containment atmosphere to the environment via the steel containment vessel. The simplified flow chart of the system based on the design is shown in Figure.1.[10]

Surrounding the containment, there are the steel containment vessel and the concrete shield building. After LOCA, MSLB or other transient accident, steam rising from break or from the In-Containment Refueling Water Storage Tank (IRWST) injects to containment and be condensed at the inside surface of the steel vessel. The passive containment cooling system is composed of following major components: air baffle, located between the steel containment vessel and the concrete shield building, which defines the cooling air flow path; air inlets composed of three rows of holes in the concrete shield building, and air exhaust which is also incorporated into the shield building structure. In order to enhance the heat transfer, the passive containment cooling water storage tank (PCCWST) which is incorporated into the shield building structure above the containment, and a water distribution system, mounted on the outside surface of the steel containment vessel, which functions to distribute water flow on the containment.

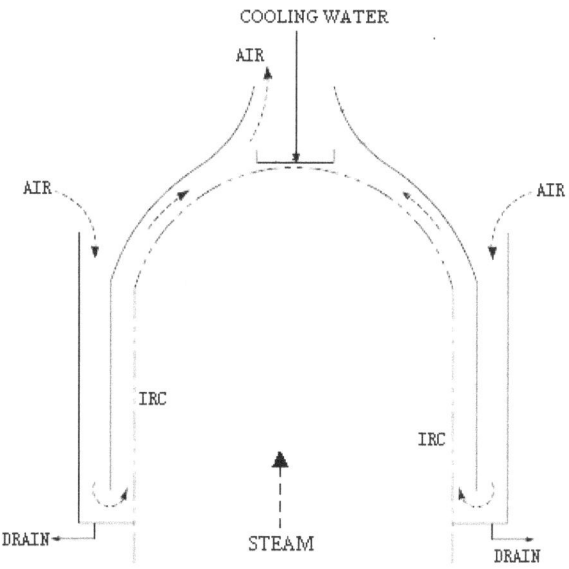

Fig.1 Flow chart of passive containment cooling system

2.2. Thermal-hydraulic model analysis

The steam injected to the steel vessel is the heat source in our analysis, and the heat is transferred to the steel wall when the steam is condensed, then the air in the air flow path will be heated by the hot steel wall and bring the heat to the heat sink—atmosphere. The thermal- hydraulic model is based on the conservative equations (mass, momentum, energy and constituent) to describe the natural circulation inside and outside the steel vessel, and the heat and mass transfer capacity is described mainly by the experiential correlations [11] which are fitted based on experiment data:

$$G = k_G \cdot M_A \cdot (p_{Ai} - p_{AG}) \quad (1)$$

$$Q = \alpha \cdot A \cdot \Delta t \quad (2)$$

Here formula (1) is the mass transfer correlation and formula (2) is the heat transfer correlation, G is the condensing or evaporating mass flux, k_G is the mass transfer coefficient, M_A is the molecular weight of gas A and p_{Ai} and p_{AG} are the partial pressure of gas A at the interface and at the bulk gas mixture respectively. Q is the heat transfer amount, α is the heat transfer coefficient, A is the heat transfer area and Δt is temperature difference between the surface of the steel vessel and the flux. Mass transfer coefficient (k_G) and heat transfer coefficient (α) are fitted based on experiment data.

The T-H performance of the system is described by the conservative equations and the correlations above. The conservative equations are deduced from the heat transfer theory and can be considered no uncertainty existing. The model uncertainty is mainly aroused by two parts: model hypothesis including boundary conditions, initial conditions and some hypothesis to simplify the physical model as well as the correlations describing the physical process. Some of uncertainty in the model hypothesis can be reflected by the input parameters' probabilistic distributions. Here we focus on the uncertainty of correlations, and heat and mass transfer correlations are fitted from the experiment data, which are two of the important factors influencing the output of the model, while the differences of coefficients (k_G and α) between experiment data and calculation values are always within 20%.

The function of the passive containment cooling system in AP1000 is to decrease the temperature and pressure in the containment, and one of the main failure modes of containment is overpressure, so the output of the T-H model is pressure in the containment.

3. RESULTS

Since the uncertainties of correlations are within 20%, we calculate the pressure in the containment with 5%, 10%, 15% and 20% correlation uncertainties. In our analysis, input parameters are set as their design values, some parameters such as diameter of steam source, steam temperature, steam mass flow rate are related to the accident, in this paper MSLB is analyzed as an example because this accident is one of the most challenging ones to the containment. In the T-H model, the heat and mass transfer correlations are used for heat transferred from mixture of air and steam to the steel vessel and steam condensed at the inside surface of the vessel, and heat transferred from steel vessel to the air in the air flow path and cooling water evaporated at the outside surface of the vessel. The heat and mass transfer correlations are in accordance for the same location, that is, the uncertainty of heat and mass transfer are highly correlated so we use the same uncertainty for the two correlations, and the 81 combination states of the correlations with 9 values of heat and mass transfer coefficients (0, ±5%, ±10%, ±15% and ±20%) as well as the results are listed in Table1. From the results, it can be seen that the difference between the max and minimum values of the pressure in the containment is 0.013MPa induced by the uncertainties of thermal-hydraulic model.

Since the design value of containment pressure is 0.5 MPa, the safety margin is:

$$\Delta P = 0.5 - P_{containment} \quad (MPa) \tag{1}$$

Here, ΔP is the safety margin and $P_{containment}$ is the pressure in the containment calculated by the thermal-hydraulic model.

Table 1: Correlation Uncertainties and results

Uncertainties of correlations		Pressure in the containment (MPa)
Heat and Mass transfer (inside surface of the vessel)	Heat and Mass transfer (outside surface of the vessel)	
0	0	0.4062
-20%	-20%	0.4138
-20%	-15%	0.4131
-20%	-10%	0.4124
-20%	-5%	0.4118
-20%	0	0.4113
-20%	5%	0.4108
-20%	10%	0.4103
-20%	15%	0.4099
-20%	20%	0.4096
-15%	-20%	0.4125
-15%	-15%	0.4117
-15%	-10%	0.4110
-15%	-5%	0.4104
-15%	0	0.4098
-15%	5%	0.4093
-15%	10%	0.4089
-15%	15%	0.4085
-15%	20%	0.4081
-10%	-20%	0.4112
-10%	-15%	0.4105
-10%	-10%	0.4097
-10%	-5%	0.4091
-10%	0	0.4085
-10%	5%	0.4080
-10%	10%	0.4075
-10%	15%	0.4071
-10%	20%	0.4067
-5%	-20%	0.4101
-5%	-15%	0.4093
-5%	-10%	0.4086
-5%	-5%	0.4079
-5%	0	0.4073
-5%	5%	0.4068
-5%	10%	0.4063
-5%	15%	0.4059
-5%	20%	0.4055
0	-20%	0.4091
0	-15%	0.4082
0	-10%	0.4075
0	-5%	0.4068
0	5%	0.4057
0	10%	0.4052
0	15%	0.4048
0	20%	0.4044
5%	-20%	0.4081
5%	-15%	0.4072
5%	-10%	0.4065
5%	-5%	0.4058
5%	0	0.4052
5%	5%	0.4047

5%	10%	0.4042
5%	15%	0.4037
5%	20%	0.4033
10%	-20%	0.4072
10%	-15%	0.4063
10%	-10%	0.4056
10%	-5%	0.4049
10%	0	0.4043
10%	5%	0.4037
10%	10%	0.4032
10%	15%	0.4027
10%	20%	0.4023
15%	-20%	0.4063
15%	-15%	0.4055
15%	-10%	0.4047
15%	-5%	0.4040
15%	0	0.4034
15%	5%	0.4028
15%	10%	0.4023
15%	15%	0.4019
15%	20%	0.4014
20%	-20%	0.4056
20%	-15%	0.4047
20%	-10%	0.4039
20%	-5%	0.4032
20%	0	0.4026
20%	5%	0.4020
20%	10%	0.4015
20%	15%	0.4010
20%	20%	0.4006

The variations of safety margin uncertainties along with the uncertainties of correlations are shown in Fig.2 and Fig.3. In the figures, the horizontal ordinate is the uncertainty of correlations, while the longitudinal coordinates is the uncertainty of the safety margin. From the figures it can be seen that the uncertainty of safety margin induced by uncertainties of correlations is within ±10%.

Fig.2 shows the uncertainty of safety margin variation along with the uncertainties of correlations inside and outside respectively, from which we can see that when uncertainty of correlations inside is zero the uncertainty of safety margin induced by uncertainty of correlations outside is from -3% to 2%, while the uncertainty of safety margin induced by uncertainty of correlations inside is from -6% to 4% when uncertainty of correlations outside is zero. From the results it can be seen that the amplitude of safety margin variation induced by the uncertainties of correlations is lower than the variation amplitude of correlations itself, the reason can be analyzed as: the heat transfer correlation influences the total heat transfer amount, it influences the temperature in the containment directly and then affects the pressure in the containment, the heat transfer coefficient is lower, the temperature and pressure in the containment are higher. While the influence of mass transfer correlation can be analyzed from two aspects: on the one hand, if the mass transfer capacity decreases, the amount of steam condensed reduces and then more steam will retain in the containment and the pressure in the containment will increase; on the other hand, the energy by which the air in the containment will be heated is mainly from the latent heat when steam is condensed, so if the amount of steam condensed decreases, the heat to the containment will be lower and then the temperature and pressure increasing velocity will decrease. The influence of correlations' uncertainties on the system pressure is synthesized result of the above two ways.

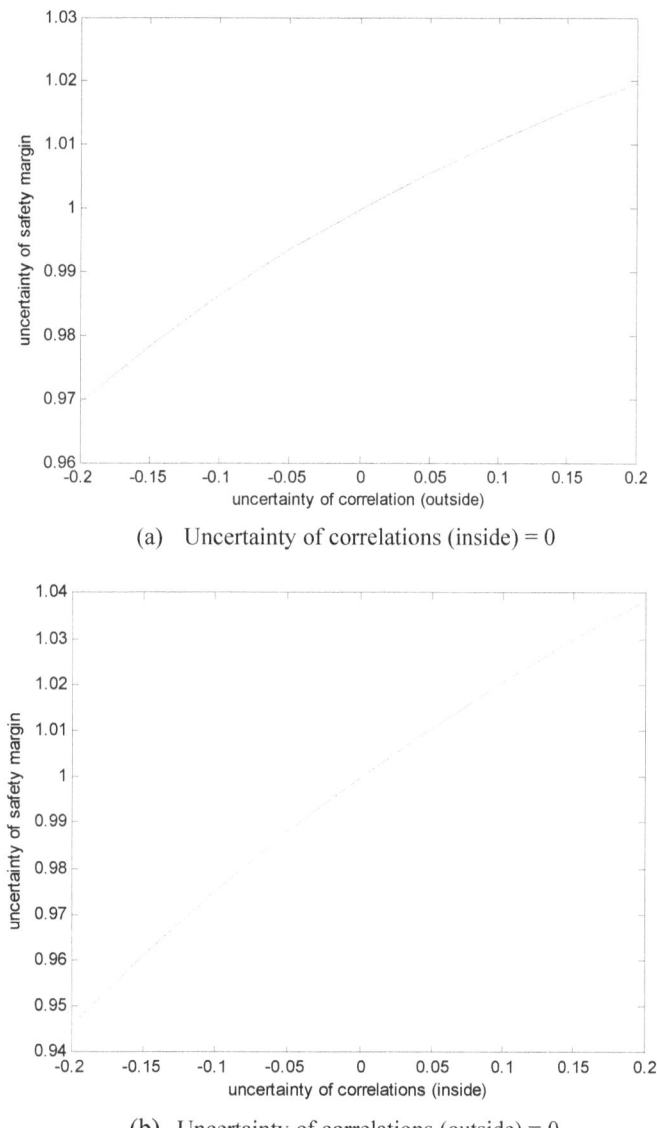

(a) Uncertainty of correlations (inside) = 0

(b) Uncertainty of correlations (outside) = 0

Figure.2 Influence of correlations' uncertainties

From the results it can be also seen that influences of uncertainties of heat and mass transfer correlations at inside and outside of the containment are difference, Fig.3 shows the influence of correlations inside and outside when the other one are -20% and 20% respectively, the influence of uncertainties of correlations inside is more important than the influence of uncertainties of correlations outside, since the correlations inside affect the heat and mass transfer synthetically as described above, while for correlations outside, mass transfer correlation affects the cold water evaporating process primarily, and heat transfer process is affected by more complex factors such as air speed and so on.

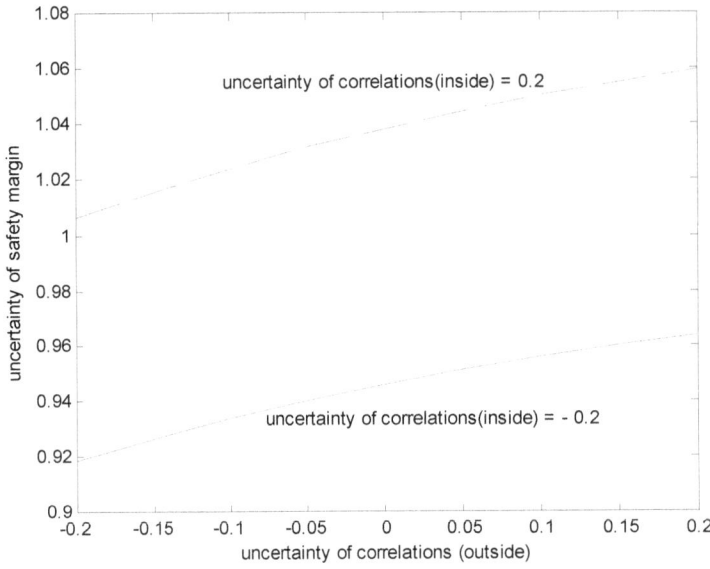

(a) Influence of correlations' uncertainties (outside) at different uncertainties of correlations for inside

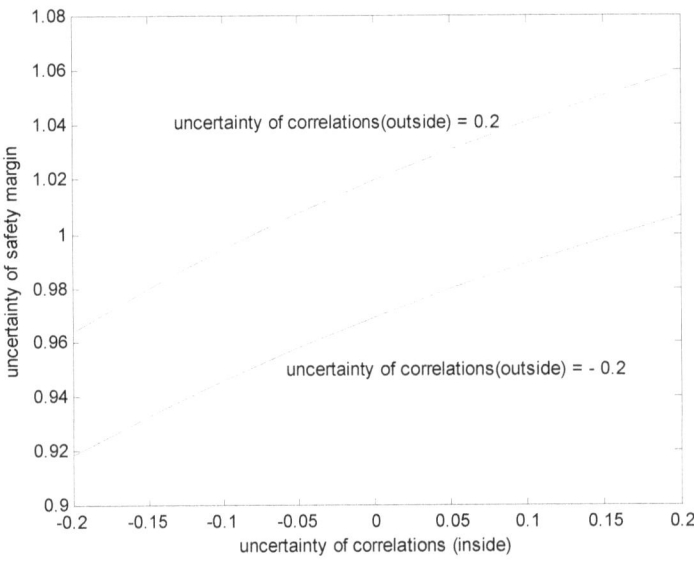

(b) Influence of correlations' uncertainties (inside) at different uncertainties of correlations for outside

Figure.3 Influence of correlations' uncertainties

4. CONCLUSIONS AND DISSCUSSION

In this paper we analyze the effect of correlation uncertainties on the system thermal-hydraulic performance, and the uncertainty of safety margin induced by uncertainties of correlations is about 10% for MSLB accident in AP1000, which may influence the system failure probability and should be considered in the system reliability analysis. From the results it can be concluded that the influence of the T-H model uncertainty on the system reliability is affected by thermal- hydraulic performance, so it will be affected by other factors such as accident (e.g. MSLB, LOCA or other accidents), uncertainties of input parameters, equipment state and so on. Since the T-H model is a complicated non-linear system and there are numerous factors influencing the system characteristic, so the uncertainties of correlations may need to be considered in the system reliability model combined with the uncertainties of input parameters and the equipment fault.

Acknowledgements

This work was supported by The National Natural Science Foundation of China [grant number: 51206042, 91326108]

References

[1] T.L.Schulz. *"Westinghouse AP1000 advanced passive plant"*, Nuclear Engineering and Design, 236, pp1547-1557, (2006).

[2] J. Oh, W. G. Michael. *"Methods for comparative Assessment of Active and Passive Safety Systems with respect to Reliability, Uncertainty, Economy, and Flexibility"*, Proceedings of 9th Conference on Probabilistic Safety Assessment and Management (PSAM9). Hongkong, China, 2008.

[3] E. Zio, N. Pedroni. *"Estimation of the functional failure probability of a thermal-hydraulic passive system by Subset Simulation"*, Nuclear Engineering and Design, 239, pp580-599, (2009).

[4] L.Burgazzi, V. Kopustinskas, VL.Lumia, et al. *"Recommendations for the integration of passive system reliability in PSA"*. Report, the 5th EURATOM framework programme 1998-2002 key action nuclear fission, 2004.

[5] Y. YU, J.J. TONG, T. LIU, et al. *"A MODEL FOR PASSIVE SYSTEM RELIABILITY ANALYSIS"*, Proceedings of 18th International Conference on Nuclear Engineering (ICONE18), Xi'an, China, 2010.

[6] Y. Y, T. LIU, J. J. TONG, et al. *"Variance Decomposition Sensitivity Analysis of a Passive Residual Heat Removal System Model"*, Social and Behavioral Sciences, 2, pp7772-7773, (2010).

[7] Y. Yu, T. LIU, J.J. TONG, et al. *"Multi-Experts Analytic Hierarchy Process for the Sensitivity Analysis of Passive Safety System"*, Proceedings of the 10th International Probabilistic Safety Assessment & Management Conference(PSAM10), Seattle, USA, 2010.

[8] E. Zio, F. D. Maio, J. J. Tong. *"Safety margins confidence estimation for a passive residual heat removal system"*, Reliability Engineering and System Safety. 95, pp828-836, (2010).

[9] Foret LJ. *"AP1000 Probabilistic Safety Assessment"*, Report, Westinghouse Electric Company LLC, Pittsburgh, PA, USA. Report no. APP-GW-GL-022, DCP/NRC1548, 2003.

[10] J. G. Sun, T. H. Chien, J. Ding, et al. *"Validation of COMMIX with Westinghouse AP-600 PCCS Test Data"*, Proc. of the 21th Water Reactor Safety Information Meeting, Bethesda, MD, 1993.

[11] A.L.Hines. *"Mass Transfer and Applications"*, Englewood Clifs: Prentice-Hall, Inc., 1985.

Sensitivity Analysis and Failure Damage Domain Identification of the Passive Containment Cooling System of an AP1000 Nuclear Reactor

Francesco Di Maio[a], Giancarlo Nicola[a], Yu Yu[b] and Enrico Zio[a,c]

[a] Energy Department, Politecnico di Milano, Via Ponzio 34/3, 20133 Milano, Italy
[b] North China Electric Power University, 102206 Beijing, China,
[c] Chair on System Science and Energetic Challenge, European Foundation for New Energy, Electricite de France, Ecole Centrale, Paris, and Supelec, Paris, France

Abstract: The paper presents an application of a variance decomposition method for the sensitivity analysis of the thermal hydraulic (TH) model of the Passive Containment Cooling System (PCCS) of an Advanced Pressurized Reactor (AP1000). The Loss Of Coolant Accident (LOCA) is considered as the most representative accident for identifying the Failure Damage Domain (FDD) of the PCCS with respect to the individual and grouped inputs most affecting the final pressure at the end of the accidental transient.

Keywords: Passive Systems Reliability, AP1000, Variance Decomposition, Failure Damage Domain.

1. INTRODUCTION

The extension of nuclear safety considerations to severe accidents and the increased safety requirements have led to a growing interest in passive systems for the safety of the future nuclear reactors. As a result, all innovative reactor concepts make use of passive safety features, to a large extent in combination with active safety and operational systems [1],[2]. Passive systems are addressed as a resource for nuclear safety improvement because of their characteristics of simplicity, reduction of human interaction and reduction or avoidance of external electrical power and signals input [2].

On the other hand, passive safety systems are affected by uncertainties that have to be properly considered to guarantee their reliability by design [3]. In fact, passive systems rely only on natural forces, (such as gravity, natural circulation, compressed gas and other physical principles) for which the classical concepts of reliability analysis does not make sense as for the pumps, fans, diesels, chillers, or other devices used in active safety systems [4]. For example, to activate safety passive systems, usually only few fail-safe valves are required to open: in case of loss of power, they automatically open by stored energy (e.g. compressed gas or batteries). Anyway, although passive safety systems are significantly simpler than active ones because they comprise significantly fewer components (with straightforward benefits on the number of tests, inspections, and maintenance activities to be planned), uncertainty due to the lack of knowledge on the physical principles driving their performance, makes passive safety systems also exposed to potential failures for which they have been designed.

Quantification of failure probability is one main goal of the system safety assessment and is usually achieved with the support of numerical models simulating the behavior of the real system. Sensitivity analysis has been widely used in engineering design to help the designers understanding the behavior of a model and make informed decisions regarding where to spend the engineering effort. Sensitivity analysis is used both in deterministic design and design under uncertainty to quantify how much the output of a model depends on the inputs and ranking variables importance [5]. These characteristics make sensitivity analysis particularly suited for both the design and the safety assessment of passive systems. Various qualitative or quantitative approaches have been developed for performing sensitivity studies, e.g. Analytic Hierarchy Process (AHP), first-order differential analysis, response surface methodology, Fourier Amplitude Sensitivity Test (FAST) and Monte Carlo sampling [5],[6]. The AHP is a qualitative method based on the consultation of multiple experts, asked to express their judgments on the relative importance of parameters to determine the overall hierarchy with regards to the defined top goal. First-order differential analysis is a quantitative local approach that uses a finite difference approximation of small output variations around the nominal best estimate values to identify the critical parameters [7]. The Response surface methodology consists in approximating the model function $f(X)$ by a simple and faster mathematical model from a database of computations. FAST is a global, variance-based sensitivity analysis method based on the principle that a model (a function) can be expanded into a Fourier series and the Fourier coefficients and frequencies can be used to estimate the mean and variance of the model, and the partial variance of individual input parameters of the model [8]. Monte Carlo sampling consists of drawing samples of the basic variables according to their probability density functions and, then, feeding them into the performance function to retrieve the output probability density function.

In this paper, we use the variance decomposition method [9] for performing the sensitivity analysis of a specific lumped thermal hydraulic (TH) model with the aim of quantifying the effects on the model output of the variability of not only single inputs but also groups of inputs, thus including also their interactions. The TH model simulates the behavior of a Passive Containment Cooling System (PCCS) when a Loss Of Coolant Accident (LOCA) occurs in an Advanced Pressurized reactor (AP1000). PCCS is an innovation used in AP1000 reactors design, aimed at improving safety [10]. PCCS operation is based on natural circulation, so that physical process failure (i.e., the actual conditions are such that natural circulation cannot be established or maintained at the time of the LOCA) becomes the important failure mode[11]. To analyze this we propose to use the outcomes of sensitivity analysis for identifying the Failure Damage Domain (FDD), a concept already used in the evaluation of risk-informed safety margins [12] and, more generally, to compactly represent the final state of a system as a function of the most important parameters which drive its response [13].

The paper organization is as follows. For self-consistency and completeness, in Section 2, the Monte Carlo method for uncertainty propagation and the Variance Decomposition method for sensitivity analysis are briefly recalled. In Section 3, the main characteristics of the AP1000 reactor design are given (Subsection 3.1) and the accident scenario considered is described (Subsection 3.2). In Section 4, the model for the long term PCCS pressure calculation is described. In Section 5, the results of the sensitivity analysis are provided and used for FDD identification. Finally, some conclusions are drawn in Section 6.

2. THE VARIANCE DECOMPOSITION METHOD FOR SENSITIVITY ANALYSIS

For simplicity of illustration, and without loss of generality, let us consider a model m whose output value y depends only on the values x_1 and x_2 of two uncertain input parameters X_1 and X_2, viz:

$$y = m(x_1, x_2) \tag{1}$$

No hypotheses are made on the structure of the model.

Monte Carlo is a global method for uncertainty analysis, which simply consists in drawing random samples of the uncertain input parameters values from their probability density functions and evaluating the model output for each set of sampled values.

Operatively, consider a set of s realizations of the two input parameters drawn from the assigned pdfs $f_{x_1}(x_1)$, $f_{x_2}(x_2)$, respectively:

$$\bar{x}^j = \lfloor x_1^j, x_2^j \rfloor \quad j=1,2,\ldots s \tag{2}$$

The model is evaluated for each of the s independently generated vectors $\bar{x}^j$, $j=1,2,\ldots,s$, to obtain a corresponding set of output values:

$$y^j = m(x_1^j, x_2^j) \quad j=1,2,\ldots s \tag{3}$$

Such set represents an independent random sample of size s of the distribution of the output y and can be analyzed using classic statistical techniques for uncertainty analysis [5],[14].

The dependence of the value of the output variable (Y) on the value of one of the two input variables, e.g. X_1, can be approximated by the expected value of Y with respect to the other variable X_2, conditioned on X_1 being equal to a given value x_1:

$$y^*(x_1) = E_{X_2}(Y|x_1) = \int m(x_1, x_2) f_{X_2|X_1}(x_2|x_1) dx_2 \tag{4}$$

where $f_{X_2|X_1}(x_2|x_1)$ is the conditional probability density of X_2 given X_1. Note that, since X_1 is fixed at x_1, y^* depends only on the variable X_2.

To evaluate how the uncertainty in the input propagates to the output of the model, the variance of the distribution of the output variable Y is decomposed as follows (see [9],[15] for further details):

$$Var[Y] = Var_{X_1}\left[E_{X_2}(Y|X_1)\right] + E_{X_1}\left[Var_{X_2}(Y|X_1)\right] \quad (5)$$

where X_1 has been indicated explicitly as subscript of the variance and expectation operators to highlight that these are applied with respect to such variable.

The sensitivity relevance of X_1 can be associated to its contribution to the output variance, i.e. the term $Var_{X_1}\left[E_{X_2}(Y|X_1)\right]$ in (5). Quantitatively, it is then customary to take the following measure as an index of the importance of the variable X_1 with respect to its contribution to the uncertainty in the output Y:

$$\eta_1^2 = \frac{Var_{X_1}\left[E_{X_2}(Y|X_1)\right]}{Var[Y]} \quad (6)$$

An operative procedure based on Monte Carlo sampling for estimating the index of importance of X_1 according to the definition (6) may be summarized as follows [16]:

1. Sample a random population of s values of X_1 $\{x_1^1, x_1^2, ..., x_1^s\}$

2. For each value x_1^j, sample r values x_2^k, $k=1,2,...,r$ from the conditioned distribution $f_{X_2|X_1}(x_2|x_1^j)$.

3. Evaluate r output values $y^{jk} = m(x_1^j, x_2^k)$; each of these values is an element of an output matrix of order (s,r).

4. For each row $j=1,2,...,s$ of the matrix, evaluate the estimate

$$\hat{y}*(x_1^j) = \frac{1}{r}\sum_{k=1}^{r} y^{jk} \cong E_{X_2}[Y|x_1^j] \quad (7)$$

5. Estimate the expected value of Y

$$\bar{y} = \frac{1}{s}\sum_{j=1}^{s} \hat{y}*(x_1^j) \cong E[Y] \quad (8)$$

6. Estimate the variances

$$\hat{V}_{X_1}\left[E_{X_2}(Y|x_1)\right] = \frac{1}{s-1}\sum_{j=1}^{s}\left[\hat{y}*(x_1^j) - \bar{y}\right]^2 \quad (9)$$

$$\hat{V}[Y] = \frac{1}{sr-1}\sum_{j=1}^{s}\sum_{k=1}^{r}(y^{jk} - \bar{y})^2 \quad (10)$$

7. Estimate the index of importance

$$\hat{\eta}_1^2 = \frac{\hat{V}_{X_1}\left[E_{X_2}(Y|x_1)\right]}{\hat{V}[Y]} \quad (11)$$

The extension of the procedure to more than two input variables and to the sensitivity analysis of groups of input variables considered simultaneously is straightforward [9],[15].

The main advantages of the variance decomposition method for performing sensitivity analysis is that it does not impose any limitative hypothesis on the structure of the model as it is, for example, with the regression methods. Moreover, it allows a straightforward evaluation of the sensitivity importance of groups of variables and not only individual ones. On the other hand, this flexibility is paid by a computational burden larger than that of other methods, e.g. regression-based methods, due to the need of computing the model output several times ($s \cdot r$) for the different input values sampled from the respective probability distributions.

3. BASIC ELEMENTS OF THE AP1000 REACTOR DESIGN

3.1. General Aspects

The Westinghouse AP1000 is a 1117 MWe (3415 MWth) pressurized water reactor (PWR), with layout simplification achieved through large operating margins and extensive implementation of passive safety systems for reduction of corrective maintenance actions in case of accident. The passive safety systems include passive Residual Heat Removal System (RHRS) and Passive Containment Cooling System (PCCS). The PCCS provides the safety-related ultimate heat sink for the plant. It cools the containment following an accident, so that the pressure is effectively controlled within the safety limits of 0.4 MPa. During an accident, heat is removed from the containment vessel by the continuous, natural circulation of air, supplemented by evaporation of the water that drains by gravity from a tank located on top of the containment shield building by means of three redundant and diverse water drain valves. The steel containment vessel provides the heat transfer surface through which heat is removed from inside the containment and transferred to the atmosphere. In addition, even in case of failure of water drain, air-only cooling is supposed to be capable of maintaining the containment below the failure pressure [10]. Figure 1 shows the PCCS of the AP1000 [Westinghouse Electric Company promotional image].

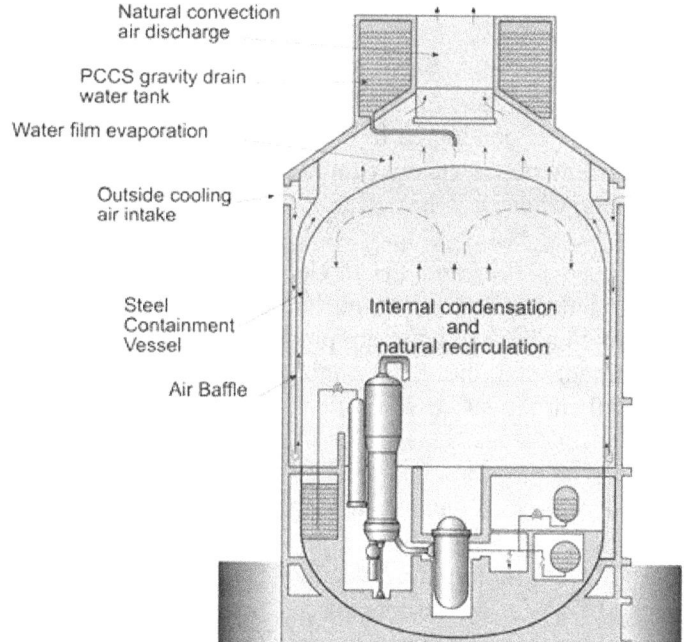

Figure 1 AP1000 Passive Containment Cooling System [Westinghouse Electric Company]

For the analysis of the functional performance of PCCS of the AP1000, TH analysis is carried out for safety assessment [17][18]. WGOTHIC [18] has been developed by Westinghouse as a conservative lumped parameter model for heat transfer with non-condensed gas, circulation and stratification for TH response analysis following i) loss of coolant accident (LOCA) ii) steam line break (SLB) accident, and containment integrity analysis.
In this paper, we aim at the identification of the FDD after a LOCA; the simplified TH model used has been developed by [19].

3.2. LOCA

The LOCA is a most dangerous accident in Pressurized and Boiling Water Reactors (PWR and BWR, respectively), whereby the stored energy of the high pressure, high temperature coolant is released to the containment by rupture of an exposed pipe. Thus, it is to be considered among the design basis accidents for AP1000 reactor design.

In general a LOCA scenario develops as follows [20]:
1) A double-ended "guillotine" pipe break in a primary coolant line allows the coolant flowing out from both ends;
2) Coolant flashes into steam due to the large amount of stored energy and is discharged rapidly into the containment building;
3) Reactor trip is automatically triggered by the protective system to assure continued sub-criticality of the reactor core;
4) The Emergency Core Cooling Systems (ECCS) cools down the core and prevents excessive decay heat-driven damage to its structures;
5) Radioactivity in the coolant is retained by the containment structure with natural deposition processes and active removal systems, eventually reducing the overall levels of radioactivity;
6) RHRS maintains ECCS effectiveness and reduces containment pressure.

A TH code for simulating a LOCA is typically divided into four phases: 1) blowdown, that includes the accident initiation (when the reactor is in a steady-state full power operation condition) to the time at which pressure equalizes to the containment pressure; 2) refill, which includes the time from the end of the blowdown to the time when the ECCS refills the vessel lower plenum; 3) reflood, which begins when water starts flooding the core until when it is completely quenched; 4) post-reflood, which starts after the core quenching and energy is released to the Reactor Coolant System (RCS) by the RCS metal, core decay heat, and the steam generators that maximize the containment pressure.

4. THERMAL HYDRAULIC MODEL

In the post-reflood phase, the steam produced in the RCS is cooled at the internal face of the steel containment vessel, and then the heat is conducted by the vessel and transferred to the air in the air channels, (see Figure 1). Cold air enters the channel through the three rows of air inlets and flows down to the bottom of the channels, where it is heated by the steel vessel up to the air diffuser to the environment.

In this paper, a steady state, lumped parameters TH model is used to analyze the effect of the air temperature and reactor power on the PCCS function. The parameters of the TH model used for calculating the PCCS capability of condensing the steam produced, and their distributions, are listed in Table 1. If the steam cannot be condensed, the vapor cumulates in the containment and results in an overpressure accidental scenario: then, the success criteria for the PCCS is set at Pcontainment < 0.4 MPa.

The selection of the distributions of the parameters in Table 1 is based on expert judgment and literature review [3],[21]. Three distributions have been used: seasonal, normal and uniform. Seasonal relates to the external air temperature T_{inlet} and pressure P_{air} variability, as inferred by historical data collected by a representative Chinese Automatic Weather Station (CAWS) in different months. Normal distributions, e.g. for the LOCA steam temperature, T_{steam}, are listed as truncated distributions with mean μ and support equal to 4σ where σ is the standard deviation. For uniform distributions, e.g. for the steam mass flow rate G, the supports from "Lower value" to "Upper value" are reported.

Figure 2 shows the distribution of the steady state containment pressure values obtained from 10000 runs of the TH code, with parameters values randomly sampled from the distributions of Table 1; the total computational time is 1894 s on a laptop machine powered by an Intel core2duo P7550 dual core processor running at 2.26 GHz. A value of 0.55 MPa is automatically assigned to the pressure, when it exceeds the safety limit of 0.4 MPa.

Table 1: List of parameters distributions

	Parameter	Description	Unit	Type of distribution	Lower value	Upper value		
1	G	Steady state LOCA mass flow rate	kg/s	uniform	6	11		
2	T_{inlet}	External air temperature	°C	seasonal	2	39		
3	P_{air}	Pressure of inlet air	MPa	seasonal	0.09837	0.1010965		

	Parameter	Description	Unit	Type of distribution	Mean value, μ	Standard Deviation, σ (% of μ)	$\mu-4\sigma$	$\mu+4\sigma$
4	T_{steam}	LOCA steam temperature	°C	normal	250	5	200	300
5	P_{steam}	LOCA steam pressure	MPa	normal	0.1	5	0.08	0.12
6	$\rho_{primary}$	Water density in primary circuit	kg/m³	normal	666.7	2	613.36	720.04
7	$P_{primary}$	Pressure of primary circuit	MPa	normal	15.5	2	14.26	16.74
8	V	Containment volume	m³	normal	58333	1	55999.68	60666.32
9	t	Containment wall thickness	m	normal	0.04455	0.5	4.37E-02	4.54E-02
10	D	Containment diamenter	m	normal	39.62	0.5	38.83	40.41
11	H	Containment height	m	normal	34.12	0.5	33.44	34.8
12	W	Width of air buffle outside containment	m	normal	0.92	0.5	0.9	0.94
13	H_1	Height of the download in air buffle	m	normal	38.11	0.5	37.35	38.87
14	H_2	Height of the upload in air buffle	m	normal	59.89	0.5	58.69	61.09
15	D_3	Diameter of the air outlet	m	normal	9.75	0.5	9.56	9.95
16	H_3	Height of the air outlet	m	normal	6	0.5	5.88	6.12
17	D_4	Diameter of uphead	m	normal	39.62	0.5	38.83	40.41
18	H_4	Height of uphead	m	normal	11.47	0.5	11.24	11.7
19	d	Diffusive coefficient (water)	m²/s	normal	2.55E-05	20	5.10E-06	4.59E-05
20	λ	Heat conduction of the wall	W/(m K)	normal	54	5	43.2	64.8

	Parameter	Description	Unit	Type of distribution	Lower value	Upper value
21	K	Air channel rugosity	-	uniform	0.00285	0.00315
22	f_1	Friction factor of corner	-	uniform	0.475	0.525
23	f_2	Friction factor of inlet	-	uniform	0.9025	0.9975
24	f_3	Friction factor of pipeup	-	uniform	0.1425	0.1575
25	f_4	Friction factor of pipeout	-	uniform	0.1425	0.1575
26	f_5	Friction factor of pipecold	-	uniform	0.1425	0.1575

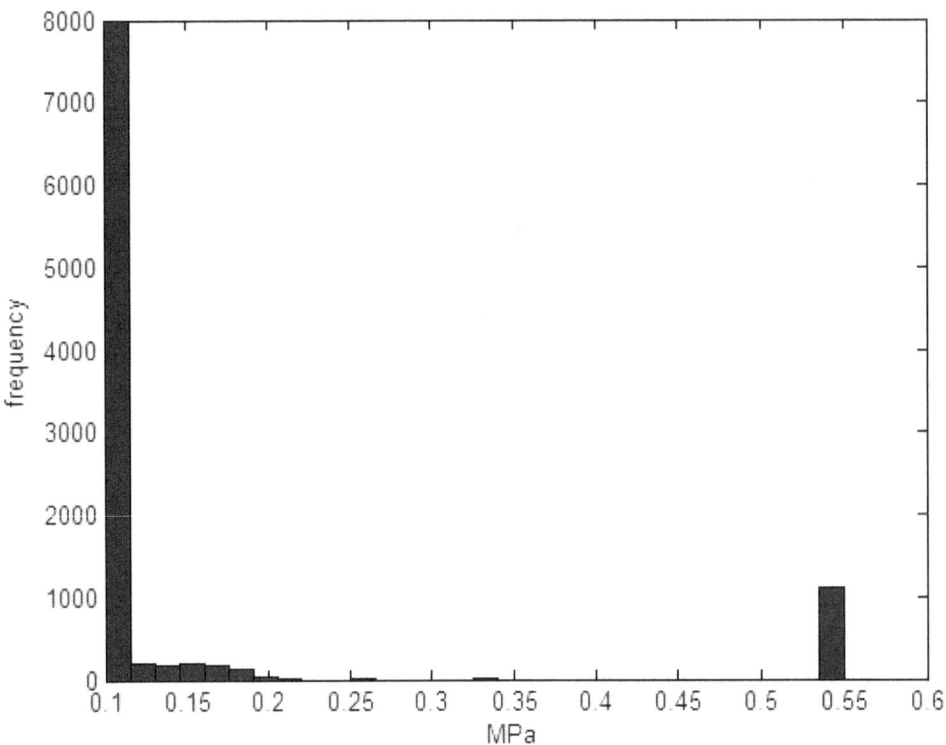

Figure 2 Histogram of the steady state containment pressure from 10000 runs of the TH code.

5. FDD BY VARIANCE DECOMPOSITION SENSITIVITY ANALYSIS

The majority of input sampled vectors lead the system to succeeding in maintaining the pressure within the limit of 0.4 MPa, but there is a not negligible probability of exceedance of approximately 10%.

5.1. Variance Decomposition Sensitivity Analysis

For variance decomposition sensitivity analysis, in our case, we select $s=90$ and $r=140$ so that 12600 runs of the TH code are performed for each one of the 26 parameters listed in Table 1, for a total of 327000 simulations (total computation time is 58251 s). The η^2 importance indexes of all the parameters are reported in Figure 3. The importances of G and T_{inlet} are clearly predominant and the importances of the other parameters are negligible.

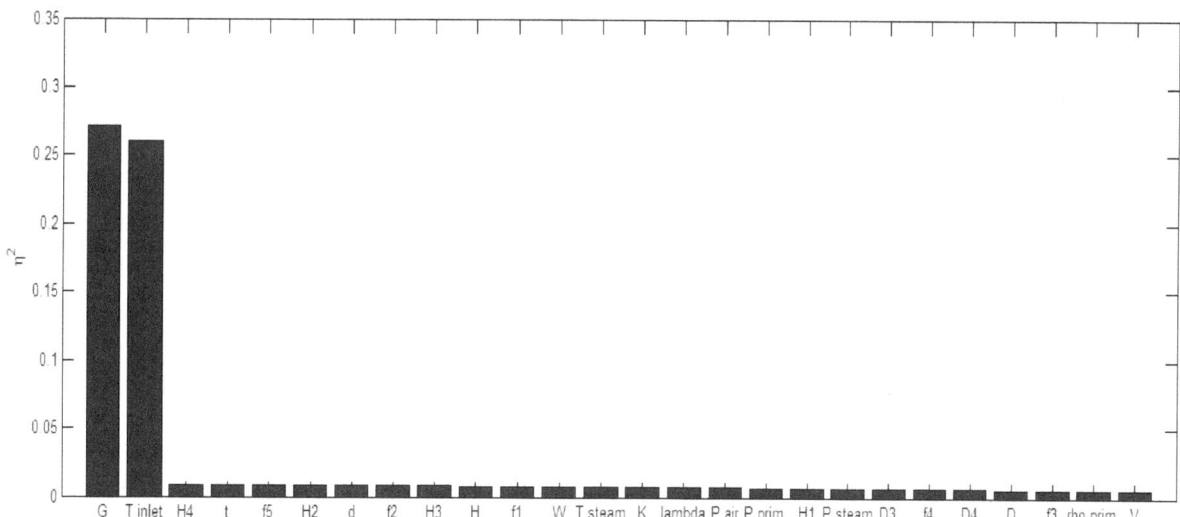

Figure 3 η^2 of the 26 parameters.

A group analysis is also performed, in which the importance index η^2 of groups of parameters is computed. Four groups have been considered: external conditions, primary coolant conditions, system geometry, materials properties and friction factors. For each group, $(s \cdot r)$ simulations with $s=90$, $r=140$, (50400 in total) have been performed.

The specific groups are:

a. External conditions: T_{inlet}, P_{air}
b. Primary coolant conditions: G, P_{steam}, T_{steam}, $\rho_{primary}$, $P_{primary}$,
c. System geometry: V, t, D, H, W, H_1, H_2, D_3, H_3, D_4, H_4
d. Material properties and friction factors d, λ, K, f_1, f_2, f_3, f_4, f_5

The results of the group variance decomposition (Table 2), again show a clear predominance of T_{inlet} and G in determining the variance of the steady state containment pressure: in fact, the groups a and b have values of η^2 greatly larger than groups c and d.

Table 2 η^2 for groups of inputs

Group	η^2
External conditions	0.2935
Primary coolant conditions	0.2648
System geometry	0.0078
Material properties and friction factors	0.0095

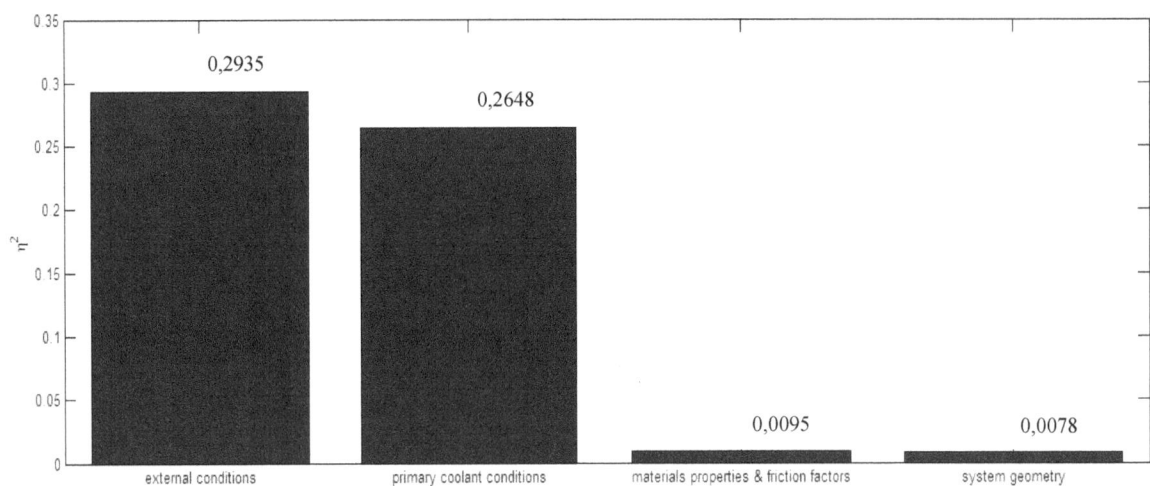

Figure 4 η^2 of the four groups of parameters.

We can conclude that under the assumptions of Table 1, G and T_{inlet} are by far the most important parameters for obtaining the PCCS response in terms of steady state containment pressure. This result aligns with our prior expectations for what regard the predominance of G and T_{inlet} with respect to the other parameters. In fact, not only G and T_{inlet} are directly linked to the energy entering (G) and leaving (T_{inlet}) the PCCS but they have also, by far, the largest uncertainties, as reported in Table 1. The other input parameters have low uncertainties due to better knowledge, and their effects on the output are modest even when sampled at maximum or minimum values of their range.

On the other hand, the finding that G and T_{inlet} are almost equally important (i.e. equally responsible for the output variability) is an information difficult to suppose a priori of the sensitivity analysis. because of the different distributions and the different relations (also nonlinear) with the model output.

5.2. Failure Damage Domain

An intuitive way to gather together all the information resulting from the previous analysis is the FDD map. This map resumes the information provided by both the MC uncertainty propagation (Figure 2) and the variance decomposition analysis (Figures 5 and 6). It can be useful for i) further sensitivity analysis without TH calculations, ii) safety margins visualization, iii) identification of mitigation strategies, iv) discussion with the regulator [13].

Figure 5 shows the FDD in the plane of the two most important parameters T_{inlet} and G, representing the failure probability of the PCCS after a LOCA.

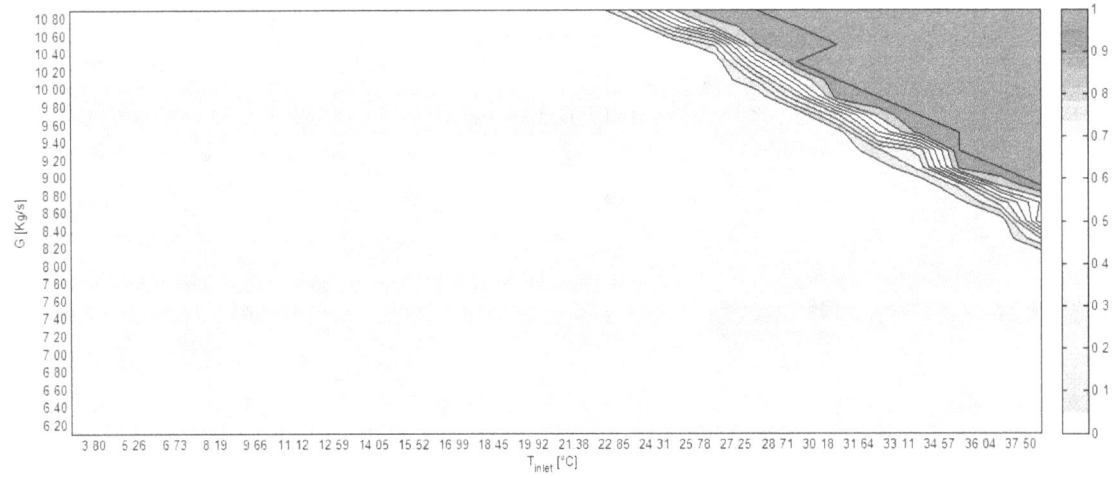

Figure 5 Failure damage domain map

6. CONCLUSIONS

In this work, we have performed a Monte Carlo sampling uncertainty propagation, variance decomposition sensitivity analysis and FDD identification with respect to a TH code that calculates the post reflood steady state pressure after a LOCA in the AP1000 passive pressure containment. The sensitivity analysis has highlighted that among all the input parameters of the code, G and T_{inlet} are by far the most important. The results of the analysis have been used to generate the FDD, which gives useful safety insights. Computational time is an aspect of primary importance in sensitivity analysis and might be a limiting factor when resorting to variance decomposition. Thus, future work will tackle this problem by developing an innovative sensitivity method that limits the need of repeating several model solutions in correspondence of different sampled input values for properly mapping the input-output relationship.

References

[1] F. J Mackay, G. E. Apostolakis, P. Hejzlar, "*Incorporating reliability analysis into the design of passive cooling systems with an application to a gas-cooled reactor*", Nuclear Engineering and Design 238 (1), 217–228, (2008).

[2] A. K. Nayak, M. R Gartia, A. Antony, G. Vinod, R. K. Sinha, "*Passive system reliability analysis using the APSRA methodology*", Nuclear Engineering and Design 238, 1430–1440, (2008).

[3] E. Zio, N.Pedroni, "*How to effectively compute the reliability of a thermal–hydraulic nuclear passive system*", Nuclear Engineering and Design Volume 241, Issue 1, Pages 310–327, (2011).

[4] W.E. Cummins, M.M. Corletti, T.L. Schulz, "*Westinghouse AP1000 Advanced Passive Plant*", Westinghouse Electric Company, LLC. Proceedings of ICAPP '03 Cordoba, Spain, May 4-7, Paper 3235, (2003).

[5] A. Saltelli, K. Chan, E. M. Scott eds., "*Sensitivity Analysis*", John Wiley & Sons, (2000).

[6] Y. Yu, T. Liu, J. Tong, J. Zhao, F. Di Maio, E. Zio, A. Zhang, "*Multi-Experts Analytic Hierarchy Process for the Sensitivity Analysis of Passive Safety Systems*", Proceedings of the 10th International Probabilistic Safety Assessment & Management Conference, PSAM10, Seattle, June, (2010).

[7] E. Zio, "*Computational Methods for Reliability and Risk Analysis*", World Scientific Publishing, (2009).

[8] S. Fang, G. Z. Gertner, S. Shinkareva, G. Wang, A. Anderson "*Improved generalized Fourier amplitude sensitivity test (FAST) for model assessment*", Statistics and Computing Volume 13, Issue 3 , pp 221-226, (2003).

[9] M. D. McKay, "*Variance-Based Methods for Assessing Uncertainty Importance in NUREG-1150 Analyses*", LA-UR-96-2695, Los Alamos National Laboratory, pp. 7-27, (1996).

[10] T. L. Schulz, "*Westinghouse AP1000 advanced passive plant*", Nuclear Engineering and Design, vol. 236, pp 1547-1557, (2006).

[11] E. Zio, F. Di Maio, S. Martorell, Y. Nebot, "*Neural Networks and Order Statistics for Quantifying Nuclear Power Plants Safety Margins*" Proceedings, European Safety & Reliability Conference (ESREL), Valencia, Spain, (2008).

[12] E. Zio, F. Di Maio, J. Tong, "*Safety Margins Confidence Estimation for a Passive Residual Heat Removal System*", Reliability Engineering and System Safety, Vol. 95, pp. 828–836, (2010).

[13] V. Rychkov, "*Failure domain approach to characterize safety margins. Probabilistic Safety Analysis perspective*", EDF R&D IDPSA workshop, NURETH-15, Pisa, 12 May, (2013).

[14] F. Di Maio, J. Hu, P. Tse, M. Pecht, K. Tsui, E. Zio, "*Ensemble approaches for clustering health status of oil sand pumps*", Expert Systems with Applications, Volume 39, Issue 5, Pages 4847–4859, (2012).

[15] M. D. McKay, "*Evaluating Uncertainty in Stochastic Simulation Models*", Proceedings of SAMO '98, Venice, April 19 -22, pp. 171–175, (1998).

[18] R. P. Ofstun, J. H. Scobel, "*Westinghouse Containment Analysis Methodology (WCAP-16608-NP, Class3)*", PA: Westinghouse Electric Company LLC, (2006).

[16] F. Cadini, E. Zio, F. Di Maio, V. Kopustinskas, R. Urbonas, "*A neural-network-based variance decomposition sensitivity analysis*" International Journal of Nuclear Knowledge Management, Vol. 2, No. 3, pp. 299-312, (2007).

[17] J. Woodcock, T. S. Andreychek, L. Conway, "*WGOTHIC Application to AP600 and AP1000 (WCAP- 15862, Class3)*", PA: Westinghouse Electric Company LLC, (2004).

[19] Y. Yu, S. Wang, F. Niu, "*Thermal–hydraulic performance analysis for AP1000 passive containment cooling system*", Proceedings of the 21th International Conference on Nuclear Engineering ICONE21, July 29- August 2,, Chengdu, Sichuan, China, (2013).

[20] F. C. Rahim, M. Rahgoshay, S. K. Mousavian, "*A study of large break LOCA in the AP1000 reactor containment*", Progress in Nuclear Energy, Volume 54, Issue 1, Pages 132–137, (2012).

[21] L. Burgazzi, "*Evaluation of uncertainties related to passive systems performance*", Nuclear Engineering and Design 230 , pp. 93–106, (2004).

The Development of a Demonstration Passive System Reliability Assessment

Matthew Bucknor*[a], David Grabaskas[a], and Acacia Brunett[a]

[a] Nuclear Engineering Division, Argonne National Laboratory, Argonne, IL 60439, U.S.

Abstract: In this paper, the details of the development of a demonstration problem to assess the reliability of a passive safety system are presented. An advanced small modular reactor (advSMR) design, which is a pool-type sodium fast reactor (SFR) coupled with a passive reactor cavity cooling system (RCCS) is described. The RELAP5-3D models of the advSMR and RCCS that will be used to simulate a long-term station blackout (SBO) accident scenario are presented. Proposed benchmarking techniques for both the reactor and the RCCS are discussed, which includes utilization of experimental results from the Natural convection Shutdown heat removal Test Facility (NSTF) at the Argonne National Laboratory. Details of how mechanistic methods, specifically the Reliability Method for Passive Systems (RMPS) approach, will be utilized to determine passive system reliability are presented. The results of this mechanistic analysis will ultimately be compared to results from dynamic methods in future work. This work is part of an ongoing project at Argonne to demonstrate methodologies for assessing passive system reliability.

Keywords: Passive System Reliability, PRA, advSMR, SFR, RMPS

1. INTRODUCTION

Recently, significant resources have been invested in the development of small modular reactors (SMRs) based on current light water reactor technology and on non-light water designs, known as advanced small modular reactors (advSMRs). These new designs offer both financial and technological advances over the current fleet of nuclear reactors. Relatively low capital costs due to compact designs, shorter construction timeframes, lower power output, and the ability to take advantage of passive safety systems can potentially increase the number of installations of SMRs globally. While these designs represent a step forward for the nuclear industry, they provide a challenge to the current regulatory framework due to the increased dependence on passive safety systems whose reliability may not be easily quantified using conventional reliability methods. This difficulty arises as a result of the nature of passive systems, which can fail functionally without a physical component failure.

In FY13, a review was conducted of the available techniques for modeling the reliability of a passive system using natural circulation [1]. The results of this survey returned three distinct classifications of passive system reliability methodologies. The first technique involved conservative bounding analyses and margin assessment. The second was the mechanistic methods of such projects as the Reliability Method for Passive Systems (RMPS) [2] developed in Europe. Lastly, simulation-based or dynamic techniques present a promising option for the most phenomenologically consistent analyses, but such techniques are still in their infancy.

The next step in this project is to demonstrate several of the surveyed methodologies by analyzing an example passive safety system. The system of choice is the reactor cavity cooling system (RCCS), which is a derivative of the reactor vessel auxiliary cooling system (RVACS) from such plant designs as the PRISM sodium fast reactor (SFR) [3]. An advanced small modular reactor (advSMR), which is a 100 MWe pool-type SFR design, will be analyzed to evaluate the reliability of the RCCS during a station blackout (SBO) scenario. The focus of this analysis is the use of reliability estimate techniques

---

* Corresponding author email: mbucknor@anl.gov

to address large uncertainties and the ability to provide a meaningful comparison of competing designs using risk-based information.

This paper details the goals of the analysis and the development of the demonstration problem including descriptions of the models and techniques utilized to analyze passive system reliability. A sister paper "Dynamic Methods for the Assessment of Passive System Reliability" [4] presents the specifics of the dynamic methodology, while the focus of this paper will be on the demonstration problem details and mechanistic modeling techniques.

2. PLANT DESIGN

The focus of this analysis is on passive safety systems reliability assessment and not on a specific reactor. For that reason a generic advSMR SFR pool-type design will be utilized in this analysis. The design characteristics of the generic advSMR are listed in Table 1. In addition to these features, the reactor design has intermediate heat exchangers (IHX) which serve as the connection between the hot and cold sodium pools. During normal operation, the IHX is utilized to transfer heat from the hot sodium pool to the secondary side of the plant for electrical power generation. However, in an accident scenario, the IHX can be utilized (if AC power is available) to reject heat via a heat sink (such as the condenser) on the secondary side of the plant. The design also incorporates a guard vessel which surrounds the primary vessel and an RCCS which is based on the design of the RCCS of the General Atomics Modular High Temperature Gas cooled Reactor (GA-MHTGR) [5]. The design will, however, be modified to accommodate the physical size of the advSMR used in this analysis.

Table 1: Design characteristics of the generic advSMR.

Characteristics	
Power rating	250 MWth/100 MWe
Primary coolant	Sodium
Primary system type	Pool
Fuel type	Metallic
Primary coolant flow rate	~ 1270 kg/s
Coolant pump type	Electromagnetic
Number of coolant pumps	4
Primary vessel height	10 m
Core inlet temperature	~ 400°C
Core outlet temperature	~ 550°C

The RCCS uses natural convection to drive air from the outside environment through a cold downcomer and into a lower plenum where the air then flows through hot riser ducts that surround the reactor pressure vessel and line the concrete containment structure. The air in the hot riser ducts is heated through a combination of radiation and convection before being exhausted back to the outside environment. The system is designed to remove decay heat, but because it is completely passive (no baffle or damper operation is required) it also functions during normal reactor operation. A plan view of the RCCS from the GA-MHTGR is shown in Figure 1 [5].

Figure 1: Plan view of the GA-MHTGR cavity and the RCCS [5].

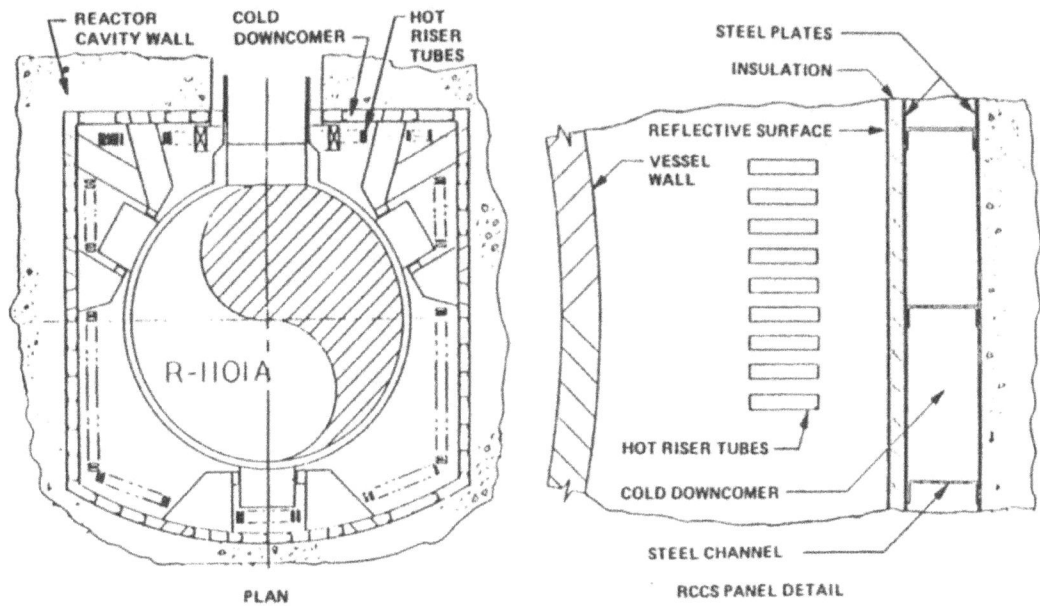

3. MODELING

The advSMR and the RCCS will be modeled together in RELAP5-3D [6]. This allows for the coupled treatment of the heat transfer effects of the natural circulation in the sodium pool and the natural circulation of air in the RCCS. Heat will transfer from cold sodium pool to the walls of the primary vessel. As the primary vessel temperature increases, heat will transfer (via radiation) to a guard vessel and then to the hot riser ducts of the RCCS that surround the primary vessel inside the containment wall. The heat is then convected from the hot riser duct walls to the air passing through each riser. The air that surrounds the vessel and the outer surfaces of the hot riser tubes will not be considered in the heat transfer calculation, as it is assumed that radiation from the guard vessel to the hot riser tubes will be the dominant mechanism of heat transfer. This assumption is consistent with previous analysis of the RCCS [7].

A basic representation of a RELAP5-3D nodalization diagram of an advSMR is shown in Figure 2. While the actual model is much more detailed, this diagram is provided as an overview of the major components that are included in the RELAP5-3D model. In this model, sodium is pumped via an electromagnetic (EM) pump from the cold pool to the inlet plenum before entering the core region. Sodium flows out of the core and into the outlet plenum before passing though the IHX system (not shown in the diagram) and back into the cold pool. This is the assumed flow pathway during normal operation and during accident scenarios where natural circulation flow will occur. Also not shown is an argon filled gap that exists between the outer wall of the primary vessel and a guard vessel. This gap will be modeled in the final RELAP5-3D model.

Figure 2: Simplified nodalization diagram of a generic pool-type advSMR RELAP5-3D model.

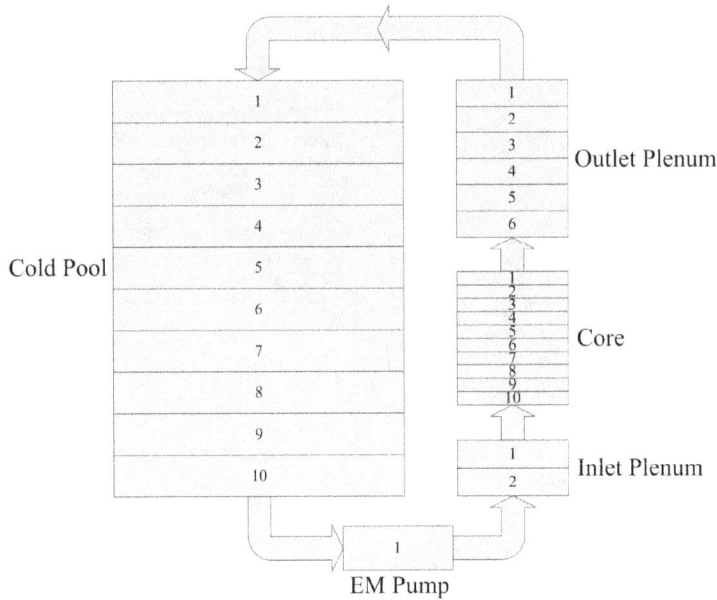

During steady-state simulations, 250 MW of power will be added to the system in the core region. Approximately the same amount of power will be removed from the system via heat structures connected to the cold pool that will be used to simulate the IHX system. The remaining power will be removed from the cold pool via heat structures which represent the guard vessel, the hot riser tubes, and the inside wall of the containment.

During the SBO scenario progression, core power will either be reduced over time to simulate generation of decay heat for protected accidents, or it will remain at operating power until negative reactivity feedback, modeled via integral reactivity feedback coefficients, ultimately reduces core power for unprotected accidents. In the SBO scenario, it is assumed that AC power from the grid will not be regained within 72 hours of the initiating event due to a massive grid failure. However, during the initial 24 hour timeframe, onsite AC power may be restored utilizing FLEX equipment, which is portable backup equipment stored both onsite and offsite as outlined in the FLEX strategy [8].

An existing RELAP5-3D model of the RCCS will be utilized in this analysis. The existing model, of which a nodalization diagram is shown in Figure 3, is from previous GA-MHTGR analyses [7]. This model will be modified appropriately for the physical size difference between the GA-MHTGR vessel and the advSMR vessel utilized in this analysis. The diameters of the primary vessels of the GA-MHTGR and of the advSMR are similar such that it is assumed that the view factors utilized in the analyses of the GA-MHTGR for radiative heat transfer will not require recalculation. However, the height of the GA-MHTGR primary vessel is taller (approximately 22 m) than that of the advSMR (10 m), and therefore the height of the heated section of the hot riser tubes will be modified accordingly. It is also important to note that radiative heat transfer between an axial node of the guard vessel and an axial node of the hot riser duct will only be calculated for nodes at the same elevation. This treatment of the radiative heat transfer is consistent with previous analysis of the RCCS [7] and greatly simplifies the calculation of view factors that are manually input into RELAP5-3D and also reduces the length of time required for each simulation to be performed.

Figure 3: Nodalization diagram of the RCCS RELAP5-3D model [7].

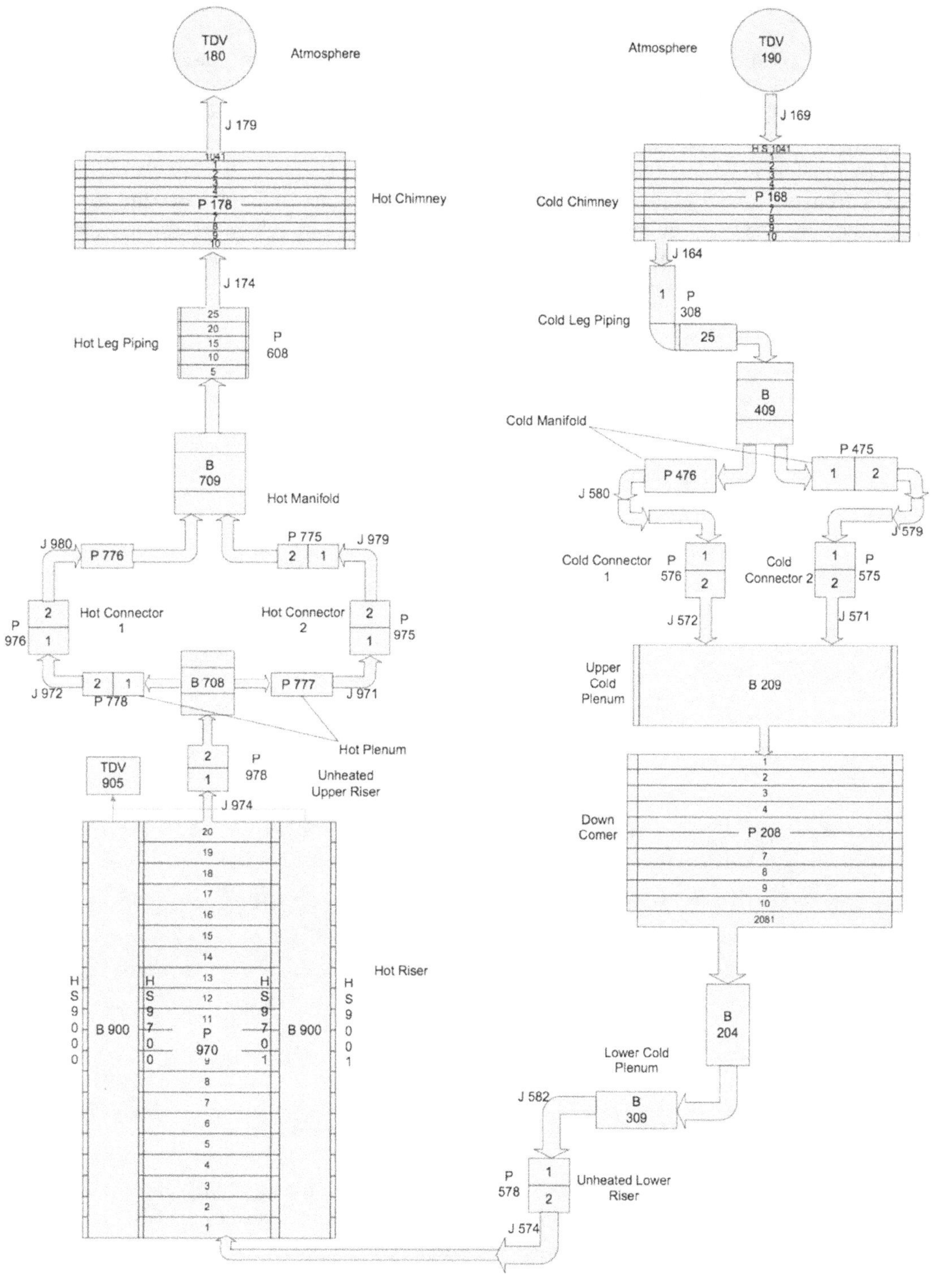

4. BENCHMARKING/VALIDATION

To benchmark the conditions in the RELAP5-3D model of the reactor during steady-state operation and long-term SBO scenarios, SAS4A/SASSYS-1 [9] will be utilized. Ideally, validation would be preferred, but benchmarking is the best available option for this analysis as experimental facilities are not available. SAS4A/SASSYS-1 has been historically utilized for SFR analyses and various aspects of the code have been validated [10,11,12,13]. The ability of SAS4A/SASSYS-1 to model both steady-state and severe accident conditions make it a viable option for benchmarking in this analysis.

To validate the conditions in the RCCS, data from experimental tests conducted at the Natural convection Shutdown heat removal Test Facility (NSTF), located at the Argonne National Laboratory, will be utilized. Originally, the NSTF was constructed to provide full scale simulation of the reactor vessel auxiliary cooling system (RVACS) for the PRISM design [3], but recently the facility has undergone renovation. The current geometry, shown in Figure 4 [14], represents a half-scale model of the RCCS from the GA-MHTGR design. In this test facility, air is drawn into the lower plenum (located in the pit) due to the chimney effect. The air then passes through hot riser ducts located in the heated test section and into the upper plenum. Two chimneys are connected to the upper plenum which allows the heated air to be released to the environment.

Figure 4: Layout of the NSTF [14].

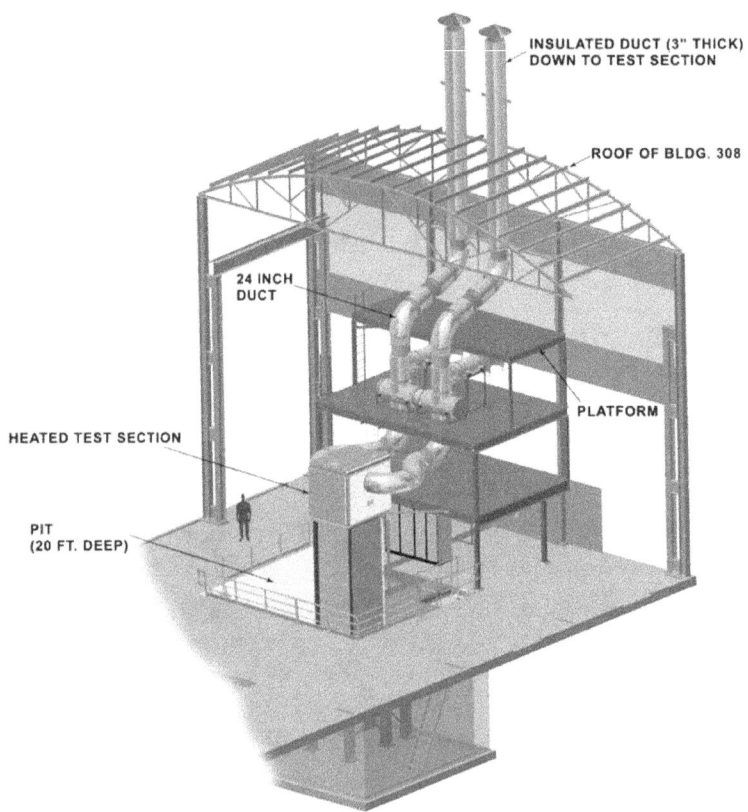

5. RELIABILITY METHODS & UNCERTAINTY ANALYSIS

This section describes two methods that will be utilized to analyze the reliability of the passive RCCS. Section 5.1 describes the first method which is mechanistic and also describes the SBO event tree that will be utilized in the analyzed scenario. Section 5.2 will briefly describe the use of a dynamic analysis which is presented in more detail in a sister paper, titled "Dynamic Methods for the Assessment of Passive System Reliability" [4]. Finally, a description and characterization of the uncertainties associated with both the advSMR and the RCCS are provided in Section 5.3.

5.1. Mechanistic Methods

The mechanistic methods used in this analysis are based on the RMPS methodology [2]. This methodology provides a systematic approach for assessing the reliability of passive safety systems. The RMPS methodology can be divided into three main phases: a preprocessing and model development phase, a simulation and propagation phase, and an analysis/post-processing phase [15]. Table 2 provides details on the main steps of each phase.

Table 2: Three phases of the RMPS methodology.

Phase 1	• Identify a system, its intended mission, and possible failure modes • Select or develop a best-estimate code to perform the analysis • Identify important system variables and characterize their associated uncertainty
Phase 2	• Conduct sensitivity analysis • Determine failure probability of the system using best-estimate code • Assess the impact of the uncertainties of the important system variables
Phase 3	• Incorporate the failure probability into the PRA

For this analysis, the results from the RMPS methodology will be represented in an event tree. An example SBO event tree, generated in SAPHIRE [16], is shown in Figure 5, which is based on the SBO event tree from the PRISM Preliminary Safety Information Document (PSID) [17]. For this initiating event, it is assumed that both off-site and on-site power (the PRISM design did not include a safety related diesel generator) are lost. For top events whose frequencies in Figure 5 are not unity, (these events are discussed later) the event frequency will be determined using data from in the NRC's "Industry-Average Performance for Components and Initiating Events at U.S. Commercial Nuclear Power Plants," NUREG/CR-6928 [18] and the PSID for PRISM [17].

In the SBO event tree, seen in Figure 5, the first top event relates to the reactor protection system (RPS) sending a signal to the Reactor Shutdown System (RSS) to SCRAM the reactor. The signal transmission is either received successfully or not. If the transmission of the signal is successful, the RSS inserts control rods into the core region. This event is labelled in the event tree as "Enough Control Rods Inserted." The PRISM PSID defines "enough" as the successful insertion of at least one out of six control rods. Because a generic reactor/core design is being utilized in this work, a complete reactivity analysis is not being performed and it is assumed that all control rods are either inserted successfully (a protected scenario) or unsuccessfully (an unprotected scenario).

Note that the event tree has two points labelled as A. In both branches containing point A, a sufficient number of control rods are not inserted to successfully SCRAM the reactor. The only difference in the two branches relates to the signal transmission from the RPS to the RSS. In the upper branch, the signal transmission to SCRAM the reactor was successful, but the RSS failed to insert enough control rods. In the lower branch the signal transmission was unsuccessful; therefore the control rods were not inserted. From the point where the branches are labelled A, they are identical and the lower branch has been removed from the figure for simplicity. The probabilities of success and failure for the two branches are identical from point A onward, however the end state probabilities will differ due to actions occurring prior to point A.

Following the "Enough Control Rods Inserted" event is the "Pump Trip" event. The EM pumps in the primary system require AC power for operation; therefore, when power is lost, the probability of pump trip is unity. The next top event is pump coastdown. EM pumps have almost no inertial force. If power is lost to the pump, the flow rate decreases so rapidly that essentially no pump coastdown occurs. To generate pump coastdown behavior, a separate unit consisting of a flywheel and synchronous machine are utilized. During normal operation, the flywheel and synchronous machine operate in parallel to the EM pump. If power is lost, the inertial energy of the flywheel is converted to electrical energy by the synchronous machine which is used to power the EM pump. This enables the EM pump to generate coastdown behavior similar to that of a mechanical pump. For this top event, five possibilities will be considered, because there are four pumps. Either four out of four pumps coastdown normally, or three out four pumps coastdown normally, and so on until the final possibility which is zero out of four pumps coastdown normally is considered. Successful pump coastdown is modeled in this fashion because the success criterion for sufficient pump coastdown is unknown due the generic reactor design utilized in this analysis.

Figure 5: Example SBO event tree.

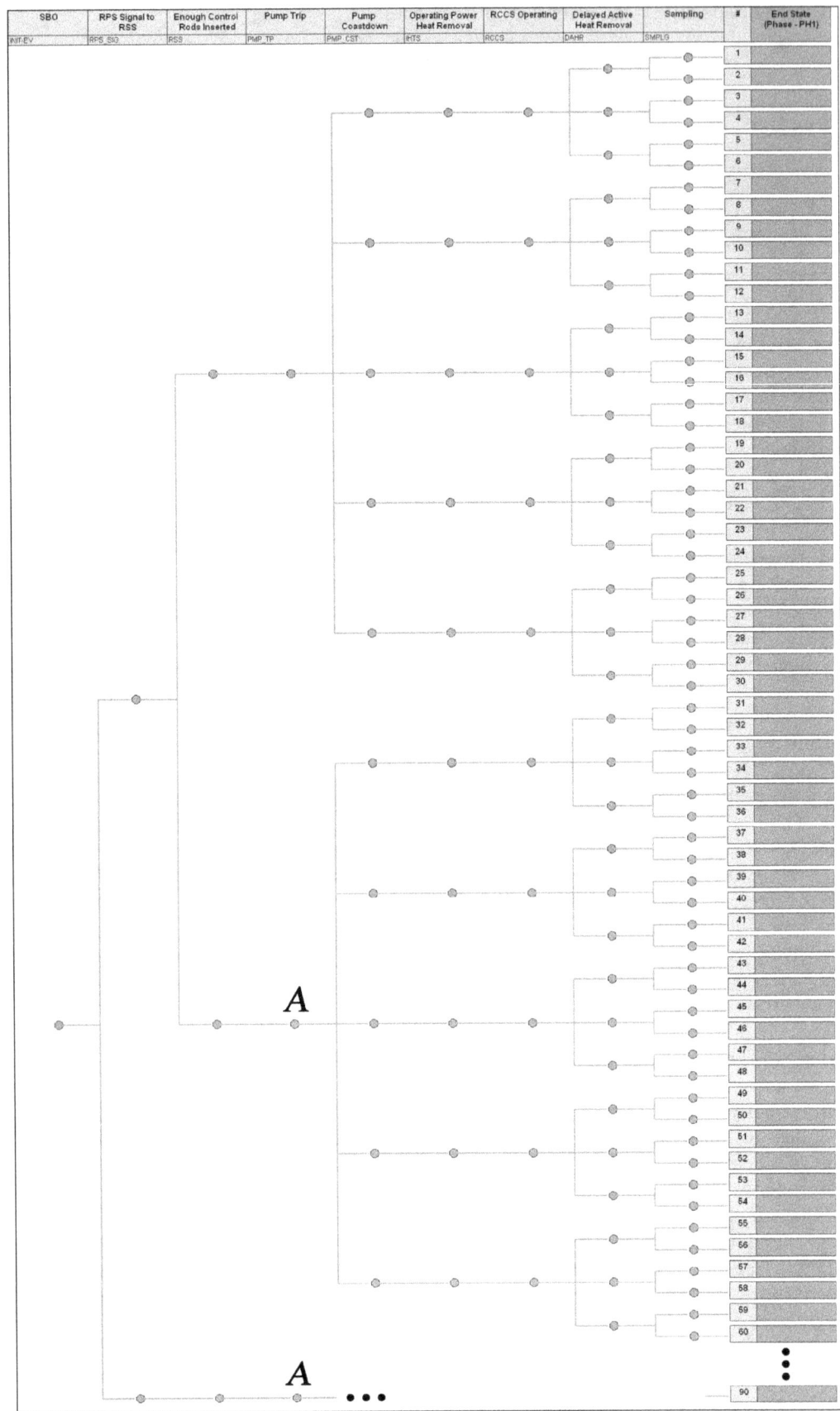

The "Operating Power Heat Removal" failure probability is unity because the initiating event is an SBO. The probability of the RCCS operating is unity because the system is passive and an SBO would not necessarily affect the structures associated with RCCS. However, in FY15, the loss and repair of single or multiple RCCS chimneys due to extreme external events, such as seismic or wind, will be considered. The next event is "Delayed Active Heat Removal" in which three possible scenarios will be investigated. The timeframe over which these scenarios occur is between 4 hours to 24 hours after the initiating event, which is roughly based on the anticipated NRC rulemaking as a response to Fukushima [19]. It is assumed that any active heat removal during this time frame will be powered by FLEX equipment as AC power from the power grid is assumed to take longer than 24 hours to restore in the SBO scenario. The first possibility is that cooling via the IHX system is not activated within 24 hours of the SBO. This could be due to unavailability or failure of the FLEX equipment, or the utility opting to allow only the passive system to operate even though FLEX equipment is available. The second and third possibilities are that heat removal via the IHX is activated through the use of FLEX equipment relatively early (9 hours after the initiating event) or relatively late (19 hours after the initiating event) in the simulation. The selection of these two representative time frames will be discussed in more detail in Section 5.3.

The final top event in the tree is labelled "Sampling". In this branch of the analysis, the success and failure probabilities resulting from the propagation of input uncertainties are assessed. For example, the uncertainty of the value for primary vessel emissivity is represented by some distribution (based on existing literature and engineering judgment). Monte Carlo random sampling will be used to create samples from the uncertainty distribution. Each sample value will be simulated in the best-estimate code (RELAP5-3D) and the scenario progression will be prescribed by a certain path through the event tree. This simulation will either result in a safe plant state or a failed plant state (which would depend on the definition of those states). If this process were repeated N times, then the sampling success probability would be the number of simulations that lead to a success (as defined by the scenario success criteria) divided by N and the sampling failure probability would be one minus the sampling success probability. This method of incorporating the simulation results from the sensitivity analysis into the event tree was utilized in the pilot application of Risk Informed Safety Margin Characterization (RISMC) to a total loss of feedwater event [20].

5.2. Dynamic Methods

The most important drawback of a traditional event tree analysis is the inability to explicitly account for time within the event tree. This can potentially lead to phenomenological inconsistencies, particularly when uncertainties are considered that can change the order of events. Another drawback of a traditional event tree analysis is that it relies heavily on the PRA analysts' judgment or on expert elicitation to determine the timing and ordering of events in the event tree. Recent developments in dynamic methods, such as dynamic event trees (DETs) [21,22,23], allow for time to be treated explicitly and provide a way to perform a consistent and mechanistic uncertainty analysis.

In this analysis, the time dependent variable (delayed active heat removal activation time) will be treated utilizing DETs. The methodology for performing this analysis is outlined in a sister paper, titled "Dynamic Methods for the Assessment of Passive System Reliability" [4].

5.3. Uncertainty Analysis

The uncertainty analysis performed as part of this work will include uncertainties that pertain to the plant as well as uncertainties that pertain to the RCCS. Table 3 contains a list of uncertainties that have been identified for inclusion in this research. Engineering judgment along with information from the PRISM PSID [17] was utilized to characterize the distributions that represent the uncertainties associated with each variable. These values represent a starting point for the analysis and may change based on new information or data from experiments conducted at NSTF.

Table 3: Uncertainty characterizations.

Uncertainty	Characterization	Comments
RCCS		
Ambient temperature	$U(-30.0, 45.0)$	Assume conservative bounds, °C.
Primary vessel emissivity	$N(0.77, 0.035)$	Mean and bounding percentiles from [17].
Primary vessel thermal conductivity	$N(1.0, 0.0125)$	Scaling factor, varies up to ±2.5% of mean.
Guard vessel emissivity	$N(0.77, 0.035)$	Mean and bounding percentiles from [17].
Guard vessel thermal conductivity	$N(1.0, 0.0125)$	Scaling factor, varies up to ±2.5% of mean.
Duct surface roughness	$\ln N(3.45, 0.70)$	Assume large range of uncertainty due to weathering, bounding values of 10 µm, 100 µm.
Plant		
Initial power level	$N(1.0, 0.025)$	Scaling factor, varies up to ±5% of mean.
Decay heat curve	$N(1.0, 0.025)$	Scaling factor, varies up to ±5% of mean.
Active heat removal actuation time	$U(4.0, 24.0)$	Time frame of interest for this analysis, hours.
Pump coastdown failure	$P(1/4) = 1.000$ E-4 $P(2/4) = 3.267$ E-6 $P(3/4) = 1.108$ E-6 $P(4/4) = 8.420$ E-7	Failure per pump considered to be several orders of magnitude larger than value utilized in [17]. Failure of multiple pumps treated using alpha factor model with staggered testing for common cause failures.

The uncertainty associated with the ambient temperature is characterized by a uniform distribution whose bounds are -30°C and 45°C. This wide range of temperatures accounts for various climates as well as seasonal changes. The mean values for primary vessel and guard vessel emissivities are consistent with values specified in the PRISM PSID [17]. The PSID also provided a conservative value for vessel emissivity that was assumed to a 95th percentile value in this analysis. These two values were utilized to characterize normal distributions which represent the uncertainty associated with the emissivities of the vessels. The thermal conductivity of the primary vessel and guard vessel are entered into RELAP5-3D as temperature dependent functions. The uncertainty associated with the thermal conductivities is assumed to be due to normal variations in manufacturing of the vessels and would not vary greatly over the lifetime of the material. These uncertainties are treated with normally distributed scaling factors, assuming that the temperature dependent functions represent mean values of normal distributions and that the scaling factors can deviate by as much as ±2.5% of the mean values. The normal distributions representing these uncertainties are characterized by the 5th percentile value being 97.5% of the mean value and the 95th percentile being 102.5% of the mean value. A scaling factor will be sampled prior to each simulation and utilized to adjust the temperature dependent thermal conductivity of each vessel. The uncertainty associated with surface roughness of the air ducts of the RCCS are also analyzed as part of this work. The surface roughness of these ducts may vary greatly over the lifetime of the facility due to accumulation of particulate material. For this analysis, the uncertainty associated with the surface roughness of the ducts is represented with a lognormal distribution with a 5th percentile vale of 10 µm and a 95th percentile value of 100 µm. These values may be adjusted to reflect experimental results from the NSTF.

The uncertainty in initial power level and decay heat will be treated with scaling factors assuming that the values may vary ±5% from the mean values. The normal distributions representing these uncertainties are characterized by the 5th percentile value being 95% of the mean value and the 95th percentile being 105% of the mean value. The active heat removal actuation time uncertainty is characterized by a uniform distribution from 4 hours to 24 hours, which is the time frame of interest for this analysis. The probabilities for failure of a certain number of pumps to coastdown normally are provided in Table 3. These probabilities were determined using the alpha factor model with staggered testing for common cause failures [24]. As was the case in the PRISM PSID, common cause failures (those not explicitly modeled in the analysis) are assumed to dominate the likelihood of failure. Unlike the PRISM PSID, a much higher value (1.0E-4 per demand) for failure of a single pump to coastdown normally is utilized in this work, as the analysis performed in the PSID was criticized for having unjustifiably (due to lack of data and details) low failure probabilities [25].

6. CONCLUSION

This paper described the preliminary development of a demonstration problem to access the reliability of a passive system utilized to reject decay heat during an SBO scenario for an advSMR. Preliminary RELAP5-3D models of both the advSMR and the RCCS were presented along with a brief description of the codes and facilities that will be used to benchmark and validate the models. Details on the uncertainty analysis were provided along with a description of the RMPS methodology. An example SBO event tree was provided which incorporates a sampling branch in which the percentages of simulations that lead to an undesirable plant state are reported.

Acknowledgements

Argonne National Laboratory's work was supported by the U.S. Department of Energy, Assistant Secretary for Nuclear Energy, Office of Nuclear Energy, under contract DE-AC02-06CH11357.

References

[1] D. Grabaskas and T. Sofu, *"Review of Existing Approaches for Modeling the Reliability of Passive Systems,"* ANL-SMR-5, Nuclear Engineering Division, Argonne National Laboratory, Argonne, IL, (2013).
[2] M. Marques, et al., *"Methodology for the Reliability Evaluation of a Passive System and its Integration into a Probabilistic Safety Assessment"*, Nuclear Engineering and Design, vol. 235, pp. 2612-2631, (2005).
[3] W. Kwant and C.E. Boardman, *"PRISM – Liquid Metal Cooled Reactor Plant Design and Performance"*, Nuclear Engineering and Design, vol. 136, pp. 111-120, (1992).
[4] A. Brunett, M. Bucknor, and D. Grabaskas, *"The Development of a Demonstration Passive System Reliability Assessment,"* to be published in Proceedings of the 12th International Conference on Probabilistic Safety Assessment and Management (PSAM12), June 22-27, Honolulu, HI, (2014).
[5] *"Preliminary Safety Information Document for the Standard MHTGR,"* HTGR-86-024, Vol. 1, Amendment 13, U.S. Department of Energy, (1992).
[6] The RELAP5-3D Code Development Team, *"RELAP5-3D Code Manual,"* INEEL-EXT-98-00834, Rev. 4, Idaho National Laboratory, Idaho Falls, ID, (2012).
[7] S. Lomperski, W. D. Pointer, C. P. Tzanos, T. Y. C. Wei and A. R. Kraus, *"Air-Cooled Option RCCS Studies and NSTF Preparation"*, ANL-GenIV-179, Nuclear Engineering Division, Argonne National Laboratory, Argonne, IL (2011).
[8] Nuclear Energy Institute, *"The Way Forward: U.S. Industry Leadership in Response to Events at the Fukushima Daiichi Nuclear Power Plant,"* Nuclear Energy Institute, Institute of Nuclear Power Operations, Electric Power Research Institute, Washington, D.C., Atlanta, GA, Palo Alto, CA, (2012).
[9] Editor: T. H. Fanning, *"The SAS4A/SASSYS-1 Safety Analysis Code System: User's Guide,"* ANL/NE-12/4, Nuclear Engineering Division, Argonne National Laboratory, Argonne, IL, (2012).
[10] H. U. Wider et al., *"Status and Validation of the SAS4A Accident Analysis Code System"*, Proceedings of the LMFBR Safety Topical Meeting, European Nuclear Society, Lyon, France, July 19 23, 1982.
[11] D. J. Hill, *"SASSYS Validation Studies"*, Proceedings of the International Topical Meeting on Safety of Next Generation Power Reactors, American Nuclear Society, Seattle, WA, May 1 5, 1988.
[12] D. J. Hill and R. A. Wigeland, *"Validation of the SASSYS Core Radial Expansion Reactivity Feedback Model"*, Trans. Am. Nucl. Soc., 56, 380, (1988).
[13] D. J. Hill, *"SASSYS Analysis of EBR-II SHRT Experiments"*, Trans. Am. Nucl. Soc., 55, 421, (1987).
[14] Argonne National Laboratory, Nuclear Engineering Division. [Online] Available: http://www.ne.anl.gov/capabilities/rsta/nstf/index.shtml [Accessed: 2 February 2014].

[15] D. Langewisch, G. Apostolakis, and M. Golay, "*The NRC-MIT Collaborative Agreement for Advanced Reactor Technology, Task 2 – Uncertainties in Passive Safety System Performance*", NRC Accession Number: ML083450601, (2008).

[16] C.L. Smith and S.T. Wood, "*Systems Analysis Programs for Hands-on Integrated Reliability Evaluations (SAPHIRE) Version 8,*" NUREG/CR-7039, U.S. Nuclear Regulatory Commission, Prepared by Idaho National Laboratory, (2011).

[17] General Electric, "*PRISM Preliminary Safety Information Document*", GEFR-00793, UC-87Ta, San Jose, CA, (1987).

[18] S.A. Eide, et al., "*Industry-Average Performance for Components and Initiating Events at U.S. Commercial Nuclear Power Plants,*" NUREG/CR-6928, U.S. Nuclear Regulatory Commission, Washington, D.C., (2007).

[19] E.M. Blake, "*The Response to Fukushima Daiichi in the United States*", Nuclear New, vol. 57, no. 2, February, American Nuclear Society, La Grange Park, IL, (2014).

[20] R. R. Sherry, J. R. Gabor, and S. M. Hess, "*Pilot Application of Risk Informed Safety Margin Characterization to a Total Loss of Feedwater Event*", Reliability Engineering and System Safety, vol. 117, pp. 65-72, (2013).

[21] A. Hakobyan, T. Aldemir, R. Denning, S. Dunagan, D. Kunsman, B. Rutt and U. Catalyurek, "*Dynamic Generation of Accident Progression Event Trees*", Nuclear Engineering and Design, vol. 238, pp. 3457-3467, (2008).

[22] M. Kloos and J. Peschke, "*MCDET: A Probabilistic Dynamics Method Combining Monte Carlo Simulation with the Discrete Dynamic Event Tree Approach*", Nuclear Science and Engineering, vol. 153, pp. 137-156, (2006).

[23] K. Hsueh and A. Mosleh, "*The Development and Application of the Accident Dynamic Simulator for Dynamic Probabilistic Risk Assessment of Nuclear Power Plants*", Reliability Engineering and System Safety, vol. 52, pp. 279-296, (1996).

[24] A. Mosleh, et al., "*Guidelines on Modelling Common-Cause Failures in Probabilistic Risk Assessment,*" NUREG/CR-5485, INEEL/EXT-97-01327, U.S. Nuclear Regulatory Commission, Washington, D.C., (1998).

[25] U.S. Nuclear Regulatory Commission, "*Preapplication Safety Evaluation Report for the Power Reactor Innovative Small Module (PRISM) Liquid-Metal Reactor,*" NUREG-1368, (1994).

Visual Monitoring Path Forecasting for Digital Human-Computer Interface in Nuclear Power Plant and its Application

Hu Hong[a*], Zhang Li[a], Jiang Jian-Jun[b], Yi Can-Nan[a], Dai Li-Cao[b], Chen Qin-Qin[a]

[a] Ergonomics and safety management Institute, HuNan Institute of Technology, Hengyang, China
[b] Human Factors Institute, University of South China, Hengyang, China

Abstract: The operators sometimes can not judge next possible monitoring object which would lead to monitoring delay or transfer error in the monitoring digital human-computer interface (DHCI) parameter information process in nuclear power plant(NPP). For this purpose, the Markov process based forecasting path dynamic plan (FPDP) method which included forecasting path model, forecasting path plan algorithm and the calculation method of transfer path success probability was proposed. Then the monitoring transfer behavior of the operators when SGTR(Stream Generator Tube Rupture) occurred abruption accidents is analyzed based on the method proposed in this paper, taking the DHCI as the source node of monitoring task of t time, the transfer path to next monitoring object was obtained successfully to improve the efficiency of monitoring and to minimize the risk of monitoring error, which will also contribute to the analysis of the driving mechanism of operators' monitoring activities, to train simulated for monitoring behavior, and to optimize the digital man-machine interface.

Keywords: Nuclear Power Plant(NPP); Digital Human-Computer Interface; Monitoring Transfer; Forecasting Path; Markov Process

1 FOREWORD

Today, many NPPs are using or plan to use digital control system (DCS). For example, DCS has been successfully used in Jiangsu Tianwan NPP[1], also, the running NPPs of Hongyanhe and Lingdong, and NPPs under construction, that are Shandong Haiyang and Zhejiang Sanmen, all will employ DCS.
DCS, on the basis of computer calculation, is characterized by digital information display, highly integrated human-computer elements, multilevel data processing and automation operation from the view of human-computer interface. So it is more advanced than traditional control system with traditional simulation technique. There is enormous important information, ie parameter information, warning massage, procedures and operating information should be obtained and analyzed to make a decision through monitoring for DHCI in DCS, so operators' monitoring has been becoming the key step for the whole monitoring execution, also it is completely different with the traditional one[2].
DCS brings great challenge to operators in that monitoring error will cause a consequent series of errors in state assessment, response plan and manipulation. Statistics show that 60%-90% system errors and 50%-70% nuclear power accidents are caused by human errors[3].
With regard to operators' monitoring process, this paper proposes FPDP method to predict the next monitoring object according to the current system state so that operators are able to select and reach the next valid monitoring object more rapidly and exactly, it will provide much helping with reducing human errors, improving monitoring efficiency, and also contributing to analyze the driving mechanism of operators' monitoring activities and to optimize the DHCI.

2 FPDP MODEL IN MONITORING PROCESS

FPDP in monitoring process refers to how to rapidly and exactly obtain the next monitoring object each time after monitoring current state or parameter information for NPP system. Investigation and interviews with operators show that there is logical correlations existed between current monitoring object and the next one, namely, the selection of the next monitoring objects depends on the current

---

* Email: fengzhisu16@163.com

system state or parameter information. That is, operators' monitoring transfer only relates to the current state or parameter information, so Markov process is available to simulate. As the whole monitoring task process is changeable, namely, the monitoring path vary with operators' monitoring transfer, this paper proposes path selection algorithm and the calculation method of Markov transfer path success probability. Forecasting path with high relevance will be selected out of all possible paths as the next monitoring object, a dynamic plan model (Fig.1) is applied to describe the process.

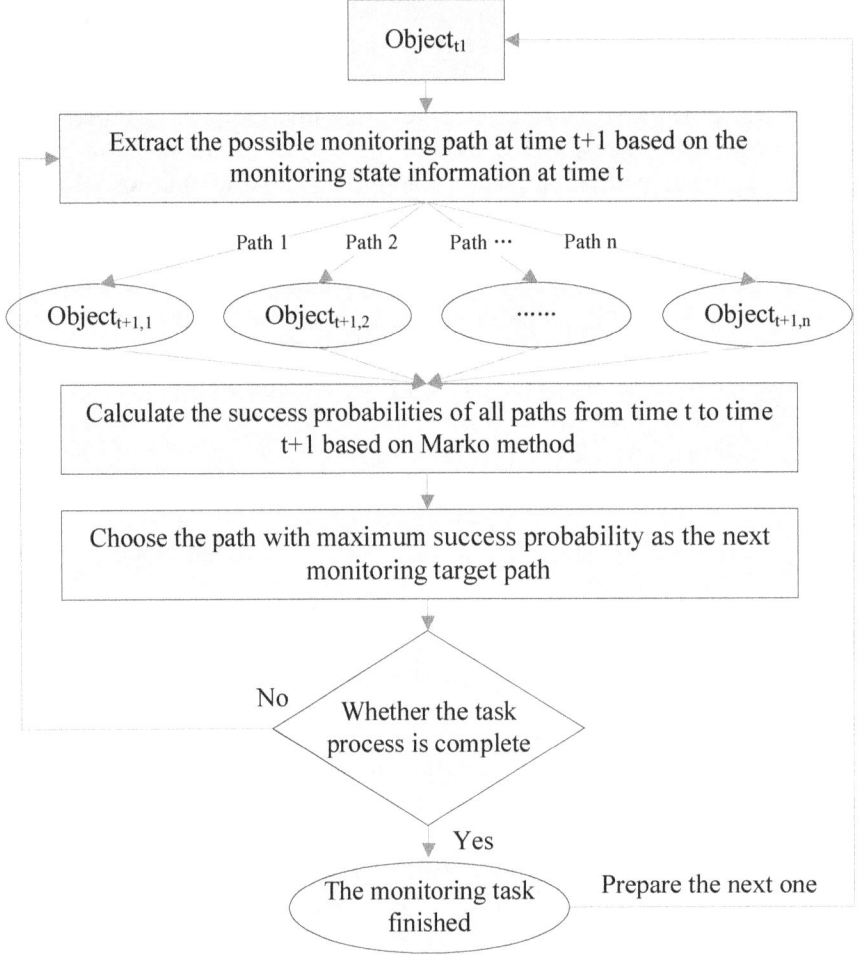

Notations:
$Object_t$: the state information of the monitoring object at time t
$Object_{t+1,n}$: the state information of the would-be monitoring object n at time t+1
Fig.1 Flowchart of FPDP model in monitoring DHCI of NPP

3 FPDP ALGORITHM IN THE MONITORING PROCESS ABOUT DHCI OF NPP

Fig.1 shows that FPDP algorithm searches effective plan through multi paths, obtains next possible monitoring object with maximum success probability and optimal way by dynamic plan method based on current monitoring object. The FPDP algorithm is specified as follows:
At t time, start from the initial object $Object_t$:
(1)Ini_Filter(t,t+1,task_info): screen out information being irrelevant to the task information of current time t and next time t+1, namely, the preliminary selection;
(2)Next_Obj_Pro(t,t+1,n): take monitoring task information at time t as the source point, count the success probability of n possible monitoring paths at time t+1;
(3)Values(t+1,n): file the success probabilities of n possible paths of time t+1 in Values array;
(4)Max(t+1,n): search the maximum success probability out of Values array;
(5)Transfer_route(t+1): Find the path with maximum success probability, and consider it as the next monitoring object;
(6)End(): If the monitoring task is finished, prepare for the next one and transfer to (1);otherwise to (2) and continue monitoring.

4 KEY ALGORITHM OF FPDP IN THE MONITORING PROCESS ABOUT DHCI OF NPP

Algorithm proposed in Chapter 3 shows that there are two key steps in the algorithm: Ini_Filter(t,t+1,task_info) and Next_Obj_Pro(t,t+1,n). We will interpret them in this part.

4.1 Ini_Filter(t,t+1,task_info) Algorithm

As operators have to face tremendous information in the monitoring process, it is essential to sieve out the irrelevant part to simplify search path and to reduce the exaggerated calculation time.

How to sieve out irrelevant part? As we know, tasks (information) with high similarity are relevant, while the others with low similarity are irrelevant. So the key step for operators' screening process is to evaluate tasks' similarity. Generally, operators judge the degree of similarity according to the index pertinence or the semantic similarity or the attribute resemblance of tasks. For example, Gao Ting discussed the grey correlation degree of customers' scarification under business model based on index system[4]; Zheng Yuhua adopted risk index to research the relevance of petroleum engineering projects[5]; Cui Qiwen studied similarity based on semanteme[6]. Inaccuracy would appear when we calculate the tasks' similarity, it would become impossible to extract similarity factors, so the calculation of similarity or association is not dealt in this study to keep the theme. This paper applies multi-branched model to screen.

(1) Build_tree(n): Set up a tree structure with n nodes in database. It is a process that should be analyzed by experts with repeated training and studying. The steps of building tree are: First, mark every tree to differ tasks, cause each tree has its own tree structure. Every node should be marked with its number and task feature, the root node denotes the task source. Second, start from the root node, if found that the current node information and some next task information is correlated, we choose the next correlated task node as the child node of current node. Third, take the child node founded in step forward as parent node and then try to find its child node, the search process would be circled in the same way introduced above until there is no relevant child node anymore. Thus a tree structure is formed, but it still need be repeated trained, studied and revised to form an expert system.

The computer pseudo-code expressions of above process are specified as below:

```
Initializing: name(tree), identifying(root_node);
Begin
 While(){
 For (i=first_adject_node;i<=total_adject_node;next_adject_node)
    If between current_node and adject_nodei is relative   Current_node_edge->= adject_node_i;
 }
  For(i=root_node;i<=all_nodes;i++)
If judge_expert(i,adject_each_node) not relative
      Delect(i,adject_edge_i);
     Else if tree_not_edge(i, adject_edge_i)
       Add(i,adject_edge_i);
End.
```

(2) Search(tree,relative_edge): Find out all of child nodes to the current task and mark them in terms of tasks' type and features. About the method of marking, two arrays are applied to respectively restore the information of current node and its child node sequences. The computer pseudo-code expressions of this process are specified as follows:

```
   begin
   Call(tree,number);
i=1; refers to the root node being the current node at the beginning
     For(j=first_child_node;j<=last_child_node;next_child,j++)
       {
   If exist_edge(current_node,child_i)
     { Crrent_array[i]=current_node_information_i;
```

```
            Child_array[j]=child_node_information_j;}
        }
                i++;
    }
End.
```
(3) build_calculation_path(parent_node,child_node): Get the information of the parent node and child nodes orderly from Crrent_array and Child_array to form the calculation path based the algorithm listed above, then we can achieve the success probability of every path.

4.2 Next_Obj_Pro(t,t+1,n) Algroithm

This part aims to calculate transfer path success probability from current node (parent node) to next possible object (child node). In monitoring DHCI, operators' monitoring transfer is influenced by relevant information of DHCI and monitoring manipulation, so it could be assumed that operator's monitoring is data-driven. If licensed operators have had enough knowledge to monitor, it can be concluded that the next monitoring state generally only relates to the current monitoring state, and that the monitoring transfers merely base on the current information, so operators' monitoring transferring has strong Markov property.

It is generally acknowledged that Markov consists of quadruples (X, A, P, R) , where X stands for states set, A for paths set, P for transfer probability between two states, R for expectation value[7]. The mathematical formulation of Markov is[8,9]: as to any integer n and any extraneous variable xi, if:

$$P(X_1 = x_{i1}, X_2 = x_{i2}, \cdots, X_n = x_{im}) > 0 \qquad (1)$$

then,

$$P(X_{n+1} = x_{in+1} | X_1 = x_{i1}, X_2 = x_{i2}, \cdots, X_n = x_{in}) = P(X_{n+1} = x_{in+1} | X_n = x_{in}) \qquad (2)$$

As to the transfer way from current state to the next, there are only two factors considered in this paper: one is the task information state, while the other is the decision process of operators. Transfer successful or not is decided by these two factors. So it is consistent with Multiplication Principle of probability theory, the transfer path error probability can be defined as:

$$\text{failure\_Tranfer\_path\_probility}(t,t+1) = P(\text{rask}_{t+1}|\text{rask}_t, \text{rask}_{t-1}, \cdots, \text{rask}_1) * p(\text{task}_{t+1}|\text{decision}_t, \text{decision}_{t-1}, \cdots, \text{decision}_1) \qquad (3)$$

Where rask_t is the current task information state at time t, rask_{t+1} is the information of possible task at time t+1, decision_t is decision process of the operators' selecting the path in monitoring. Eq. (3) indicates that the transfer process is mainly affected by current task state and operators' decision upon selecting the path in monitoring. The current task state reveals the physical properties of the system at that moment, while decision of selecting path is about the operators' mental activities. According to Eq. (2), we find that the transfer-influenced factors at time t+1 are only concerned with factors at time t, so Eq. (3) can be simplified as:

$$\text{failure\_Tranfer\_path\_probility}(t,t+1) = P(\text{rask}_{t+1}|\text{rask}_t) * p(\text{task}_{t+1}|\text{decision}_t) \qquad (4)$$

thus the transfer path success probability is:

$$\text{Succ\_Tranfer\_path\_probility}(t,t+1) = 1 - P(\text{rask}_{t+1}|\text{rask}_t) * p(\text{task}_{t+1}|\text{decision}_t) \qquad (5)$$

How to calculate the success probability according to Eq.(5)? We can see from Eq.(5) that if we know the methods to calculate $P(\text{rask}_{t+1}|\text{rask}_t)$ and $p(\text{task}_{t+1}|\text{decision}_t)$, we can easily get the value. So we will discuss their calculations followed.

(1) $P(\text{rask}_{t+1}|\text{rask}_t)$ Algroithm

As mentioned in the previous analysis, $p(\text{task}_{t+1}|\text{decision}_t)$ refers to the error probability caused by the transfer process of task information state from time t to time t+1, Δt denotes the time space between two states, in terms of relevant study[2], the error probability of Δt is obtained:

$$\begin{pmatrix} p(task_i(t)=0) \\ p(task_i(t)=1) \end{pmatrix} = \begin{pmatrix} e^{-F_P(t)\Delta t} & 0 \\ e^{F_P(t)\Delta t}-1 & 1 \end{pmatrix} \begin{pmatrix} p(task_i(t-1)=0) \\ p(task_i(t-1)=1) \end{pmatrix} \qquad (6)$$

Where $task_i(t)=0$ denotes the current normal state, $task_i(t)=1$ the current abnormal state, while $F_p(t)$ the monitoring error probability of $task_i(t)$.

For convenience of calculation, Eq. (6) can be simplified as listed in Fig.1.

Table 1 Calculation formulae of transferring failure ratio between two consecutive states

P{ task$_j$(t)| task$_j$(t-1) }		task$_j$(t-1)	
		0	1
task$_j$(t)	0	$e^{-F_P(t)\Delta t}$	0
	1	$e^{F_P(t)\Delta t} - 1$	1

(2) p(task$_{t+1}$|decision$_t$) Alogroithm

The term p(task$_{t+1}$|decision$_t$) refers that the operators' monitoring transfer to next object depends on operators' current decision error, namely, the value is the error probability at time t, so, p(task$_{t+1}$|decision$_t$) and p(decision$_t$) are mathematically equivalent, that is:

$$p(task_{t+1} | decision_t) \Leftrightarrow p(decision_t) \qquad (7)$$

The decision process is decisive to transfer in that decision error will consequently lead to transfer error and this process is mainly influenced by the physical properties of task information and operators' individual factors. This decision process influenced by multi-factors can be simplified as[7]:

$$P(decision_t)=P(decision_t|task\ character,\ human\_factors) \qquad (8)$$

Actually, Eq. (8) is a calculation under multi-conditions. According to relevant studies[7], the conditional probability with many parent nodes can be solved by conditional probability with single parent node, and the expressions are listed below:

$$P(n = S_{N_i} | M_1 = S_{M_1,P_1}, M_2 = S_{M_2,P_2}, \ldots M_K = S_{M_K,P_K}) = \lambda \sum (\prod_{j=1}^{j=k} P(N = S_{N_i} | M_j = S_{M_j,P_j})) \qquad (9)$$

Where, $\lambda = \sum_{i=1}^{n} P(N = S_{N_i} | M_1 = S_{M_1,P_1}, M_2 = S_{M_2,P_2} \ldots M_K = S_{M_K,P_K}$

How to judge whether a factor could be the decision influencing factor? Based on the author's study and experience in NPPs, analysis of the interviews with the operators, other experts' studies and the characteristics of this research itself [10,11,12,13], we conclude 6 factors being taken into consideration (Table 2).

Table 2 The correlation factors that influence operators' decision

affecting factors	Variables	affecting factors	Variables
Operators' knowledge and experience	d1	Operators' training level	d2
Task complexity	d3	Decision support system	d4
Stress level	d5	Time stress	d6

Toward to the factors in Table 2, combine Eqs. (8) and (9), then the further derivation expression of P(decisiont) is as following:

$$p(decision_t) = \sum_{i=1}^{i=6} [\sum (\prod_{j=1}^{k} p(decision_t | d_j))] \qquad (10)$$

Where i refers to the factor count, j refers to the state of factors.

5 CASE STUDY

This paper take SGTR accident of NNP as an example and its 3K00118YMA DHCI as the information source node at time t in the monitoring process. The laboratory equipments used in the experiment are eye tracking system (Tobii), virtual workstation, and accident simulation screen(developed by soft Visual studio.net). In terms of the FPDP method constructed above, the specific process in this case are:

(1) Build_tree(n)：set up tree structure of the monitoring path based on FPDP method;
(2) Take the DHCI of 3K00118YMA as the initial monitoring object at time t;
(3) According to Search(tree,relative_edge) algorithm, find out all of child nodes screen to the parent node screen(3K00118YMA) from the monitoring path tree of SGTR, then the parent node screen and the next possible child node screens will be obtained (Fig.2).

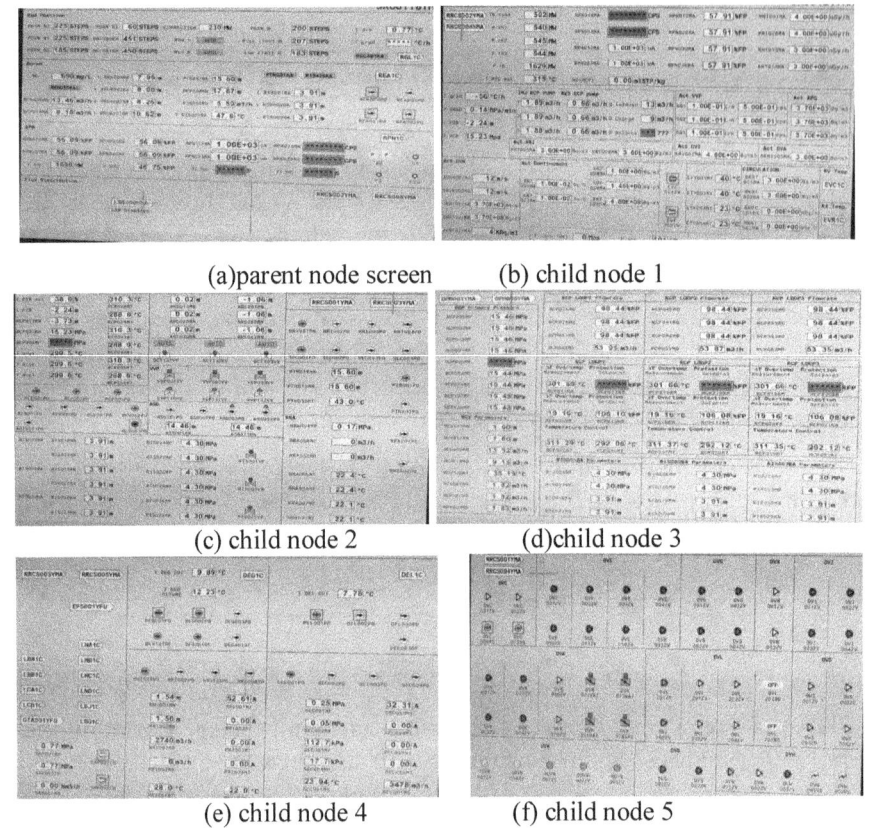

Fig.2 The father node and its child nodes selected in SGTR accident

(4) Get the tree structure of the parent node and the child nodes according to build_calculation_path(parent_node,child_node) (Fig.3).

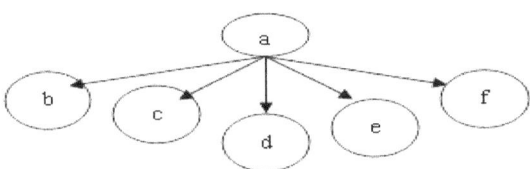

Fig.3 The father node and its child nodes selected in SGTR accident

Where the letter a in Fig.3 stands for the screen (a) in Fig2, in a similar way, letter b stands for screen (b) in Fig.2, ect.
(5)Next_Obj_Pro(t,t+1,n) algroithm
For ease of description, the transfer path success probability is separated into 2 parts: P(raskt+1|raskt) and P(decisiont). The follows are the respective calculations of them:
P(raskt+1|raskt) algroithm
When accident happens, the current task state and the next task state are abnormal, Eq. (12) can be obtained according to Table 1:

$$p(task_j(t)\,|\,task_j(t-1)) = e^{F_p(t)\Delta t} - 1 \qquad (12)$$

Parameters of parent node and each child node are obtained by experimental analysis and listed in Table 3:

Table 3 Parameter values of parent node and some child nodes obtained from experiment

$a \to b$		$a \to c$		$a \to d$		$a \to e$		$a \to f$	
$F_p(t)$	Δt	$F_p(t)$	Δt	$F_p(t)$	Δt	$F_p(t)$	Δt	$F_p(t)$	Δt
0.0008	29	0.0001	20	0.005	47	0.003	39	0.001	35

We can obtain the P(raskt+1|raskt) value of node pair (the parent node with a child node) according to Eq. (12) and Table 3(See Table 4).

Table 4 The results of the $P(rask_{t+1}|rask_t)$ value of node pair (the parent node with a child node)

Transferring path	$a \to b$	$a \to c$	$a \to d$	$a \to e$	$a \to f$
$P(rask_{t+1}|rask_t)$	0.023	0.002	0.265	0.124	0.035

① $P(decision_t)$ algroithm

Among the 6 decision affecting factors in Table 2, d1,d2,d5 and d6 are unrelated to the execution of the task but to the operators' themselves, that is, though it is irrelevant to the path, there would be different states for them in monitoring. In terms of [9] and [14], we can achieve the values of d1,d2,d5and d6 of different states, Table 5 can be obtained based on Eq. (8).

Table 5 The error probability of factor d1,d2,d5and d6, and their sum of products of different state

d_1			j=1	j=2	j=3	d_2	j=1	j=2	j=3
			0.25	0.002	0.001		0.01	0.02	0.001
Decision	j=1	0.1	0.025	0.00004	0.00001		0.003	0.0004	0.00003
	j=2	0.02	$P(decision/d_1)$= 0.02505				$P(decision//d_2)$= 0.00343		
	j=3	0.01							
	d_5		j=1	j=2	j=3	d_6	j=1	j=2	j=3
			0.03	0.01	0.003		1.0	0.1	0.01
Decision	j=1	0.1	0.003	0.0002	0.00003		0.1	0.002	0.0001
	j=2	0.02	$P(decision//d_5)$= 0.00323				$P(decision//d_6)$= 0.1021		
	j=3	0.01							

Note: j=1,2,3 refers to the state bejing high, middle and low.

According to the actual situation, factor d3 and d4 only have one sort of state, but their values vary with DHCI. Combined with expert judgment and actual experience in NPP, the values can be obtained in accordance with different transfer paths, the results are listed in Table 6 as follows:

Table 6 The error probability of factor d3 and d4 in different paths

Transferring path	$a \to b$	$a \to c$	$a \to d$	$a \to e$	$a \to f$
$P(decision/d_3)$	0.06	0.001	0.9	0.6	0.2
$P(decision/d_4)$	0.001	0.005	0.1	0.08	0.04

In accordance with Table 5, Table 6 and Eq. (11), the values of $P(decision_t)$ of each transfer are listed in Table 7.

Table 7 The error probability of $P(decision_t)$

Transferring path	$a \to b$	$a \to c$	$a \to d$	$a \to e$	$a \to f$
$P(decision_t)$	0.1948	0.13981	1.13381	0.8138	0.3738

③ The results of Succ_Tranfer_path_probility(t,t+1)
Based on Tables 4 and 7, Eq. (5), the success probability of each path is listed in Table 8:

Table 8 The successful probability of each transferring path

Transferring path	$a \to b$	$a \to c$	$a \to d$	$a \to e$	$a \to f$
Succ_Tranfer_path_probility(t,t+1)	0.996	0.998	0.7	0.899	0.987

(6) Max(t+1,n)
Select the maximum transfer probability from Table 8, we can find that the maximum is 0.998.
(7) Transfer_route(t+1)
Based on step (6), we can know that the next transfer path would be (a) → (c), that is, the parent node screen (a) transfers to the child node screen (c)(child node screen 2 in Fig.2)
(8) END().
According to the FPDP method for DHCI of NPP proposed in this paper, operator's transfer path was obtained successfully through monitoring 3K00118YMA in the SGTR accident. The forecasting result is in consistent with path transferring (frame (a) transfers to the frame (c) in Fig.2) of responding manipulation instruction given by operators in an actual SGTR accident of a NPP, as well as the transferring path achieved by video investigation and eye tracking system analysis when operators handle with SGTR accident in simulator. Therefore, the FPDP algorithm proposed in this paper is available.

6 CONCLUSION

This paper proposed FPDP method for the monitoring transferring process of DHCI of NPPs, which consists of plan mode, execution flow and key algorithm and so on. Then, the method is applied to analyze SGTR accident of digital NPP and the predicted monitoring transferring path which is consistent with actual situation. Thus the method proposed herein does helpful to decrease operator's monitoring errors, which will also contributes to analyze the driving mechanism of operators' monitoring activities, to train simulated for monitoring behavior, to optimize the digital man-machine interface, to analyze the monitoring behavior on other fields: radar, intelligent robot, aerospace, and so on.
However, there exists some limitations as to this method, for example, we only select 6 factors that influences operators' decision, undoubtedly, there are more than 6 factors that would influence decision; also the value of the factors adopted from the existed data or experts judgments; meanwhile, the formation of the tree structure still need improving to reduce the complexity of inquiry.

Acknowledgements

The research presented in this paper is supported in part by National nature Science Foundation (Nos. 70873040, 71071051)of China, research projects of Ling Dong Nuclear Power Company Ltd. (KR70543) of China, and Innovation Ability Construction Projects based on the new Industry- Academy- Research Cooperation of HuNan Province（2012GK4101）.

References
[1] Yu J B. *"The Characteristics of the Digital Contorl System in Nuclear Plant"*, Modern Science, 22, pp. 48-49, (2008).
[2] Jiang J J, Zhang L, Wang Yq, Peng Y Y, Zhang K, He W. *"Markov Reliability Model Research of Monitoring Process in Digital Main Control Room of Nuclear Power Plant"*, Safety Science, 49, pp. 843-851,(2011)

[3] Jiang J J, Zhang L, Wang Y Q, Yang D X , Zhang K, He W. *"Association Rules Analysis of Human Factor Events Based on Statistics Method in Digital Nuclear Power Plant"*,afety Science, 49, pp. 946-950,(2011)

[4]Gao T. *"Gray Correlation Assessment of Customer Satisfaction on c2c E-business Model"*, Modern Business Trade Industry, 24, pp. 368-369(2011)

[5]Zheng Y H, Luo D K. *"Gray Correlation Assessment of the Risk of Petroleum Engineering Project Based on Utility Theory"*,Project Management Technology, 9, pp. 100-103(2011)

[6]Cui Q W, Xie F. *"An Improved Computational Method for Method for Conceptual Semantic Similarity in Domain Onthlogy "*, Computer Applications and Software, 29, pp. 173-175(2012)

[7] Luiz Guilherme Nadal Nunes, Solon Venaˆncio de Carvalho,Rita de Ca´ssia Meneses Rodrigues. *"Markov decision process applied to the control of hospital elective admissions"*, Artificial Intelligence in Medicine, 47, pp. 159—171(2009)

[8]Ran J J, Zhao Y J,Liang C. *"The Application of the Prediction of Precipitation State i Based on Weighted Markov Chain"*,Yellow River, 28,pp. 32-34(2006)

[9] Jiang J J, Zhang L, Wang Y Q, Peng Y Y, Zhou C, Zhang K. *"Research on Cognitive Reliability Model for Main Control Room Considering Human Factors in Nuclear Power Plants"*, Nuclear Power Engineering, 33, pp. 66-73(2012)

[10] Shen S H. Smidts C, Mosleh A. *"A methodoloty for collection and analysis of human error data based on a cognitive model:IDA"*, Nuclear Engineering and Design, 172, pp. 157-186(1997)

[11]Massimo B. *"Assessment of human reliability factors:A fuzzy cognitive maps approach"*, International Journal of Industr'ial Ergonomics, 37,pp. 405–413(2007)

[12]Y.H.J. Chang, A. Mosleh. *"Cognitive modeling and dynamic probabilistic simulation of operating crew response to complex system accidents. Part 2:IDAC performance influencing factors model"*, Reliability Engineering and System Safety, 92,pp. 1014-1040(2007)

[13]Y.H.J. Chang, A. Mosleh. *"Cognitive modeling and dynamic probabilistic simulation of operating crew response to complex system accidents. Part 4: IDAC causal modelof operator problem-solving response"*, Reliability Engineering and System Safety, 92,pp. 1061-1075(2007)

[14] Mahadevan S, Zhang R X, Smith Natasha. *"Bayesian Network for System Reliability Reassessment"*,Structural Safety, 23,pp. 231-251(2001)

Individual Differences in Human Reliability Analysis

Jeffrey C. Joe[a*] and Ronald L. Boring[a]
[a] Idaho National Laboratory, Idaho Falls, ID, USA

Abstract: While human reliability analysis (HRA) methods include uncertainty in quantification, the nominal model of human error in HRA typically assumes that operator performance does not vary significantly when they are given the same initiating event, indicators, procedures, and training, and that any differences in operator performance are simply aleatory (i.e., random). While this assumption generally holds true when performing routine actions, variability in operator response has been observed in multiple studies, especially in complex situations that go beyond training and procedures. As such, complexity can lead to differences in operator performance (e.g., operator understanding and decision-making). Furthermore, psychological research has shown that there are a number of known antecedents (i.e., attributable causes) that consistently contribute to observable and systematically measurable (i.e., not random) differences in behavior. This paper reviews examples of individual differences taken from operational experience and the psychological literature. The impact of these differences in human behavior and their implications for HRA are then discussed. We propose that individual differences should not be treated as aleatory, but rather as epistemic. Ultimately, by understanding the sources of individual differences, it is possible to remove some epistemic uncertainty from analyses.

Keywords: HRA, Individual Differences, Human Performance, Crew Performance.

1. INTRODUCTION

The nominal model of human error in human reliability analysis (HRA) typically simplifies the variability in how operators respond to initiating events. In the context of procedure use, human success and failure are often treated dichotomously in HRA methods, thereby implying that there is one path to modeling human error (i.e., the operator either used the procedure correctly or s/he did not). Furthermore, individual differences in operator behavior, and crew-to-crew variability in performance, are typically treated as aleatory (i.e., random) uncertainty. Given this treatment as an aleatory factor, most HRA methods address this issue by using a measure of central tendency to represent the characteristics of an "average" operator or crew.

However, there is a considerable amount of literature that shows individuals and teams vary significantly in their performance, and that there is observable and reliable relationship between this variability in performance and antecedent factors. For example, one of the key findings in the international HRA empirical study [1, 2] was that there was variability in how crews responded in given plant upset scenarios. Given the same scenario, the same indicators, the same procedures, and the same training, crews varied in their decisions on what actions to carry out. Braarud and Kirwan [3] further noted that variability in crew performance in the international HRA empirical study was closely related to task complexity in that greater variability in crew performance was seen in more complex scenarios than in simpler ones. Heimdal [4] also noted research showing an interaction between task complexity and procedure adherence. Crews that followed procedures verbatim were faster at simple tasks, but slower on complex tasks. Crews that operated in a culture where verbatim compliance was not required tended to be faster and more accurate on complex tasks, but slower on simple tasks. This finding implies that crew-to-crew variability in performance on simple and complex tasks can vary significantly simply as a function of procedure usage and adherence. Heimdal's findings also show that crew-to-crew variability in performance is present even in simple

---

[*]Corresponding Author: Jeffrey.Joe@inl.gov

tasks, thereby demonstrating that the assumption in HRA that crews will perform the same on routine tasks is not always correct.

Toquam, Macaulay, Westra, Fujita, and Murphy [5] studying nuclear power plant crews in Japan also observed performance differences between nuclear power plant crews, and noted that three primary antecedent factors contributed to variability in performance between teams. These antecedent factors were task characteristics (e.g., how routine and simple versus unusual and complex is the task the team must perform), team member characteristics (e.g., intelligence, personality types, specific cognitive abilities, etc.), and team dynamics, or team characteristics, such as group cohesion and communication practices. Additionally, Gertman, Haney, Jenkins, and Blackman [6] found that the emotional stability of individual crewmembers' personalities was related to their future performance. Research by Itoh, Yoshimura, Ohtsuka, and Matsuda [7] showed that perceptual speed and memory (i.e., cognitive abilities) are also related to individual nuclear power plant crewmember performance, and as a result, overall team performance. Similarly, many studies show that group cohesion or lack thereof, can have an effect on team performance (for a review, see Evans and Dion [8]).

Recent work by Massaiu, Hildebrandt, and Bone at the Halden Reactor Project [9] also found that behavioral differences can arise in crews operating nuclear power plants, particularly in crew decision-making during emergency operation. They developed the Guidance Expertise Model (GEM), which explains crew behavior, and differences in crew behavior, in terms of the interaction of internal resources and external resources during different control modes (i.e., general or macrocognitive behaviors). Specifically, the two control modes in GEM are (1) narrowing and (2) holistic view, and roughly correspond to times in which the operators are performing procedure driven tasks, and times when operators are trying to ascertain what the bigger picture is for the emergency situation, respectively. Furthermore, whether the crew is engaged in either the narrowing or holistic view control mode depends on external resources (e.g., the availability and relevance of procedures), and internal resources, such as experience and teamwork (i.e., the ability for teams of people with different personalities and proclivities to communicate and work together effectively). With this model guiding their research, they confirmed that even under the same simulator trial and with the same emergency procedures, nuclear power plant crews varied in their performance as a function of the availability and interaction of external and internal resources.

Though this is not a study on variability in individual or crew performance, research by Galyean [10] noted similar issues in HRA in terms of handling individual differences. The focus of that paper was on the use of performance shaping factors (PSFs) that HRA uses to modify a human error probability (HEP) for a given event. Galyean noted that most HRA methods conflate differences between people and differences in the environment in their PSF taxonomies. For example, his analysis noted many HRA methods use a workload PSF, but that the effect size of workload on a nominal HEP is a function of the person performing the task, the task, and the environment. That is, a task that is known to increase workload, such as performing a complex mathematical calculation, likely depends on differences in cognitive ability, such that the overall workload level of those with higher cognitive ability is affected less than those with lower cognitive ability. As such, the workload PSF conflates influences on the nominal HEP that are attributable to differences between people and the differences in the task being performed. To remedy this issue, Galyean proposed a set of three orthogonal PSFs: population capacity, organizational environment, and event specifics, whereby population capacity is the PSF that directly considers differences related to the individual.

Overall, all of the research summarized highlights an important area of human performance that is rarely considered in HRA: individual differences in operator behavior (e.g., responses to initiating events). While HRA methods include uncertainty in quantification, there is the inherent assumption that well trained crews will tend to vary little and that any differences between crews are simply aleatory (i.e., random). While this assumption holds true in some areas (e.g., the ability to carry out well defined actions for understood phenomena), complex situations that go beyond training and procedures will garner differences in responses according to individual operators' understanding and decision making. Control room phenomena – including the communication patterns between operators, the rigidity of the control room command structure, crew openness to new ideas, and safety culture – are areas that can vary considerably across crews and plants. These differences are just a few of the many important antecedents that contribute to individual differences in behavior, and these antecedents and their effects on operator performance must be considered within HRA in order to achieve valid analyses.

This paper explores examples of individual differences taken from operational experience and the psychological literature. The impact of these individual differences in human performance and their implications for HRA is then discussed. We propose that individual differences should not be treated as aleatory, but rather as epistemic – an area of modeling uncertainty caused by lack of knowledge. Ultimately, by understanding the sources (i.e., antecedents) of individual differences, it is possible to remove some epistemic uncertainty from analyses.

2. PSYCHOLOGICAL RESEARCH ON INDIVIDUAL DIFFERENCES

One common feature of the studies on nuclear power plant crews and individual operators in Section 1 is that the antecedent factors contributing to individual differences in behavior originate from one of three general categories: (a) personality traits, (b) characteristics of the environment, or (c) the interaction of both. Examples of personality traits identified as contributing to individual differences in behavior include intelligence [5] and emotional stability [6]. Examples of environmental characteristics include task complexity [3, 5], and availability and relevance of procedures [9]. Examples of the interaction of the person and environment include the finding that crew-to-crew variability in their performance on simple and complex tasks depends on how they use and follow their procedures [4]. We also argue that team dynamics, such as group cohesion and communication practices, are a kind of interaction between the person and the environment, in that a person is interacting with someone else, who is part of their working environment.

This concept of manifested behavior being a function of both the person (e.g., their innate traits and abilities) and the environment is not new. Kurt Lewin made this observation in his book, *Principles of Topological Psychology* [11], where he presented his equation $B = f(P,E)$, which is a simple heuristic formula for 'behavior is a function of the person in their environment'. In the context of individual differences in behavior, Lewin's equation posits that differences arise when different people interact with the ever-changing environment. Given this, one implication of Lewin's equation is that differences in the observed behavior of two different people can occur even when both people are in the same environment, because individual differences can be due to differences between people. Thus, the fact that individual differences can arise solely because of differences between people explains why in the international HRA empirical study [1, 2], despite the fact that all nuclear power plant crews were given the same scenarios and procedures to follow (i.e., the same environment), there were differences observed in their behaviors. This section elaborates on some of the seminal findings in psychology on the nature of individual differences, and in particular, the observable and reliable relationship between antecedent factors, such as personality traits and differences in abilities, and subsequent differences observed in behavior.

2.1. Personality Psychology

The field of psychology has a long and storied history, whereby experimental and clinical psychologists have created a vast scientific literature that has both spanned the breadth and plumbed the depths of human behavior. One of the paradoxical findings of psychology is that some aspects of human behavior are generalizable to all, some aspects of human behavior are common to a subset of the population, and some aspects are unique to a person. As Murry and Kluckhohn [12] put it: "Every man is in certain respects (a) like all other men, (b) like some other men, (c) like no other man."

Because psychology provides this particular insight into what makes humans (a) similar (i.e., nomothetic), (b) similar only with certain unique groups, and (c) unique (i.e., ideographic), some have criticized psychological research as "schizophrenic" in nature in the sense that it appears to draw contradictory conclusions about the nature of human behavior. Others, however, simply view these three foci of psychological research as complementary endeavors. In fact, we argue it is not possible to understand what is nomothetic about humanity, or most any other tangible object for that matter (e.g., pumps, valves, turbines, steam generators, reactor vessels, containment structures, etc.) without understanding what is ideographic, and vice versa. An understanding of one requires knowledge of the other. The discovery of individual differences in a person's behavior, or the ways in which a person is like some other people, could not have occurred without understanding what is nomothetic about that person as well.

As with medicine, where physicians can specialize in different sub-fields (e.g., oncology, pediatrics, neurology, etc.), psychology also has areas of specialization. Personality psychology, more so than cognitive, social, or clinical psychology, has focused on understanding the full range of ways in which people are all alike, like some (but not others), and like no one else. Personality psychology currently studies these three foci through a number of different theoretical lenses. Some of the major theoretical perspectives include the psychoanalytic (i.e., Freudian) approach, behavioral (i.e., Skinnerian) approach, biological approach, dispositional/trait approach, humanistic approach, and the cognitive approach.

While there is value in researching personality through all of the different theoretical lenses, for purposes of understanding individual differences in HRA, special attention should be given to the cognitive approach to personality psychology, primarily because most of the second-generation HRA models (e.g., SPAR-HRA [13], MERMOS [14], and CREAM [15]) are based on a cognitive psychological understanding of human behavior. That is, a central premise of these second-generation HRA models is that in order to model human behavior, it is important to understand how humans cognitively represent and process information. Furthermore, while the terminology varies between cognitive theories and models of personality, the common features among all of them are that information that is external to the individual is cognitively represented and processed in stages, and that stable personality traits of the individual can influence or bias the way in which information is processed. The first stage is when the information is detected or noticed (i.e., information is collected). In the next stage, cognitive effort is applied to interpret or make sense of the information. It is within these first two stages that information is recast into a mental representation, which is a structured way of reorganizing the information to make that information easier to comprehend, as well as store and recall in memory. The third stage involves the human making a decision about what course of action to take, and the final stage is usually described in terms of the human performing the behavior or action.

It is important to note in these cognitive information processing models within personality psychology that antecedent factors (e.g., personality traits, perceptual abilities, etc.) influence an individual's information processing, and that each stage of information processing represents the opportunity for ideographic variability. At the perceptual stage, physiological differences (e.g., visual acuity or color blindness) or experience (e.g., learned perceptual biases) may shape the way different people take in sensory information. At the cognitive stage, knowledge, temperament (e.g., emotive vs. analytical), and experience shape the way decisions are made, and the way decision options are weighted (e.g., near-term vs. long-term strategy). Finally, the individual may even approach action in response to decisions in different manners as a function of some aspect of their personality (e.g., reserved vs. impulsive).

Said in more specific theoretical terms, the cognitive approach to personality psychology has as a fundamental premise the notion that humans create mental representations of, and process information in stages, and that individual differences in behavior can arise from: (1) differences in the external environment (e.g., variability in external stimuli), and (2) the subtle, but meaningful, antecedent cognitive factors that lead to differences in the ways in which people represent and process information. Two well-known and broadly accepted cognitive theories of personality and individual differences are Kelly's Personal Construct Theory [16] and Mischel and Shoda's Cognitive-Affective Systems Theory [17]. Both theories posit that people have mental representations that help people detect, process, organize, and then act upon information, but that personality traits and other antecedent factors lead to individual differences in the formation of people's mental representations. That is, no two people are exactly identical in their genetic make-up, cognitive abilities, and their life experiences, and as a result, even in the same situation, with the same training, people can process incoming information differently. Some examples of potential differences at each stage of information processing are listed below.

Two people in the same situation, with the same training may nevertheless:
1. Detect, notice, and/or attend to different bits of information that are part of the more complex feature set of a given situation
2. Have differing abilities to recall information stored in their respective memories that is needed to aid in comprehending what the bits of information are
3. Evaluate the meaning and importance the various bits of information differently
4. Have different attitudes towards decision making (i.e., one may be risk-averse, and the other risk-seeking)
5. Have different preferences for engaging in some actions and not others (e.g., one may prefer habitual behaviors over novel behaviors)

In short, the cognitive approach to personality psychology explains individual differences as a function of differences in people's mental representations of information. As Burger [18] put it, "A Christmas tree can remind one person of his or her religious values, another of family and seasonal joy, and a third of sad memories from childhood." (pg. 473). These different interpretations of a Christmas tree are a function of different people imparting a different meaning or symbolic significance to it. The cognitive approach to personality psychology would ascribe these differences in meaning or symbolic significance to differences in how people cognitively organize and represent the salient information they can recall from memory that is activated once a person sees a Christmas tree.

2.2. Individual Difference Effects in Information Processing

This section summarizes a number of research studies that demonstrate how antecedent factors (e.g., personality traits, cognitive abilities, etc.) affect information processing; specifically during the processing stages of detecting, processing, organizing, and then acting upon information. One antecedent factor that has been shown to have a robust effect across a number of stages of information processing is cognitive ability (i.e., "intelligence"). First, it should be clear that there is general consensus within the field of psychology, and in society in general, that there are individual differences in cognitive ability. Given this, Stanovich and West [19] studied the relationship between cognitive ability and the ability to, "Evaluate the quality of an argument independent of one's feelings and personal biases about the proposition at issue" (pg. 351), and found that those with greater cognitive ability were better able to suppress their biases when evaluating argument quality versus those with lesser cognitive ability. Additionally, later work by Stanovich and West [20] as cited in Rachlinski [21] found that differences in cognitive ability affected the propensity to make certain kinds of cognitive errors, such as the conjunction fallacy (i.e., Kahneman [22]), and the ability to solve complex problems (e.g., the Wason card selection task). Research by Rachlinski [21] also reported that those with greater cognitive ability made fewer cognitive errors and were less susceptible to using faulty reasoning when solving the complex problems.

There are also a number of other research studies that have looked at how antecedent factors affect the decision-making stage of information processing. Research by Lauriola, Levin, and Hart [23] examined individual differences in decision-making as a function of a person's tolerance for uncertainty. In this research, decision-making under uncertainty is defined as having two distinct subcomponents: decisions under ambiguity, and risky decision-making. Risky decision-making is akin to the kinds of decisions studied by Kahneman and Tversky [24]: where the probabilities of all possible outcomes were known, but the outcome itself was unknown. Decisions under ambiguity are decisions that people must make when the probabilities of outcomes are unknown, and the outcome is also unknown. What Lauriola, Levin, and Hart [23] discovered was that there is a stable dispositional (i.e., personality) trait that explains differences in decision-making where uncertainty is high. First, they were able to assess that some people are generally more risk seeking and others are generally more risk-averse across a number of situations. Then, they were also able to demonstrate that when risk seeking and risk-averse people were presented the same ambiguous situation (where uncertainty was high), that those who were risk seekers consistently made riskier decisions than those who were risk avoiders.

3. IMPLICATIONS OF INDIVIDUAL DIFFERENCES FOR HRA

3.1. General Implications for HRA

Given the literature reviewed in Section 2, there are a number of general implications for the treatment of individual differences in HRA, which are summarized here. First, it is interesting to note that the orthogonal PSFs developed by Galyean [10] share some similarities with Lewin's equation, $B = f(P,E)$ [11], in that the PSF population capacity is similar to P, and the PSF organizational environment is similar to E. The model by Massaiu, Hildebrandt, and Bone [9] (i.e., GEM) is also similar to work by Lewin in that crew behavior is a function of their personalities/internal resources and environment/external resources, and that differences in behavior can arise from either the person, the environment, or the interaction of person and environment. The fact that [9] found crew-to-crew variability as a function of the interaction between internal and external resources gives credence to the notion that factors such as personality traits and task complexity are antecedent factors that contribute to subsequent variability in crew performance.

The research by Stanovich and West [19, 20] and Rachlinski [21] demonstrates how individual differences in cognitive abilities can have a significant and meaningful impact on behaviors, specifically the rate and kinds of mental errors people can make in decision-making. Based on this research on cognitive information processing, a greater cognitive ability generally means that a person is generally better able to:

- Detect/notice/attend to more bits of information that are part of the more complex feature set of any given situation
- Process/comprehend information more quickly and accurately
- Organize/evaluate information more efficiently and accurately

Given these findings, one general implication for HRA is that while HEPs for performance on discrete tasks are typically calculated as if everyone (e.g., nuclear power plant operators) were all alike, it is quite likely that there are a number of meaningful differences between operators that may affect the operator's likelihood for making an error, and therefore affect the calculated HEP for a given task. For example, the research by Lauriola, Levin, and Hart [23] showing how differences in tolerances for uncertainty differentially affect decision-making supports the assertion that the nomothetic treatment of operators is likely an oversimplification. As such, the quantification of HEPs may be an oversimplification in that it is typically based on a single point measure of central tendency, when it is clear there is meaningful variability around that central point.

3.2. Practical Implications for HRA

There are four key areas where individual differences affect the assumptions in HRA:

1. *Human failure events (HFEs):* HFEs are those human errors that are risk significant and contribute to the overall failure likelihood of a system. In much contemporary HRA practice, HFEs are treated as a single path of operator behavior. For example, an HFE might be defined as the failure to isolate a cooling system. Assumed within defining the HFE this way is the idea that the path that led to this failure centered on the operator's failure manually to engage

a close valve switch. However, there are other possible ways such an error could occur (e.g., operator actuates the wrong valve switch). HFEs must not be artificially constrained to a single course of operator action; rather, they should consider multiple paths the operator could take. The ATHEANA HRA method [25] proposes consideration of multiple paths to various outcomes—deviation paths from the nominal path. Such an approach is, in fact, supported by evidence from the international HRA empirical study [1, 2], in which crews followed different paths either to success or failure in a scenario. It should be noted that such multiple possible paths has the opportunity to shape dependence modeling in the HRA, since one path will shape the likelihood of subsequent paths, which potentially primes particular HFEs depending on the path taken.

2. *Performance shaping factors (PSFs):* Performance shaping factors are those aspects of a situation, task, or individual that influence the likelihood of a particular behavior. Most HRA methods consider PSFs, but few consider the extent to which PSFs may manifest differently on the individual. For example, the SPAR-H method [13] treats PSFs as a set of multipliers on a nominal HEP. These multipliers are invariant across individuals. Thus, the effect of high stress is assumed to be constant across individual operators. Indeed, a sharp reduction in performance is accepted in the psychological literature to be a reflection of high stress levels (i.e., the so-called stress cliff). However, there also exists a point where stress can be considered to enhance performance (i.e., eustress). In fact, the optimal level of performance is usually at a level of moderate stress. Within HRA, there is no consideration of the individual thresholds for different levels of stress. One operator may have a high threshold of stress, while another may have a much lower threshold. In other words, given the same situation, one operator may find him/herself at the point of optimal performance, while another operator may find his or her performance sharply degraded by stress. Stress perception is individual; yet, there is no way in SPAR-H or other methods to treat the effects of stress differently across operators. PSFs should not be applied in a manner that overlooks individual differences in the effect of performance.

3. *Human error probabilities (HEPs):* A consequence of not considering individual differences is that a single probability distribution is used when calculating the HEP. In fact, given the possibility of different outcomes due to individual factors, it should be assumed that in many cases the true distribution is multimodal. There is a range of possible outcomes, which is being treated as noise or uncertainty in much of the HRA that is performed. By aligning the HEP to the actual performance range of individual operators, the HRA turns much of what is treated as aleatory uncertainty into epistemic certainty.

4. *Individual vs. crew performance:* It must be noted that individual differences can be modeled to account for both *intra* and *inter* crew differences in performances. Intra crew differences are, of course, those differences that occur because of variability in individuals, while inter crew differences are those sources of variability that occur because crews perform actions differently. There are a number of factors that influence crew dynamics, (e.g., cohesiveness and communication styles), which are simply the meta-level manifestation of the micro-level differences between individuals in the crew. The central point here is that both individuals and crews manifest ranges of behaviors. To date, much of the psychological literature has focused on differences at the individual level, but newer literature [26, 27] is looking at group behaviors such as team cognition and other aspects of team performance. This distinction between individual and crew performance may not be as crucial to HRA at this stage as acknowledging and accounting for variability in performance between individuals or crews.

While the above points emphasize the importance of considering individual differences, not everything is susceptible to individual differences. Reactor operators, for example, are screened through a rigorous curriculum for mental capacity, while the idiographic nature of their performance is minimized through extensive training. Thus, for well-trained and well-understood situations, there is no reason to assume that crews with similar capabilities, background, and training using the same control room and the identical procedures would perform significantly differently. In off-normal situations that present some degree of ambiguity or complexity in diagnosis, certain differences might be expected to surface. HRA has to date done a good job of accounting for nominal behavior in crews. As HRA matures and as it is fine-tuned to reflect a greater range of scenarios (e.g., Level 2 and Level 3 HRA), it becomes important to enhance HRA to capture the spectrum of possible outcomes, including those attributable to individual differences.

4. CONCLUSION

We have attempted to demonstrate in this paper that:

- Person-to-person and crew-to-crew variability in behavior exists
- There is an observable, reliable, and therefore systematically measureable relationship between antecedent factors (e.g., personality traits, communication practices, etc.) and variability in person and crew performance.
- This variability has meaningful and risk significant effects on operator and crew performance

This means it is not necessary to treat variability in performance (i.e., individual differences) as aleatory. Individual differences can be treated as epistemic uncertainty, and can therefore be reduced through the acquisition of more knowledge.

We are not alone in making these points. The researchers involved in the international HRA empirical study [1, 2] drew the same general conclusion from their first experiment, and proposed improvements to HRA methods. Namely, while HRA methods typically do not factor in variability in performance when estimating HEPs, they suggested that sensitivity studies assessing the epistemic effects of antecedent factors on performance variability should be performed. These sensitivity studies would vary an antecedent known to contribute to performance variability (e.g., task complexity, procedural guidance), and then estimate the range of effects it has on performance. Doing this would, "Provide additional insights into the regions of validity of the methods and identify potential improvements in the use of the HRA models." (pg. 3-116). Not only do we concur with this suggestion, we have also proposed elsewhere [27] that there are many aspects of team performance and team dynamics (e.g., communication) that can be modeled using event trees and fault trees, thereby making these factors less aleatory and more epistemic in our understanding of their effects. Specifically, we argued that: (1) errors in teamwork are different than individual errors, (2) teamwork errors contribute to plant risk in ways that are unique from individual errors, and (3) that these teamwork errors can be meaningfully modeled and quantified such that the adverse effects of their under-specification in HRA can be reduced. These are two research ideas that propose concrete and actionable steps that HRA methods can do to consider more completely the effects of variability in operator and crew performance on the assessment of human contributions to plant risk. Of course, we recognize there are many other ideas that can help address the issue of individual differences, and we are supportive of any research that helps achieve the goal of improving HRA.

Acknowledgements

INL is a multi-program laboratory operated by Battelle Energy Alliance LLC, for the United States Department of Energy under Contract DE-AC07-05ID14517. This work of authorship was prepared as an account of work sponsored by an agency of the United States Government. Neither the United States Government, nor any agency thereof, nor any of their employees makes any warranty, express or implied, or assumes any legal liability or responsibility for the accuracy, completeness, or usefulness of any information, apparatus, product, or process disclosed, or represents that its use would not infringe privately-owned rights. The United States Government retains, and the publisher, by accepting the article for publication, acknowledges that the United States Government retains a nonexclusive, paid-up, irrevocable, world-wide license to publish or reproduce the published form of this manuscript, or allow others to do so, for United States Government purposes.

The views and opinions of authors expressed herein do not necessarily state or reflect those of the United States government or any agency thereof. The INL issued document number for this paper is: INL/CON-14-31940.

The authors gratefully acknowledge the late Dana Kelly, who championed early discussions on individual differences in HRA. Regrettably, Dr. Kelly passed away before we had the opportunity to complete the paper with his invaluable input. We dedicate this paper to his memory.

References

[1] E. Lois, V. Dang, J. Forester, H. Broberg, S. Massaiu, M. Hildebrandt, P. O. Braarud, G. Parry, J. Julius, R. Boring, I. Mannisto, and A. Bye, *"International HRA Empirical Study – Phase 1 Report: Description of Overall Approach and Pilot Phase Results from Comparing HRA Methods to Simulator Performance Data,"* (NUREG/IA-0216, Vol. 1). U.S. Nuclear Regulatory Commission, 2009, Washington, DC.

[2] S. Massaiu, A. Bye, P.O. Braarud, H. Broberg, M. Hildebrandt, V. Dang, E. Lois, and J. Forester, *"International HRA Empirical Study, Overall Methodology and HAMMLAB Results,"* In A. B. Skjerve & A. Bye (Eds.) *Simulator-Based Human Factors Studies Across 25 Years.* Springer-Verlag, 2011, London, UK.

[3] P.O. Braarud and B. Kirwan, *"Task Complexity: What Challenges the Crew and How Do They Cope,"* In A. B. Skjerve & A. Bye (Eds.) *Simulator-Based Human Factors Studies Across 25 Years.* Springer-Verlag, 2011, London, UK.

[4] J. Heimdal, *"Operational Culture Literature Review,"* (HWR-901). Halden Reactor Project, 2008, Halden, Norway.

[5] J. Toquam, J. Macaulay, C. Westra, Y. Fujita, and S. Murphy, *"Assessment of Nuclear Power Plant Crew Performance Variability,"* In M. Brannick, E. Salas, & Prince, C. (Eds.) *Team Performance Assessment and Measurement.* Lawrence Erlbaum Associates, 1997, Mahwah, NJ.

[6] D. Gertman, L. Haney, J. Jenkins, and H. Blackman, *"Operational Decision-Making and Action Selection Under Psychological Stress in Nuclear Power Plants,"* (NUREG/CR-4040). U.S. Nuclear Regulatory Commission, 1985, Washington, DC.

[7] J. Itoh, S. Yoshimura, T. Ohtsuka, and F. Matsuda, *"Cognitive Task Analysis of Nuclear Power Plant Operators for Man-Machine Interface Design,"* In Proceedings of the Topical Meeting on Advances in Human Factors Research on Man Machine Interactions. American Nuclear Society, Nashville, TN, 96-102, (1990).

[8] C. Evans and K. Dion, *"Group Cohesion and Performance a Meta-Analysis,"* Small Group Research, 22(2), 175-186, (1991).

[9] S. Massaiu, M. Hildebrandt, and A. Bone, *"The Guidance-Expertise Model: Modeling Team Decision Making with Emergency Procedures,"* Paper presented at the 10th International Conference on Naturalistic Decision Making (NDM 2011), Orlando, FL, (2011).

[10] W. Galyean, *"Orthogonal PSF Taxonomy for Human Reliability Analysis,"* Proceedings of the 8th International Conference on Probabilistic Safety Assessment and Management, PSAM-0281, New Orleans, LA, 2006.

[11] K. Lewin, *"Principles of Topological Psychology,"* McGraw-Hill, 1936, New York, NY.

[12] H. Murray and C. Kluckhohn, *"Personality in Nature, Society, and Culture,"* Knopf, 1953, New York, NY.

[13] D. Gertman, H. Blackman, J. Marble, J. Byers, and C. Smith, *"The SPAR-H Human Reliability Analysis Method,"* (NUREG/CR-6883). U.S. Nuclear Regulatory Commission, 2005, Washington, DC.

[14] P. Le Bot, *"Human Reliability Data, Human Error and Accident Models – Illustration Through the Three Mile Island Accident Analysis,"* Reliability Engineering and System Safety, 83 (2), 153-167, (2004).

[15] E. Hollnagel, *"Cognitive Reliability and Error Analysis Method,"* Elsevier Science Ltd., 1998, Amsterdam, The Netherlands.

[16] G. Kelly, *"The Psychology of Personal Constructs,"* Norton, 1955, New York, NY.

[17] W. Mischel and Y. Shoda, *"A Cognitive-Affective System Theory of Personality: Reconceptualizing Situations, Dispositions, Dynamics, and Invariance in Personality Structure,"* Psychological Review, 102, 246-268, (1995).

[18] J. Burger, *"Personality,"* Brooks/Cole, 1997, Pacific Grove, CA.

[19] K. Stanovich and R. West, *"Reasoning Independently of Prior Belief and Individual Differences in Actively Open-Minded Thinking,"* Journal of Educational Psychology, 89 (2), 342-357, (1997).

[20] K. Stanovich, and R. West, *"Individual Differences in Reasoning: Implications for the Rationality Debate?,"* Behavioral and Brain Sciences, 23, 645–726, (2000).

[21] J. Rachlinski, *"Cognitive Errors, Individual Differences, and Paternalism,"* University of Chicago Law Review, 73 (1), 207-229, (2006).

[22] D. Kahneman, *"Thinking, Fast and Slow,"* Farrar, Straus, and Giroux, 2011, New York, NY

[23] M. Lauriola, I. Levin, and S. Hart, *"Common and Distinct Factors in Decision Making Under Ambiguity and Risk: A Psychometric Study of Individual Differences,"* Organizational Behavior and Human Decision Processes, 104, 130-149, (2007).

[24] D. Kahneman and A. Tversky, *"Choices, Values, and Frames,"* Cambridge University Press, 2000, Cambridge, UK.

[25] J. Forester, A. Kolaczkowski, S. Cooper, D. Bley, and E. Lois, *"ATHEANA User's Guide,"* (NUREG-1880), U.S. Nuclear Regulatory Commission, 2007, Washington, DC.

[26] E. Salas and S.M. Fiore, *"Team Cognition: Understanding the Factors That Drive Process and Performance,"* American Psychological Association, 2004, Washington, DC.

[27] J. Joe and R. Boring, *"Modeling and Quantification of Team Performance in Human Reliability Analysis for Probabilistic Risk Assessment,"* Proceedings of the 12th International Conference on Probabilistic Safety Assessment and Management (PSAM 12, Paper #7), (INL/CON-14-31339), 2014, Honolulu, HI.

Cultural Profiles of Non-MCR Operators Working in Domestic NPPs

Jinkyun Park[a*], and Wondea Jung[a]

[a] Korea Atomic Energy Research Institute, Daejeon, Rep. of Korea

Abstract: Traditionally, the safety of NPPs has been evaluated by the PSA (Probabilistic Safety Assessment) or PRA (Probabilistic Risk Assessment) technique that quantifies the integrated safety of a whole system. In this regard, HRA (Human Reliability Analysis) plays an important role because it should quantify the possibility of HFEs (Human Failure Events) affecting the safety of NPPs. Therefore, the provision of sufficient data that are helpful for understanding the nature of HFEs under a given accident sequence is indispensable for estimating more realistic HRA results. To address this issue, one of the technical obstacles is the cultural effect on the performance of human operators. That is, it is suspicious for an HRA practitioner to use HRA data collected from another country or organization without sufficient understanding the nature of cultural differences. In this study, as one of the practical approaches to unravel this question, the cultural profiles of non-MCR operators are investigated in detail with respect to their operational experience.

Keywords: PSA, HRA, Operational Culture, Non-MCR Operators, Operational Experience.

1. INTRODUCTION

It is well known fact the performance of human operators (or human error) is decisive for the safety of nuclear power plants (NPPs). Accordingly, it is very natural that significant efforts continue to be applied to reduce the potential for human error. In this light, one of the most disseminated approaches is to conduct human reliability analysis (HRA).

Traditionally, the safety of NPPs has been evaluated by the PSA (Probabilistic Safety Assessment) or PRA (Probabilistic Risk Assessment) technique that quantifies the integrated safety of a whole system based on the analysis of event trees (ETs) and fault trees (FTs) representing all the plausible accident sequences. In this regard, since the accident sequences can be initiated by two kinds of events such as human failure events (HFEs) and hardware failure events, HRA takes part in quantifying the possibility of those HFEs. Therefore, the provision of sufficient data that are helpful for understanding the nature of HFEs under a given accident sequence is indispensable for estimating more realistic HRA results [1, 2]. To address this issue, recent efforts largely emphasize the collection of HRA data from simulated emergencies [3-6].

Unfortunately, the collection of HRA data seems not to be easy because of several technical reasons. One of them could be the cultural effect on the performance of human operators. Gertman et al. articulated that "Culture influences the probability of a person following a specific course of action and thus may affect the probability of actions [7]." Similarly, based on the results of existing studies, Kim et al. pointed out that "In addition, if an operator has their own strategy to use a procedure, or a specific operating culture exists in the operator's organization, the strategy will affect the method used for following the procedure [8]." Accordingly, it is suspicious for an HRA practitioner to use HRA data collected from another country or organization without sufficient understanding the nature of cultural differences. Conversely say, the HRA practitioner will use HRA data with confidence, if there is a clue upholding that the profiles of different cultures are very similar or even homogeneous.

In this light, Skraaning et al. showed a striking result indicating that six different organizations comprised of 81 MCR (Main Control Room) operators from three different countries (i.e., Sweden,

---

* kshpjk@kaeri.re.kr

Korea and United States) have similar culture profiles [9]. In addition, although there are some discrepancies, Park and Jung pointed out that the cultural profiles of MCR operators working in the domestic NPPs of Rep. of Korea seem to be similar to those of non-MCR operators [10]. These results strongly imply the possibility of a cross-cultural generalizability among operating personnel who have the responsibility of NPP operations. In this end, it is indispensable to clarify the reason of discrepancies observed from the comparison between MCR operators and non-MCR operators.

In this study, as one of the practical approaches to unravel this question, the cultural profiles of non-MCR operators are investigated in detail with respect to their operational experience. As a result, it is expected that the discrepancies can be soundly explained by the difference of operational experience among non-MCR operators.

2. PREVIOUS STUDIES

It is very natural to anticipate that the behavior of human operators will be largely affected by the cultural characteristics of an organization to which they belong (e.g., organizational culture). However, in addition to the organizational culture, it is strongly expected that there could be other cultural characteristics affecting the performance of human operators. For example, Hofstede articulated that is the culture is the collective programming of the mind which distinguishes the members of one group or category of people from another [11]. In addition, he articulated that the culture has several layers distinguishable from many levels, such as a national, regional, ethnic, religious, linguistic, gender, social class and organizational level. Of them, he proposed 5 and 6 cultural dimensions to represent the national and organizational culture, respectively. Table 1 summarizes 11 cultural dimensions with their meaning. The value of each dimension can be quantified by the scores of several questionnaires developed by Hofstede [11]. More detailed information can be found from Ref. [9, 10].

In this regard, based on the review of existing literatures, Heimdal claimed that not only an organizational culture but also a national culture can affect the behavior of human operators [12]. In other words, it is assumed that not only the organizational culture but also the national culture should be considered in parallel in order to properly understand the behavior of human operators working in a certain organization. From this assumption, Skraaning et al. compared the cultural profiles of 81 MCR operators based on the Hofstede's cultural dimensions [9]. All of the participants are working in Westinghouse 3-loop PWRs (Pressurized Water Reactors) in three different countries (i.e., Sweden, Rep. of Korea and United States of America). As a result, although there are some discrepancies, it was observed that both the national and the organizational culture profiles of MCR operators seem to be very similar. These results are very interesting because it is generally anticipated that the cultural profiles of human operators are probably different along with their nationalities (i.e., the national culture) as well as organizations (i.e., the organizational culture).

One plausible explanation for these results would be that MCR operators in three different countries share similar values and norms probably acquired from the operational experience of NPPs [9]. That is, it is possible to assume that the cultural profiles of MCR operators could be comparable to those of non-MCR operators. This is because, although working places are different, they are likely to share similar knowledge and expertise acquired from: (1) the operation of Westinghouse 3-loop PWRs, and (2) similar education and training contents. In order to scrutinize this expectation, the cultural profiles of MCR operators working in the domestic NPPs of the Rep. of Korea are additionally compared with those of non-MCR operators (e.g., field operators and auxiliary operators) who are working in the identical units. In other words, if the cultural profiles of non-MCR operators are also similar to those of MCR operators, it could be good evidence supporting the existence of a homogeneous operational culture across operating personnel working in NPPs. Table 2 summarizes the age and operational experience of participants belonging to domestic NPPs, and Figure 1 shows the results of these comparisons.

Table 1. 11 Cultural Dimensions with Their Meaning; reproduced from Ref. [10]

Culture	Dimension	Meaning
National culture	PDI (Power Distance Index)	The extent to which the less powerful members of institutions and organizations within a society expect and accept that power is distributed unequally
	IDV (Individualism Index)	Individualism stands for a society in which the ties between individuals are loose
	MAS (Masculinity Index)	Masculinity stands for a society in which emotional gender roles are clearly distinct; Femininity stands for a society in which emotional gender roles overlap
	UAI (Uncertainty Avoidance Index)	The extent to which the members of institutions and organizations within a society feel threatened by uncertain, unknown, ambiguous, or unstructured situations
	Long-term Orientation Index (LTO)	Long-term Orientation stands for a society that fosters virtues oriented towards future rewards, in particular perseverance and thrift
Organizational culture	P1 (process vs. result oriented)	In a result oriented culture, people perceive themselves to be comfortable in unfamiliar situations
	P2 (employee vs. job oriented)	An employee-oriented organization takes responsibility for people's welfare, and important decisions are often made by groups or committees
	P3 (parochial vs. professional	With high professional scores, the employees' private lives are perceived to be their own business, where they are hired on the basis of their professional skills only
	P4 (open vs. closed system)	In an open culture, almost everyone fits into the organization, and it takes only a few days to feel at home
	P5 (loose vs. tight control)	Tight control cultures are cost-conscious, keep meeting times, and jokes about the company are rare
	P6 (normative vs. pragmatic)	Employees of normative cultures view their tasks toward the outside world as implementations of inviolable rules, correctly following organizational procedures

Table 2. Summary of participants

Unit	Belong to	Designation	Number of participants	Age (year) Mean	Age (year) SD*	Experience (year) Mean	Experience (year) SD*
1	MCR	MCR1	24	39.08	4.99	12.17	7.19
1	Non-MCR	Non-MCR1	31	38.20	6.17	9.24	5.93
2	MCR	MCR2	20	40.30	6.13	12.67	5.45
2	Non-MCR	Non-MCR2	21	34.10	4.24	4.93	3.13

*SD is short for standard deviation.

From Fig. 1, at a glance, it seems that organizational culture profiles obtained from non-MCR operators are almost identical with those from MCR operators. In addition, except two dimensions (i.e., MAS and UAI), the national culture profiles of non-MCR operators appear to be congruent with those of MCR operators. This implies that operating personnel probably share similar cultural profiles. In this light, it is necessary to explain why operating personnel showed some discrepancies on two dimensions. In other words, if there is a reason clarifying why operating personnel have different values in these two dimensions, then it is possible to anticipate the existence of a common operational culture that is reproducible, predictable, and in common under the context of NPP operations. For example, in the MAS dimension, it seems that MCR operators have different values compared to non-MCR operators. In contrast, in the case of the UAI dimension, except non-MCR operators working in the Unit 1 (i.e.,

Non-MCR1), even Non-MCR2 showed similar values to those of MCR operators (i.e., MCR1 and MCR2). Of them, one plausible explanation on the MAS dimension is the responsibility of operating personnel working in MCRs.

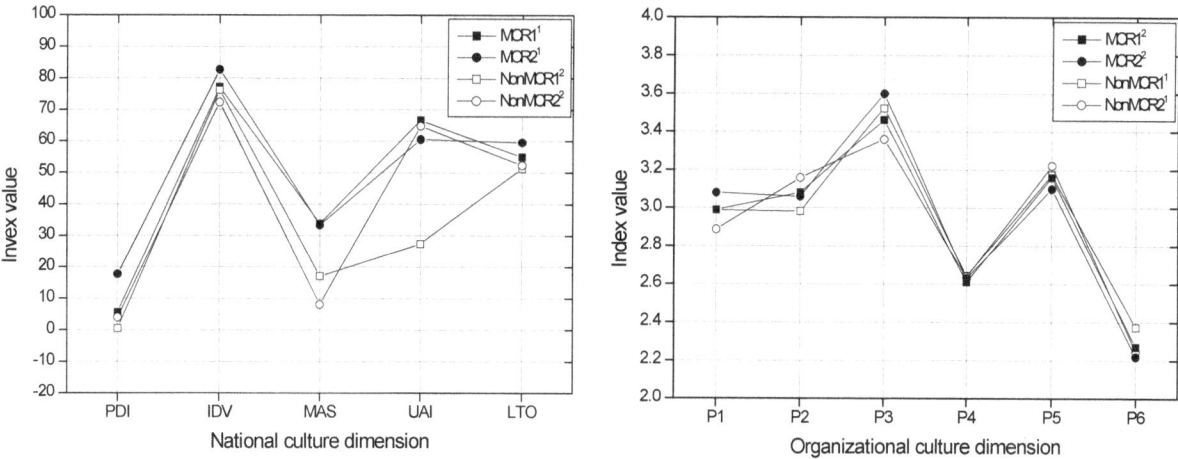

1. MCR1 and MCR2 denote operating personnel working in the MCR of the Units 1 and 2, respectively.
2. NonMCR1 and NonMCR2 designate operating personnel working as non-MCR operators in the Units 1 and 2, respectively.

Figure 1. Cultural profiles identified from two groups of operating personnel; reproduced from Ref. [10]

According to Table 1, it is expected that the Masculinity will decrease when emotional gender roles are largely overlap. This allows us to assume that the value of the MAS dimension on a certain group will increase if all the group members have distinctive and unique roles and responsibilities. Conversely, if the distinctions of both roles and responsibilities among group members are blurred, it is expected that the value of the MAS dimension will decrease. In this regard, the values of the MAS dimension rated by MCR operators should be greater than those of non-MCR operators because the former have clear roles and responsibilities with respect to the operation of NPPs [13, 14]. On the contrary, in comparison with MCR operators, the roles and responsibilities of non-MCR operators are apt to be overlapped with respect to the situation of a local field. Therefore, although other causes may exist, it is promising that the roles and responsibilities of operating personnel is one of the important factors affecting the MAS dimension across operating personnel [10].

Unfortunately, this explanation does not seem to be enough for the UAI dimension because human operators belonging to the NonMCR2 group showed higher value comparing to those who belong to the NonMCR1 group. This strongly implies that there could be other causes resulting in the discrepancies of two dimensions. For this reason, the effect of the operational experience of human operators on each cultural dimension is investigated in detail.

3. THE EFFECT OF OPERATIONAL EXPERIENCE ON CULTURAL DIMENSIONS

First of all, a total of 52 non-MCR operators are assigned into four classes based on their operational experience that represents how many years they have worked in Westinghouse 3-loop PWRs. After that, the mean values of 11 cultural dimensions are calculated with respect to the each class of operational experience. Table 4 shows the number of non-MCR operators included in each class, and Table 5 summarizes the mean values of each cultural dimension. For example, the mean values of 5 cultural dimensions (i.e., PDI, IDV, MAS, UAI, and LTO) pertaining to the national culture of non-MCR operators who have experienced the operation of Westinghouse 3-loop PWRs greater than 15 years are -25.00, 83.00, 90.00, 31.00, and 60.00, respectively. It should be noted that the values of the

5 cultural dimensions usually vary between 0 and 100, but there are times when it could be below 0 or above 100 [9].

Table 3. Number of non-MCR operators involved in the four levels of operational experience

Level	Experience	Unit 1 (ratio)	Unit 2 (ratio)
1	< 5 years	9 (29%)	12 (57%)
2	5-10 years	10 (32%)	6 (29%)
3	10-15 years	4 (13%)	3 (14%)
4	> 15 years	8 (26%)	0 (0%)

Table 4. Mean values of 11 cultural dimensions with respect to experience levels

Level	PDI	IDV	MAS	UAI	LTO	P1	P2	P3	P4	P5	P6
1	9.76	65.00	17.14	37.86	51.43	3.17	3.14	3.51	2.54	3.16	2.10
2	-0.31	80.00	-11.88	63.75	50.00	2.94	2.90	3.77	2.60	3.04	2.25
3	0.71	86.43	-8.57	20.71	51.43	2.81	3.00	3.38	2.71	3.05	2.67
4	-25.00	83.00	90.00	31.00	60.00	2.73	3.20	2.93	3.00	3.40	2.33

From Table 4, two kinds of interesting tendencies can be identified. Figure 2 will be helpful for clarifying them.

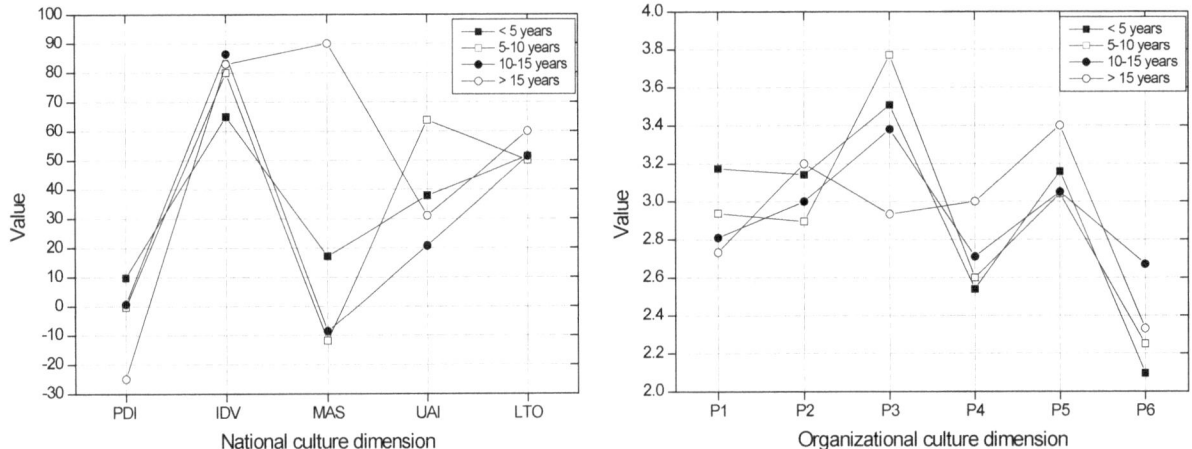

Figure 2. Cultural profiles with respect to the operational experience of non-MCR operators

The first tendency is that non-MCR operators who experienced the operation of Westinghouse 3-loop PWRs more than 15 years (i.e., *Level 4*) seem to have quite different cultural profiles comparing to others belonging to the rest three levels. For example, the value of MAS dimension observed from non-MCR operators involved in the *Level 4* is 90.00 while those of non-MCR operators belonging to the *Levels 1, 2*, and *3* are 17.14, -11.88 and -8.57, respectively. Similarly, the value of P3 dimension observed from the *Level 4* is 2.93 that is quite lower than the values gathered from the *Levels 1, 2* and *3*. This tendency could become more evident if we look at Fig. 3 in which cultural profiles corresponding to the Level 4 are not considered.

The second tendency is that, even though the cultural profiles of non-MCR operators who experienced plant operations more than 15 years are not considered, there are some discrepancies in several dimensions, such as the MAS and UAI. For example, from Fig. 3, the values of the MAS dimension observed from the *Levels 2* and *3* are quite close (i.e., -11.88 and -8.57) while that of the *Level 1* is relatively different (i.e., 17.14). In addition, it seems that the values of the UAI dimension gathered from the *Levels 1, 2,* and *3* are quite different each other (i.e., 37.68, 63.75, and 20.71, respectively).

At the same time, conversely say, the values of other dimensions are relatively close each other (e.g., the values of the LTO observed from the *Levels 1, 2* and *3* are almost identical). This tendency strongly alludes to the fact that, to some extent, the cultural profiles of a certain group could be *estimated* based on the weighted average of the cultural profiles of all group members with respect to their experience on plant operations.

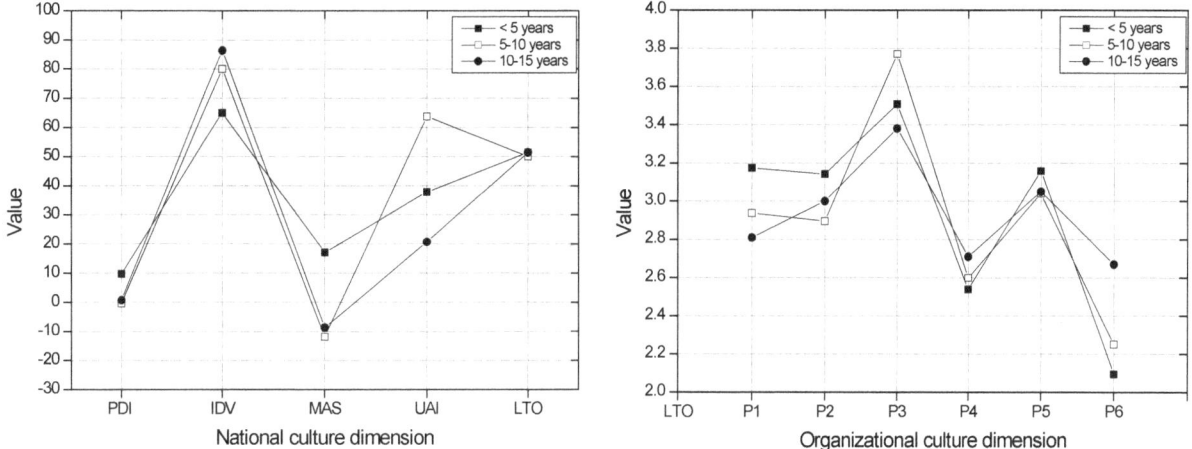

Figure 3. Cultural profiles for non-MCR operators belonging to the *Levels 1, 2* and *3*

For example, let us recall Table 3 in which the relative percentages of non-MCR operators are classified with respect to the four levels of operational experience. Based on these relative ratios, the value of the PDI dimension in the Unit 1 can be estimated by the weighted average of mean values with respect to the four levels of operational experience (refer to Fig. 4).

Level	Ratio	Level	PDI
1	29%	1	9.76
2	32%	2	-0.31
3	13%	3	0.71
4	26%	4	-25.00

$0.29 \times 9.76 + 0.32 \times (-0.31) + 0.13 \times 0.71 + 0.26 \times (-25.00) = -3.63$

Figure 4. Estimating the value of the PDI dimension in the Unit 1

In this way, all the values of 11 dimensions can be estimated. Figure 5 compares the cultural profiles of the Unit 1 with the estimated values of 5 cultural dimensions pertaining to the national culture.

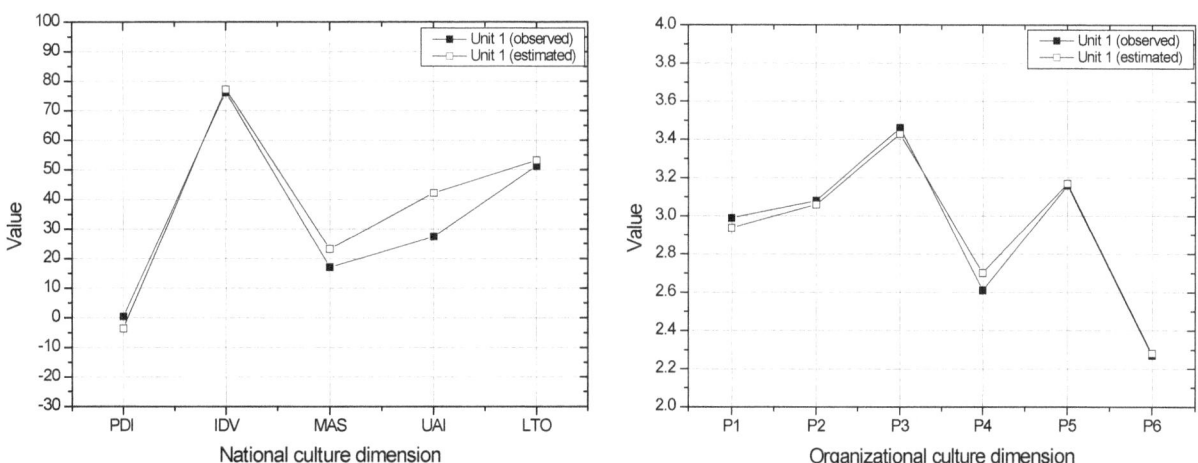

Figure 5. Cultural profiles for the Unit 1 – observed and estimated values

4. GENERAL CONCLUSION

It is evident that the amount of available data for conducting HRA is one of the decisive factors affecting the quality of HRA results. Unfortunately, the collection of HRA data is not easy because of several technical reasons including the cultural effect on the performance of human operators. In this light, several researchers empirically observed that, although there are some discrepancies, operating personnel seem to share similar cultural profiles even though they are working in NPPs located in different countries. In order to confirm this observation, it is indispensable to clarify the reason of discrepancies observed from the comparison between MCR operators and non-MCR operators.

In this study, as one of the practical approaches to unravel this question, the cultural profiles of non-MCR operators are investigated in detail with respect to their operational experience. To this end, the variations of cultural profiles are compared with respect to the four levels of operational experience. As a result, it is expected that, to some extent, the cultural profiles of a certain group could be *estimated* based on the weighted average of the cultural profiles of all group members with respect to their experience on plant operations. This means that the discrepancies could be soundly explained by the difference of operational experience among non-MCR operators. Therefore, although the result of this study is not sufficient for drawing a firm conclusion, it is possible to say that this study is able to contribute to start the very first step to scrutinize the possibility of a cross-cultural generalizability among operating personnel who have the responsibility of NPP operations.

Acknowledgements

This work was supported by Nuclear Research & Development Program of the National Research Foundation of Korea grant, funded by the Korean government, Ministry of Science, ICT & Future Planning (Grant Code: 2012M2A8A4025991).

References

[1] R. L. Boring. "*Fifty Years of THERP and Human Reliability Analysis,*" Proc. 11th Probabilistic Safety Assessment and Management / European Safety and Reliability 2012 (PSAM11/ESREL 2012), Helsinki, Finland, June 25-29 (2012)

[2] J. Chang, and E. Lois. "*Overview of the NRC's HRA data program and current activities,*" Proc. 11th Probabilistic Safety Assessment and Management / European Safety and Reliability 2012 (PSAM11/ESREL 2012), Helsinki, Finland, June 25-29 (2012)

[3] E. Lois, V. N. Dang, J. Forester, H. Broberg, S. Massaiu, M. Hildebrandt, P. Ø. Braarud, G. Parry, J. Julius, R. Boring, I. Männistö, and A. Bye. "*International HRA empirical study - Description of overall approach and pilot phase results from comparing HRA methods to simulator performance data,*" NUREG/IA-2016, Vol. 1, Washington, DC: US Nuclear Regulatory Commission, 2009 (also issued as HWR- 844, OECD Halden Reactor Project Work Report, Halden, Norway, 2008)

[4] A. Bye, E. Lois, V. N. Dang, G. Parry, J. Forester, S. Massaiu, R. Boring, P. Ø. Braarud, H. Broberg, J. Julius, I. Männistö, I., and P. Nelson. "*International HRA empirical study - Results from comparing HRA method predictions to simulator data from SGTR scenarios,*" NUREG/IA-2016, Vol. 2, Washington, DC: US Nuclear Regulatory Commission, 2011 (also issued as HWR-915 OECD Halden Reactor Project Work Report, Halden, Norway, 2010)

[5] J. Marble, L. Huafei, M. Presley, J. Forester, A. Bye, V. N. Dang, and E. Lois. "*Results and insights derived from the intra-method comparisons of the US HRA empirical study,*" Proc. 11th Probabilistic Safety Assessment and Management / European Safety and Reliability 2012 (PSAM11/ESREL 2012), Helsinki, Finland, June 25-29 (2012)

[6] Y. J. Chang, D. Bley, L. Criscione, B. Kirwan, A. Mosleh, T. Madary, R. Nowell, R. Richards, E. M. Roth, S. Sieben, and A. Zoulis. "*The SACADA database for human reliability and human performance,*" http://dx.doi.org/10.1016/j.ress.2013.07.014 (2013)

[7] D. Gertman, S. Novack, and J. Marble. "*Culture representation in human reliability analysis,*" Proc. Interservice/Industry Training, Simulation, and Education Conference (I/ITSEC).

[8] Y, Kim, J. Park, and W. Jung. "*Measuring variability of procedure progression in proceduralized scenarios,*" Annals of Nuclear Energy, vol. 49, p. 41-47 (2012)

[9] G. Skraaning, J. Park, and J. O. Heimdal. "*Cross-cultural generalizability in the nuclear domain: a comparison of culture profiles for control room operators in Swedish, Korean, and US plants,*" HWR-1027, OECD Halden Reactor Project (2012)

[10] J. Park, and W. Jung. "*Comparing cultural profiles of MCR operators with those of non-MCR operators working in domestic nuclear power plants,*" Submitted to Reliability Engineering and System Safety.

[11] Hofstede, G. "*Culture's consequences,*" 2nd ed., Saga Publication, 2001

[12] J. O. Heimdal. "*Operational culture literature review,*" HWR-901, OECD Halden Reactor Project, Halden, Norway (2007)

[13] N. Moray. "*Cultural and national factors in nuclear safety,*" In: Safety Culture in Nuclear Power Operations, Wilpert, B., and Itoigawa, N. (eds.), p. 37-59, Taylor and Francis, 2001, London

[14] J. Park. "*The complexity of proceduralized tasks,*" Springer-Verlag, 2009, Berlin

Improving Scenario Analysis for HRA

Claire Taylor
OECD Halden Reactor Project, Institute for Energy Technology (IFE), Halden, Norway

Abstract: The International Human Reliability Analysis (HRA) Empirical Study [1, 2, 3, 4] concluded that variability in predictions of human error probabilities are in part due to deficiencies in the qualitative scenario analysis for some HRA methods. The study showed that it can be difficult for HRA analysts to gain a good understanding of how a scenario is likely to unfold, what challenges it may present to operators, how operators are likely to respond, and where performance problems may occur. Although some HRA methods include guidance on qualitative scenario analysis, most methods state only that this should precede quantification, but without specifying methods for this or the depth to which the scenario analysis should go.

A study is underway at the Halden Reactor Project in Norway to investigate scenario analysis and why it is considered difficult. The study focuses on the experience of HRA analysts in their everyday work, with the goal of understanding the challenges they face. The aim of the study is to develop a practical guidance handbook for use when performing scenario analysis. The results will include good practices implemented by analysts and further recommendations for improvement.

This paper describes the plan for this study, the findings to date and how these findings will inform a further proposed study on the development of a database to support HRA.

Keywords: HRA, Scenario Analysis, Qualitative Analysis.

1. INTRODUCTION

The International HRA Empirical study [1, 2, 3, 4] generated important insights into the strengths and weaknesses of HRA methods. The study identified variability in predictions of human error probabilities, in part due to deficiencies in the qualitative scenario analysis for some HRA methods. The study showed that it can be difficult for HRA analysts to get a good understanding of how a scenario is likely to unfold, what challenges it may present to the operators, how the operators are likely to respond, and where they may experience performance problems. Although some HRA methods include guidance on qualitative scenario analysis, other methods leave it entirely to the analyst to decide which approach to use and to which depth the scenario analysis should go.

1.1 Background

Following the International HRA Empirical Study, a US HRA Domestic Study [5] was performed which replicated the International study approach (with minor differences) to evaluate HRA methods using data from a training simulator at a US plant. During the US study [5], the HRA teams involved were asked to keep "HRA diaries" documenting their approach to the analyses they performed. These diaries were subsequently reviewed for information about the methods used by the various teams, particularly during scenario analysis. Analysis of the diaries showed significant variability between analysts with respect to resource usage (i.e., time spent on the analysis), their approaches to data collection, and the analysts' own expertise in HRA.

Semi-structured interviews were conducted in December 2012 with a small number of participants from the US study to further investigate the approach they used during the study and the challenges they encountered. In the interviews, the analysts reported a number of challenges, including a lack of clear guidance for Human Failure Event (HFE) identification and awareness or interpretation of Performance Shaping Factors (PSF)s; a lack of data to validate their analyses; uncertainties about how to document information collected, especially where an expert judgment has been made; and understanding where improvements can be made after the quantification (e.g. in terms of ergonomics, procedures, etc.), based on the findings of the qualitative scenario analysis.

A workshop was held at the US Nuclear Regulatory Commission (NRC) in Rockville, Maryland, in December 2012, attended by HRA researchers and practitioners from a number of international organizations, to discuss the possibility of an international collaborative study on the use of simulator data to support HRA. A second workshop was held in Halden, Norway, in May 2013 to further develop this study proposal. At the second workshop it was agreed that the scenario analysis study would complement the larger international collaboration, as it proposes to investigate the process of scenario analysis itself to understand why it is difficult, how analysts use the available information, good practices for scenario analysis, and how analysts could make use of a data source such as an HRA database.

At the Halden workshop, a number of use cases for HRA were presented, one of which incorporated three activities: scenario familiarization (which includes defining the HFE scenario, scope and purpose of the analysis), qualitative data collection, and qualitative data analysis (which includes description and representation of how tasks are performed and human error identification). It was agreed at the workshop that these three activities generally encompass the scenario analysis part of an HRA. The discussion of this use case at the Halden workshop indicated that access to and availability of data is one of the most significant challenges for HRA analysts at this stage of the HRA, and that an HRA database may be useful in such instances.

Following the workshop, it was agreed that this scenario analysis study should focus on how typical HRA analysts gather data for their analysis, what kind of data they collect, the challenges associated with this, and how they could potentially use a database to support or supplement their qualitative data collection and analysis activities. Therefore the scenario analysis study will form the first phase of the international collaborative HRA data study as it will provide valuable insights regarding what kind of data analysts would like to be able to access. This in turn will enable the international collaborative study team to tailor the HRA database to include information that will be of practical use to HRA analysts.

2. STUDY OVERVIEW

2.1 Objectives of the Scenario Analysis Study

The objectives of the scenario analysis study are to explore the following questions:

- How do HRA analysts plan and conduct scenario analysis?
- What issues or aspects of the scenario do analysts typically focus on?
- What part of scenario analysis is the most challenging, and why?
- What kinds of data or information sources do analysts typically use (for example, event reports, databases, site visits, expert interviews)?
- How do analysts resolve uncertainties or contradictions during the analysis?
- How do analysts safeguard against biases or misinterpretation during the analysis?
- What good practices have analysts developed or implemented to overcome the challenges experienced?

The ultimate goal of the study is to develop a handbook for HRA analysts with practical guidance on how to perform scenario analysis. The handbook will contain advice on best practices and the aim is to write the handbook in collaboration with experienced analysts from the HRA community.

As noted earlier, the findings from this study will also be used as input to the international collaborative study on using simulator data for HRA, by providing real-world examples of the kinds of data that analysts need, how they can use these data, and the challenges they have experienced associated with obtaining these data.

2.2 Study Plan

The study will focus on how HRA analysts perform scenario analysis in their everyday work, to gain an understanding of the real-world issues and challenges that they face. The study comprises five key steps, as shown in Figure 1 and described below. At the time of writing this paper, the first two steps have been completed.

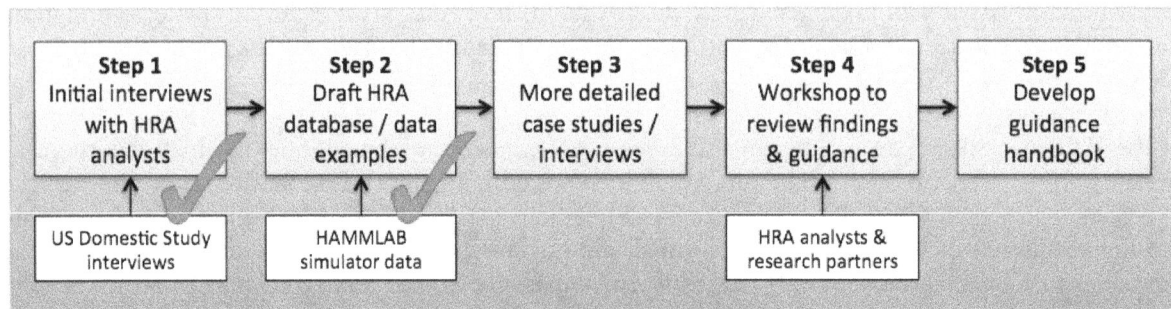

Figure 1: Scenario Analysis Study Plan

2.2.1   Step 1: Initial interviews with HRA analysts

The first step in the study was to interview HRA analysts from different organizations and/or who use different HRA methods, to gain an understanding of how they usually perform scenario analyses for HRA. As noted earlier, a number of semi-structured interviews were held in December 2012 with participants from the US HRA Domestic Study. However, the interviews revealed that not all of the interviewees perform HRA as part of their "normal" everyday job. Therefore, the interviewees chosen for this scenario analysis study have been selected on the basis that they regularly perform HRA as part of their job (i.e. approximately 50% of their time is involved in HRA).

The aim of the interviews was to understand the reality of conducting a scenario analysis for HRA; what approaches do the different analysts use; what data do they seek to collect as part of the scenario analysis; what issues and challenges do they regularly face; what good practices have they developed or adopted to overcome these problems?

2.2.2   Step 2: Develop a draft HRA database and/or HRA data examples

A small sample of data collected from the Halden HAMMLAB simulator, as shown in Figure 2, has been selected. These data examples will be provided to the HRA analysts during the Step 3 case studies to test whether they consider this data to be of practical use for scenario analysis, and to explore how they could use the data. The findings here will be used as a first input to the planned international collaborative HRA data study and will also inform the development of recommendations and guidance for the scenario analysis handbook.

The data examples are taken from a Steam Generator Tube Rupture (SGTR) scenario that was simulated in the HAMMLAB. The first example includes empirical data showing timings for establishing Feed & Bleed (F&B). The second example includes qualitative data summarizing the response of the crews, and the third example includes qualitative data describing the "crew stories", which is a detailed description of how the scenario unfolded for one crew.

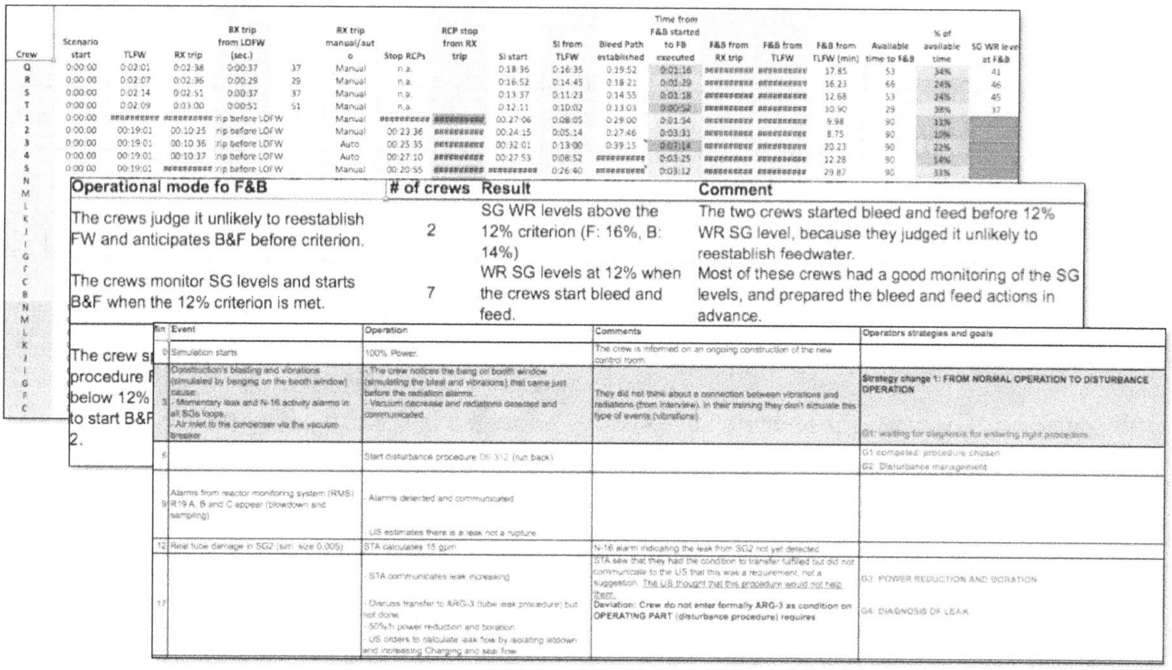

Figure 2: Data examples for testing during this study

The study plan originally considered the possibility of developing a draft HRA database, but due to the interview and case study schedules, there has not been sufficient time to develop this. However, the data examples are presented in a way that mimics what an analyst could find if conducting a search in such a database. More detail on how to present and search for this data will form part of the larger international collaborative HRA data study.

2.2.3 Step 3: Case studies with HRA analysts

During the Step 1 interviews, interviewees were asked about the possibility of taking part in a more detailed case study to be carried out in 2014. The objective of the case studies would be to shadow an HRA analyst (or team of analysts) as they perform a scenario analysis for HRA during the course of their normal work. This would provide an opportunity for observation and/or discussion of how the analysts collect, sort, analyze and document data, what problems or difficulties they encounter during this process, and the approaches or good practices that they use to overcome these.

During the case studies, the analysts would also be provided with the Halden HAMMLAB simulator data (either as a sample HRA database, or individual examples of data that could be retrieved from such a database), to evaluate whether they could use this data to inform or support the scenario analysis process. Two potential detailed case studies have been identified and, at the time of writing this paper, work is underway to make arrangements for these.

2.2.4 Step 4: Workshop with analysts and research partners

Following the case studies, the findings will be analyzed and presented at a workshop, with the HRA analysts involved in the study and the research partners involved in the proposed international collaborative HRA study (depending on availability). The aim of the workshop will be to discuss the findings from the interviews and case studies, and to develop recommendations for improvements to the scenario analysis process and for the guidance handbook.

2.2.5 Step 5: Develop scenario analysis handbook

The main deliverable from this study will be a handbook of practical guidance and recommendations to improve the scenario analysis process for HRA analysts. It is intended that this handbook will assist analysts in preparing for and conducting scenario analysis, particularly for less experienced analysts.

The handbook is intended to be independent of any particular HRA method, but where good practices or recommendations specific to a particular method are identified, these will be included for those analysts who choose to use that method.

3. A TYPICAL APPROACH TO SCENARIO ANALYSIS

As noted earlier, a use case for scenario analysis was presented at the Halden workshop in May 2013. This use case described a generic approach to scenario analysis (within the context of an overall generic approach to HRA, as shown in Figure 3) and was presented to determine whether all of the workshop participants had the same understanding of the main steps in a typical scenario analysis activity. The dotted arrows in Figure 3 demonstrate that HRA is not necessarily a linear process, and that there may be several iterations between or even within individual steps.

The feedback from the workshop participants was that this diagram adequately describes the main steps in a generic HRA. The only exception was the final step – human error reduction – referring to, for example, an impact assessment of the results of the HRA to determine whether improvements should be made to reduce the likelihood of human error. It was agreed that this step is a consequence of the overall HRA and not actually a part of the generic HRA process. However, as this part of the approach to HRA is outside the scope of this project, this issue is not discussed further here.

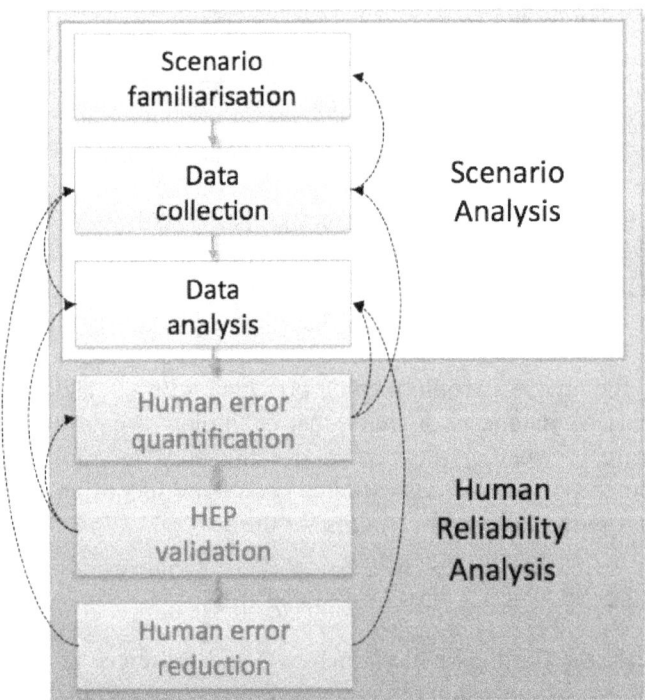

Figure 3: A typical approach to HRA and Scenario Analysis

Therefore, the steps in a typical scenario analysis are generally agreed to be as follows:

- **Scenario familiarization:** This step involves the preparation for the HRA, when the analyst explores and defines the scenario, scope and purpose of the HRA. The analyst will usually gather basic information about the scenario in order to prepare for the more formal data collection step. This can often include discussions with Probabilistic Risk Assessment (PRA) analysts and/or other subject matter experts, and a documentation review (e.g. scenario description, event reports, operator stories, operating procedures, system descriptions, plant drawings, shift logs, etc.).

- **Data collection:** This step involves a more formal process for collecting information about the HRA scenario. The information collected will usually be qualitative, but may also include quantitative data, for example, from a training simulator. The data collection step may include a walk-through or talk-through of the scenario with subject matter experts in the training simulator or on the plant, task observations, interviews with operators, process engineers, trainers or other relevant personnel and/or review of local event reports and additional plant documentation, such as checklists, job aids, etc.
- **Data analysis:** This step usually involves the description and representation (for example, in a hierarchical and/or tabular task analysis structure) of how tasks are performed for the scenario of interest. This will be based on the qualitative and quantitative information gathered during the data collection step, and will often include some human error identification. This step provides an opportunity to check for knowledge gaps in the information gathered about the scenario. The step may include some task and error analysis.

The above description of a generic approach to scenario analysis was used as to set the scene for the interviews carried out for this scenario analysis project; i.e. that scenario analysis tends to take place before quantification and tends to include primarily qualitative data collection and analysis activities.

4. FINDINGS TO DATE

4.1 Interviews with Participants in US HRA Domestic Study

4.1.1 Overview of the Domestic Study

The 2010 US HRA Domestic Study [5] was a follow-on study from the International HRA Empirical Study [1, 2, 3, 4], primarily to test the consistency and accuracy of HRA predictions amongst different analyst teams using the same HRA methods. A key difference between the International and the Domestic studies is that for the International study, the HRA teams did not have an opportunity to visit the "plant" (in this case, the HAMMLAB simulator in Halden) to observe and collect data about the scenarios under analysis; instead, data from the HAMMLAB simulator were provided to them. However, for the domestic study, the HRA teams were able to visit the plant site to observe a training scenario, take notes, review simulator logs, audio and video recordings and debriefing interviews.

The HRA teams in the domestic study also had the opportunity to interview the simulator instructors about the scenarios being analyzed and other relevant aspects, such as use of procedures, expected operator actions, etc. Therefore, the domestic study methodology more closely resembled a real-life approach to qualitative data collection for HRA. For this reason, it was considered relevant to this scenario analysis study to interview some of the participants in the domestic study to explore their experiences with aspects of the HRA such as: identification of Human Failure Events (HFEs); data collection to inform and support the analysis; challenges and limitations associated with the HRA methods used, and their "wish list" for improvements to HRA and HRA methods for future analyses.

4.1.2 Findings from Interviews

Nine participants of the domestic study were interviewed in total. The interviewees had varying levels of HRA knowledge and experience, ranging from very experienced persons who perform HRA as part of their daily work and/or have been involved in the development of HRA methods, through to persons who had no experience of HRA prior to the domestic study, and who do not perform HRA as part of their daily work. We considered it useful to interview both the experienced HRA analysts to learn from their insights into real-world analyses, and also to interview the non-experienced study participants as their insights from the domestic study would be similar to those of analysts who are new to HRA.

Scenario Familiarization

The scenario and the HFEs for analysis are usually defined by the PRA, although the definitions can sometimes be at a high level, in which case the HRA analyst will need to investigate and describe these in more detail. In other cases, the HRA analyst may be provided with a scenario description only, and will have to identify the operator actions and HFEs themselves. The definition and/or scope of the scenario and the HFEs can often change over the course of the analysis as more information about these are collected and analyzed. It is important to revisit these definitions throughout the analysis to ensure they are sufficiently described.

Data Collection

Analysts will almost always review operating procedures in the first instance, to gain an understanding of the HFEs and operation actions for the scenario being analyzed. Some interviewees stated that they will also usually discuss the scenario and HFEs with a PRA analyst to ensure they have a good understanding of the initiating events and sequence timings, etc.

All of the interviewees considered that a visit to the plant and/or the training simulator is the best way to collect information about the scenario and the HFEs. Here the analyst can get an impression of how quickly the scenario could unfold, how many and what kinds of cues and alarms the operator would expect to receive, whether there could be conflicting or simultaneous communications from multiple sources, and the impact of factors such as these on operator performance.

The interviewees with experience of visiting training simulators noted that it may not always be possible to observe the actual scenario being analyzed, however it is always useful to observe whatever scenarios are being run in the simulator at the time of the analyst's visit. Two of the more experienced interviewees stated that the purpose of observing simulator runs is not to collect empirical or statistical data about that scenario, but rather to collect qualitative information about how operating crews work; i.e. how they use procedures, how they communicate with each other, how they respond to alarms, how they work together as a team, how they interface with the system and equipment, the impacts of PSFs on their performance, etc.

Most of the analysts interviewed stated that it is important to fully understand the scenario and the HFEs before visiting the plant or training simulator to perform observations and talk with plant personnel. The analyst should first study available documentation such as system descriptions, operating procedures, training documents, event reports, and previous analyses (if these exist) etc. to familiarize and inform themselves on all aspects of the scenario. One of the interviewees noted that it is often useful to talk to the PRA team at the plant, as they will often have a better understanding of the purpose of the HRA and can help to ensure that the HRA analyst understands how the scenario fits into the overall PRA model and the factors that are likely to affect the scenario.

Data Analysis

One of the interviewees stated that most of the data analysis is done prior to visiting the plant or simulator, based on the information collected beforehand, e.g. from reviewing operating procedures and other relevant documentation, discussion with PRA analysts, previous knowledge of the plant, etc. The analyst will then review the analysis with a PRA team prior to visiting the plant or simulator, and uses the plant visit as a means of checking and confirming the data analysis. However, this approach does not appear to be typical for the other analysts interviewed, most of whom stated that they would perform some high-level analysis prior to the plant visit to inform their data collection. The majority of their analysis would be performed after the visit, when they have gathered the required data.

Main Challenges

There were two key challenges (relevant to this scenario analysis study) that were reported by some of the interviewees. For the less experienced analysts, a significant challenge is the lack of guidance on how to develop an overall understanding of the scenario, how to prepare for a site visit and interviews with plant personnel, and how to collect and interpret qualitative data. It can be difficult even for experienced analysts to correctly identify and interpret PSFs.

Another challenge is that simulator observations may not always reflect the reality of the scenario; for example, in the simulator, operators may skip or run quickly through certain steps in a procedure to quickly get to the relevant instructions, whereas in reality they would take their time and work systematically through all of the procedural steps.

4.2 Interviews with HRA Analysts for Scenario Analysis Study

At the time of writing this paper, seven semi-structured interviews have been held with individuals working in HRA (i.e. who perform HRA for at least part of their job). The interviewees come from a range of different organizations, including a nuclear regulatory organization, a power plant operating organization and a number of different consulting organizations. The interviewees also have different experiences in terms of the HRA methods that they currently use (or have used in the past) and the context within which these methods are applied for HRA.

Planning And Conducting Scenario Analysis

Most of the analysts interviewed use a similar approach for planning and conducting the scenario analysis. To familiarize themselves with the scenario and HFE(s), some analysts reported that they will contact the plant in advance to request information relevant to the scenario – this may include system descriptions, operating procedures, event reports, and information about the how the scenario is likely to unfold etc. Many analysts stated that they would usually spend some time discussing the scenario and HFE(s) with a PRA analyst to ensure they understand the context of the scenario and the operator actions that are likely to occur.

Some analysts noted that they will review previous analyses of the same or similar scenarios (if these are available), in some cases because the HRA they are currently engaged in may be an update to a previous analysis, and in other cases because the previous analyses can provide useful information about the scenario, the types of issues that may influence the scenario and about what was done previously to analyze these.

All of the analysts reported that they will visit the plant and/or simulator, but opinions varied regarding which site is more useful for data collection. In some cases, the interviewees stated that it is almost impossible to realistically reproduce the scenario at the plant (due to the fact that it is an operating plant) and so the information obtained does not reflect how a scenario would actually unfold. Rather, simulator observations were considered by these interviewees to be more useful as the scenario can be more realistically reproduced and so the analyst obtains a more accurate representation of how operators would react in that scenario.

However, almost all of the analysts reported that it can be difficult to observe the exact scenario being analyzed, and often the analyst has to make do with observation of similar scenarios or even with whatever scenario is being simulated at the time of the visit. In this case, the analyst can obtain more general information about team working, communication, use of procedures etc., rather than information specific to the analysis. A small number of interviewees noted that, as external contractors, it can be more difficult to get access to the simulator and so they cannot guarantee that they will be able to observe any simulations. Regardless, all of the interviewees stressed the importance of talking to operators (and simulator trainers, where possible) in addition to performing observations, to verbally talk through and get their opinions on the scenario.

Types of Data or Information Obtained During Scenario Analysis

All interviewees reported similar types of information or data that they seek during scenario analysis, which includes:

- The timeline and sequence of events – this was quoted by most analysts as being the most important aspect of the HRA;
- The role of the operator during the scenario – what actions the operator is expected to perform;
- Which procedures operators use, and how they use the procedures;
- Which alarms and cues operators receive, how they respond to these, and the time available and time taken for response;
- Which decisions the operators will have to make, and how they make these decisions;
- General information about how the operating team works together, and how they communicate during the scenario.

Some of the analysts noted that it may not always be possible to obtain all of the above information; sometimes the analyst must "take whatever they can get". But almost all of the interviewees specified that it is essential to get the operators' opinions on the scenario – how easy or difficult the scenario would be and whether what they are expected to do would be obvious to them (and how) – again stressing the importance of talking to the operators as part of the scenario analysis.

Key Challenges Associated with Scenario Analysis

Getting the timeline and sequence of events for the scenario was reported as one of the main challenges for scenario analysis. Many of the interviewees noted that it can be difficult, especially for less experienced analysts, to understand fully how the scenario will progress and at which points the operators are required to perform certain actions. Lack of experience and knowledge of analysts can also create challenges in terms of understanding the plant systems and how these are operated in reality, to ensure that the analysis reflects a realistic "as operated" scenario, rather than how it is modeled in the PRA. Without a good understanding of the plant response and scenario progression, the analysts may not ask the right questions to get the information that they need for a realistic analysis.

Another key challenge related to the above is ensuring that the information received is accurate and trustworthy. This can be especially difficult given the usual time constraints of the analysis and the busy schedules of operators and simulator trainers at the plant. Some analysis reported that it can be difficult to get access to the most knowledgeable people because they tend to be quite busy and/or it may be difficult to get access to a number of different people to cross-check and confirm information received.

Resolution of Uncertainties or Contradictions

Many of the analysts noted that they often have to deal with uncertainty during the scenario analysis, because they are unable to collect information about all aspects of the specific scenario that is being analyzed. One interviewee noted that there tends to be more uncertainty in the analysis if the scenario takes place over a longer time window (e.g. 24 hours or longer) or if the operating instructions are not yet fully developed because, for example, the analysis is for a new plant system.

In cases like these, the analyst must make assumptions about the scenario. This can be quite typical in HRA and is acceptable, but the analysts stated that it is important to ensure that any assumptions are clearly documented so that they are transparent to the PRA analysts and the plant. Some interviewees also recommended discussing uncertainties with operators and PRA analysts as they may be able to provide the missing information.

The interviewees generally agreed that they do not often get contradictory information during the scenario analysis, but some noted that it can happen because the person providing the information may be unaware of recent changes or may lack the requisite knowledge. In this case, they noted that it is particularly important to ensure that the analyst speaks to more than one operator and, if possible, observes the scenario in the simulator to correct the contradictory information. The analyst can also discuss this with the PRA team to determine which information source is more trustworthy. Again, this should be clearly documented for transparency.

Safeguarding Against Bias in the Analysis

There is a risk that HRA analysts (and consequently the HRA itself) may become biased as a result of the information sources received during the scenario analysis. Some of the interviewees reported that, to prevent this from happening, they treat every analysis separately and always work on the assumption that every plant is different. Some of the interviewees stated that they frequently review previous relevant analyses to see what was done before, which could bias the analyst. However, they noted that, as long as the analyst only uses the previous work to enhance their own knowledge and does not directly copy from this, then it should not introduce a significant risk of bias. It is imperative that the HRA reflects the reality of how the plant is operated now.

Many of the interviewees agreed that it is possible that operators can be biased in their opinions of how the scenario might unfold. It is not unusual for operators to think that "this situation would never happen to us" or "we would never make that mistake". The interviewees noted that it can be difficult to get operators into the right mindset to be able to understand the scenario and the potential errors that could occur. Again, the way that most analysts resolve this problem is to ensure that they interview more than one operator and/or observe the scenario in the simulator.

Good Practices

The interviewees were asked about good practices that they have developed and/or implemented when conducting scenario analysis, as listed below:

- Contact the plant in advance of a visit to ask for relevant information and documentation – this will help the analyst to familiarize themselves with the plant and the scenario to help ensure they focus on the appropriate issues when they visit the plant or simulator.
- Talk to the operators and/or the simulator trainers to understand the reality of the "as operated" scenario.
- Document assumptions and any expert judgments made during the analysis to ensure these are transparent for future readers of the analysis.
- Get operating staff and the PRA team to review the final HRA report to ensure the information underpinning the analysis is correct.
- Review relevant event reports and other documented operating experience (OPEX) reports as these often provide valuable insights into previous similar events.

Desired Improvements

Finally, the interviewees were asked about improvements that they would like to see regarding the scenario analysis phase of HRA. Most of the analysts identified a need for better guidance on how to perform scenario analysis, to reduce inter-analyst variability in terms of how the approach scenario analysis.

A small number of analysts noted that there needs to be better interaction between the PRA and HRA analysts, to ensure that HRA analysts understand the context of the HRA within the PRA and to ensure that the HRA is reported in such a way that it is useful to and provides the information needed by the PRA.

Many of the analysts also reported a need for guidance on how to collect the right information to enable a better assessment of dependency later in the analysis. Assessment of dependency is generally considered to be an area of current uncertainty within HRA, with little method guidance available on how to do this, and so it is unsurprising that this has been cited as an area for improvement.

5. CONCLUSIONS AND NEXT STEPS

A total of sixteen interviews have been carried out to date to input to this study on scenario analysis for HRA. From these interviews, it can be concluded that there is widespread agreement on the importance of qualitative scenario analysis as it underpins the remainder of the HRA. It is also clear from the interviews that the quality of the scenario analysis is often dependent on the knowledge and experience of the analyst. Furthermore, because many HRA methods do not provide explicit guidance on how to perform scenario analysis, there is inconsistency between analysts' approach to this phase of the HRA which can further impact on the quality of the overall HRA.

The interviews also reinforced the importance of going to the plant and/or simulator to collect information for scenario analysis. It is clear that scenario analysis cannot, and should not, be performed in isolation or as a desk-top exercise. The input from talk-throughs, observations and discussions with operating staff, simulator trainers, and PRA analysts is essential to ensure the HRA reflects the "as operated" reality of the scenario.

The findings from these interviews will be used to develop an initial guidance document for analysts performing HRA. This will be supplemented by the findings from some more detailed case studies which are being planned at the time of writing this paper. It is intended that a workshop will be held later in the year to review the guidance with a panel of HRA analysts, and a final version of the guidance document will be issued at the end of the year.

Acknowledgements

The author wishes to acknowledge the interviewees who have participated to date in this study.

References

[1] E. Lois, V. N. Dang, J. Forester, H. Broberg, S. Massaiu, M. Hildebrandt, P. Ø. Braarud, G. Parry, J. Julius, R. Boring, I. Männistö and A. Bye. "*International HRA Empirical Study – Pilot Phase Report – Description of Overall Approach and First Pilot Results from Comparing HRA Methods to Simulator Data*", HWR-844, NUREG/IA-0216 Volume 1. OECD Halden Reactor Project (2008)

[2] A. Bye, E. Lois, V. N. Dang, G. Parry, J. Forester, S. Massaiu, R. Boring, P. Ø. Braarud, H. Broberg, J. Julius, I. Männistö and P. Nelson. "*The International HRA Empirical Study – Phase 2 Report – Results from Comparing HRA Methods Predictions to HAMMLAB Simulator Data on SGTR Scenarios*", HWR-915, NUREG/IA-0216 Volume 2.OECD Halden Reactor Project (2010).

[3] V. N. Dang, J. Forester, R. Boring, H. Broberg, S. Massaiu, J. Julius, I. Männistö, H. Liao, P. Nelson, E. Lois and A. Bye. "*The International HRA Empirical Study – Phase 3 Report – Results from Comparing HRA Methods Predictions to HAMMLAB Simulator Data on LOFW Scenarios*", HWR-951, NUREG/IA-0216 Volume 3. OECD Halden Reactor Project (2011).

[4] J. Forester, V. N. Dang, A. Bye, E. Lois, S. Massaiu, H. Broberg, P. Ø. Braarud, R. Boring, I. Männistö, H. Liao, J. Julius, G. Parry and P. Nelson. "*The International HRA Empirical Study – Final Report – Lessons Learned from Comparing HRA Methods Predictions to HAMMLAB Simulator Data*", NUREG-2127 (HWR-373), OECD Halden Reactor Project (2013).

[5] A. Bye, V. N. Dang, J. Forester, M. Hildebrandt, J. Marble, H. Liao and E. Lois. "*Overview and Preliminary Results of the US Empirical HRA Study*". In Proceedings of the 11th International Probabilistic Safety Assessment and Management Conference (PSAM), Helsinki, Finland (2012).

Can we quantify human reliability in Level 2 PSA?

Lavinia Raganelli[a,b,*], Barry Kirwan[c]

[a] Imperial College, London, United Kingdom
[b] Corporate Risk Associate, London, United Kingdom
[c] Eurocontrol, Brétigny-sur-Orge, France

Abstract: In current safety practice in the nuclear power domain, the demand for Level Two PSA by regulatory organizations has become mandatory, and this has received greater priority after the Fukushima-Daiichi accident in Japan in March 2011. However, there are many challenges in the process of performing a Level Two PSA. Most of the challenges are related to uncertainties in the plant state in such accident scenarios. However, even assuming that it is possible to know the exact extent of damage in a selected scenario, a key question remains: "What level of detail is required for describing the human response?" In reality, damage to equipment and the exact plant status are not predictable; therefore Severe Accident Management Guidelines (SAMGs) and Emergency Operating Procedures (EOPs) offer guidelines for operator behaviour rather than specifying the procedural details of actions. In this paper the appropriate level of detail for the analysis of operator action in Level Two PSA models is discussed, as are the difficulties in conducting Human Reliability Assessment (HRA) for vaguely defined actions. It is found that most current HRA approaches for Level 2 PSA rely heavily on expert judgment, but is such expertise valid? This paper explores potential ways forward for HRA in Level 2 PSA.

Keywords: PRA, Human Factors, Uncertainty, Level 2.

INTRODUCTION

A Severe Accident scenario in a nuclear power plant could lead to loss of containment integrity and a melted core. The consequence is an uncontrolled release of radioactivity into the environment. Level Two PSA is concerned with the progression of accident sequences until the release into the atmosphere and the ability of the containment to withstand overpressurisation. In current PSA practice the modelling of Level Two conditions has assumed greater importance and major regulatory bodies have asked for an updated and improved modelling of severe accident sequences following recent accident history, in particular the Fukushima-Daiichi nuclear power plant accident in Japan on 11th March 2011, which culminated in a meltdown of three of the plant's six reactors, and a large release of radioactive material into the environment. This was the second accident of this magnitude, the first being Chernobyl in 1986 in the Ukraine. These are the only two accidents so far in civil nuclear history to reach a level of 7 on the International Nuclear Event Scale.

In a Severe Accident time-frame the operators play important roles. They need to assess plant status using the information available during the accident sequence, and they then need to operate the available auxiliary systems to maintain or reinstate core cooling to prevent core damage and mitigate plant damage, and avoid atmospheric release once the core has been damaged. The modelling of operator actions in a Level 2 time-frame has not yet been formalized. To date, in a PSA model the practice consists of introducing general, high level actions, and assigning them a probability of failure based on expert judgement. But given the potential severity of such accidents that L2 PSA models, and the given the critical role of the operators during severe accident evolution, both the validity and the utility of the assessment of the human role must come under scrutiny in terms of its validity and utility in assuring reactor safety. This paper therefore discusses the key issue in assessing the human role in severe accidents, namely the large uncertainties that are associated with Level 2 PSA models, especially how they influence human action assessment. The paper then discusses how HRA in L2 PSA is currently done, and explores ways to improve the quality, validity and utility of the process.

---

[*] l.raganelli13@imperial.ac.uk

First, Level 2 PSA itself must be outlined in terms of its objectives and its requirements, and this is discussed in the next section.

REQUIREMENTS FOR LEVEL 2 PSA AND ITS OBJECTIVES

In any process involving radioactive materials, the primary requirement is to ensure the protection of the public, the workers and the environment from the harmful effect of ionising radiation. The main safety principle when dealing with an installation containing radioactive materials is to maintain risk As Low As Reasonably Achievable (ALARA). International Atomic Energy Agency (IAEA) safety standards establish specific requirements in terms of risk assessment and risk acceptability. The standards include the requirement to carry out both a deterministic and a probabilistic assessment of risk. Overall, a PSA provides a methodology for identifying accident sequences that originate from various Initiating Events, and it allows a systematic evaluation of accident frequencies and consequences. Internationally, three levels of PSA are generally recognised:

Level 1 PSA, where plant design and operations are analysed to identify possible event sequences that could lead to core damage. Level 1 PSA provides insights into design weaknesses and into accident prevention.

Level 2 PSA, where a quantitative assessment predicts the consequences following reactor fuel damage. Level 2 PSA is concerned with the analysis of how a release of radioactive materials from the reactor core could lead to an environmental release. A Level 2 PSA provides input into assessing the importance of accident sequences leading to core damage and on the importance of mitigation of severe accidents consequences.

Level 3 PSA, where the consequences of a radioactive release outside of the reactor building are evaluated. In this case the focus is on the environmental contamination and on public and workers' health.

The benefits gained from performing a Level 2 PSA include:
- making sure that systems to mitigate consequences are in place;
- verifying the aptness and the limits of the containment for retaining radioactive materials;
- providing plant personnel with indications for action in case of severe core damage accident sequences.

The Level 2 analysis develops along two trajectories, one is containment response to fuel meltdown through Containment Event Trees (CETs), and the other is progression of the accident sequences through Accident Progression Event Trees (APETs). It is important to be aware of the limitations of modelling. Any limitation affecting the Level 1 PSA will be inherited by the Level 2 PSA. Moreover, if the starting point for a Level 2 assessment is a Level 1 PSA there might be underdeveloped sequences and portions of the Level 1 model. Level 1 PSA is mainly concerned with core damage frequencies, not with the structural safety of the containment and its related systems, thus differences in the development of a station model. If the containment integrity has not been considered in the Level 1 model, the experts need to create a containment model in the Level 1 frame before developing a Level 2 model.

Considering the logic models that need to be developed in a Level 2 scenario, it is important to remain true to the accident sequence chronology. To this end IAEA recommends [1] to divide the accident sequence into three parts:
- Phase 1: Immediate response of the plant to the plant damage state caused by the initiating event through the early period of in-vessel core damage.
- Phase 2: Late period of in-vessel core damage up to failure of the reactor pressure vessel.
- Phase 3: Long term response of the plant.

In a Level 1 analysis numerous sequences and initiating events that lead to core damage frequencies are identified. When the effectiveness of the containment integrity needs to be assessed, there is no advantage in treating every Level 1 sequence individually after core damage and until release frequency. Hence the IAEA recommends that [1]:

"Accident sequences should be grouped together into plant damage states (PDS) in such a manner that all accidents within a given plant damage state can be treated in the same way for the purposes of the Level 2 PSA."

"Plant damage states should represent groups of accident sequences that have similar accident timelines and generate similar loads on the containment, thereby resulting in a similar event progression and similar radiological source terms."

THE ROLE OF THE OPERATOR DURING SEVERE ACCIDENTS

IAEA provides international guidelines on management of accidental scenarios following core damage. The Severe Accident Management guidance gives some information on how to manage and organise recovery and mitigation when extremely unlikely events are happening. During these scenarios the role of the operator is a key one, as automatic responses of the plant control and auxiliary systems may no longer be available or reliable. The objectives of severe accident management are given as follows [2]:

- Preventing significant core damage;
- Terminating the progress of core damage once it has started;
- Maintaining the integrity of the containment as long as possible;
- Minimizing releases of radioactive material;
- Achieving a long term stable state.

As Vinh Dang et al (2009) [3] report, the actions the operators are required to perform during Level 2 accident sequences are different to routine ones, and are also different from those that they perform in known emergency conditions. The main differences are outlined below and are taken directly from the paper.
Some of the most significant differences are:
- The prescriptive character of EOPs (Emergency Operating Procedures) vs. the informative nature of SAMGs (Severe Accident Management Guidelines). EOPs represent a plan that should be followed to the extent possible while SAMGs are more akin to a set of options with informative character. This distinction is not completely unambiguous since there are a few areas with scope for the control room operators' judgment in the EOPs and, conversely, some accident measures with clear criteria within the SAMG.
- The optimal response (whether or not to implement the SAM measure) cannot be fully determined in advance. One of the reasons for the informative rather than prescriptive character of SAMG is that the uncertainties concerning accident progressions hinder the determination of the optimal response in advance. Some of these uncertainties will not be eliminated during the accident, such that the determination of the optimal response for the situation "at hand" remains subject to uncertainties.
- The responsible staff for making the decisions within the EOPs and the SAMG. The decision-making responsibility for SAMG actions lies with the head of the Emergency Response Team (ERT), who is advised by the SAMG team, the Emergency Response Organization (ERO), and, in some cases, other external experts. Some SAM measures require the agreement of the authorities.

- The need to consider radiation exposure in assessing the actions to implement the SAM measure, in terms of feasibility as well as constraints on the execution.

There are usually several choices the plant manager and the operators need to face in order to reach a long term stable state, and a few points are highlighted below, see [2] for further reading. The operators need to be aware that auxiliary and support system could be used for a different purpose than originally intended. Also the option of restoring failed equipment should be considered. All levels of personnel present at the station need to collaborate fully with any external authority or with new personnel reaching the power station. The station manager needs to be aware of how the management hierarchy changes, e.g. whether an emergency team arrives at station, and how to best secure the exchange of information between the new arrivals and the plant team.

The potential for wrong diagnosis by the operators should be minimised by providing redundancy and diversity of signals and feedback. However, the signals should not be confusing and it should be possible for the operator to detect if the signals are giving a wrong warning. In an ideal scenario the operators will have been trained through simulator exercises to respond and react to severe accident conditions. Guidelines such as SAMGs or similar, should also be provided to help taking decision. The SAMGs should always outline both advantages and disadvantages of any mitigation actions.

UNCERTAINTIES IN LEVEL 2 PSA

Uncertainties Due To Physical Model

To obtain a suitable PSA model the dynamics of a Severe Accident need to be understood. However, Severe Accidents, by definition, are extremely unlikely to happen; if the design of the station is effective the probability of its occurrence is practically eliminated ($<10^{-7}$ events per reactor year). Due to this definition, very few SA sequences have been observed so far, the most notable examples being at Fukushima-Daichi station and Chernobyl.

The lack of a sufficient number of observed events means a scarcity of data that would demonstrate the evolution of a Severe Accident. Thus, to understand the development of a Severe Accident the analysts use integrated system computer codes that portray the evolution of the accident from the molten core condition to containment failure. The codes are able to simulate possible plant scenarios, using as input physical parameters describing plant conditions.

Given the lack of data and the difficulties in validating the codes, the models are affected by high uncertainties. It is practical to divide uncertainties related to any physical process into *epistemic* and *aleatory* uncertainties. The first is due to lack of sufficient information while the second is tied to the statistical variation of recurring phenomena. Therefore, in a Severe Accident scenario aleatory uncertainties stemming from the different possible accident developments limit the accuracy of the model. At the same time epistemic uncertainty on the signal and plant feedback impacts on the effective management of the accidental conditions.

To develop an effective time-dependent model in an integral system code (MAAP, MELCOR) the engineers should identify the parameters that regulate the development of the accident sequence. The dominant parameters are related to the following:

- Degree of core damage and configuration of debris/molten core
- Core/corium coolability
- Hydrogen release into the containment
- Reactor Coolant System integrity
- Operator actions that could change the event sequence.

Practically speaking the dominant phenomena usually influence the core coolability, affecting the classical thermal hydraulic parameters: pressure, heat transfer coefficient, temperature, viscosity, etc.

Any mechanism that avoids bypass or failure of the Reactor Cooling System (RCS) is certainly influencing the outcome of an accident scenario. However, if the RCS fails, a successful containment of radionuclides could be still achieved if there are mitigating features that keep the radioactivity in the containment. So the first logical step would be to understand if there are phenomena that allow the core to be cooled inside the Reactor Pressure Vessel.

Once the dominant parameters have been identified, MAAP or similar codes can run different analyses, varying these parameters, namely the heat exchange coefficient of a partially molten core, or the size of the of the primary circuit leak. The variation range of the parameters is usually given by experimental results or theoretical models. Once a sufficient number of simulations have been run, it is possible to extrapolate uncertainties of the parameters conditioning each event in an accident sequence.

The issues in uncertainty evaluation for instrumentation feedback have not been solved. In the ASTEC (Ref) 2013 report it is stated how the instrumentation available during the Fukushima-Daichi accident did not provide sufficiently reliable feedback on cores and spent fuel ponds even weeks after the accident. Also the lack of reliable signals for longer than the first year did not allow precise location of core debris. Unreliable instrumentation renders SAMG implementation difficult, because if the plant status in not well diagnosed, it is impossible to successfully manage the accident sequence. Appropriately performing instrumentation needs to be developed. As an alternative, already-existing instruments need to be positioned diversely or redundantly to allow a better follow-up of core degradation and cooling conditions. These uncertainties influence the operator knowledge of the plant status and so they impact the probability of operators performing the correct action.

Although the uncertainties introduced in a Level 2 PSA due to inaccurate instrumentation feedback are not necessarily considered explicitly when evaluating the Human Error Probability, they could be. The impact of inaccurate instrumentation feedback on operator performance can be quantified using HRA techniques such as NARA (Nuclear Action Reliability Assessment) or SPAR-H (Standardised Plant Assessment of Risk – HRA). Both these techniques contain a specific performance shaping factor related to feedback. The maximum effect on the Human Error Probability for this factor can range from x30 (NARA) to x50 (SPAR-H). It should be noted that in the recent Empirical Benchmark of HRA techniques led by the USNRC [4] that the size of this factor was found to be appropriate. One of the scenarios used in the Benchmark included misleading instrumentation, which had a significant impact on human performance in realistic real-time simulations with licensed operators and industry EOPs.

Uncertainties the PSA model

Once a Severe Accident development is understood, the next step is to build Event Trees that cover the possible paths of the accident sequences. The ETs, as noted earlier, are Accident Progression Event Trees and Containment Event Trees. During the transition from L1 to L2 the frequencies resulting in a damaged core are grouped into Plant Damage States. Then the APETs are developed, showing the possible evolutions of an accident sequence. The evaluation of each accident sequence frequency is carried out explicitly considering the dependent probabilities of all the events occurring in the sequence. The probabilities related to each branch are conditional probabilities, thus the calculation tools need to be accurately set. In each APET (or CET) branching node the possible outcome path is dependent upon multiple parameters. Deterministic calculations need to assess the value range the parameters could assume for a certain path to be taken. The input is a probability distribution or a point value with a range of uncertainty. The analyst then chooses the best way to represent the uncertainty.

Before and after running a Monte-Carlo simulation, in order to identify the relevant parameter sensitivity studies can be carried out, it would help to select the critical input data influencing the release frequency and intensity. Sometimes, if a certain parameter is subject to high uncertainties, the easiest route is to assign it a subjective probability distribution and then verify the influence it has on the overall release frequency distribution through Monte-Carlo simulation.

CURRENT LEVEL 2 PSA MODELLING PRACTICE

Once the Level 2 PSA model has been built, the frequencies or probabilities in each event tree node need to be assigned. The probabilities result from expert judgement, software outputs, plant walk downs and simulator experiments. They can be updated during the plant lifetime if new data becomes available. The whole evaluation process needs to be recorded for future reference. In recent years, Level 2 PSA has been requested by regulatory bodies in many nations having a developed nuclear program. Internationally, for Level 2 PSA the reference models are the already cited IAEA standards and the NRC NUREG-1150 report.

In 2013, the EURATOM consortium has conducted a report relating the state of the art for Level 2 PSA performed in various European countries [5]. The scope of the study is to provide an update of differently implemented Level 2 PSAs and to encourage the reduction of such differences. Most of the level 2 models are somewhere between a full scope PSA model and a limited scope one. As a matter of fact, the different national regulatory bodies have pushed the developers towards prioritising different aspects in the modelling, and thus different levels or detail. As a result, the general IAEA guidelines and the NRC NUREG-1150 have been followed to different levels.

In the Sizewell B Level 2 PSA [6] the process of assigning probabilities to the nodes of the different trees is showed in detail. Because of the peculiarity of post melt core scenarios the branching ratios of the phenomena are not known in detail, so expert judgment is heavily relied upon. The probability of failure, and of success, assigned to each node results usually from an informed decision made by experts. The knowledge could come from different sources; multi-physics computer codes, literature, experiments, other expert judgements, other plants. Unfortunately due to imperfect recording procedures it is not always possible to trace how the choices are made, so it is difficult for subsequent analysts to form an idea on how information leading to a certain choice was gathered.

CURRENT PRACTICE IN MODELLING OPERATOR'S ACTIONS

The role played by the operators during Severe Accidents becomes part of the Level 2 PSA model, because operator actions can significantly influence the outcomes of recovery and mitigation. However the choices made by practitioners on how to include human performance in a PSA Level 2 model differ depending on the resources available and modelling choices. Thus there is a continuous exchange of information between PSA model, SAMGs and thermal hydraulic analyses.

During Severe Accidents the operators are expected to intervene and bring the station to the safest possible state. The Level 2 PSA model needs to reflect the possible choices of operators and how they achieve their goals. The dependencies between actions performed in Level 1 and those that should be performed in Level 2 also have to be explored.

The development of Severe Accident Management Guidelines (SAMGs) is strongly encouraged by IAEA and international regulatory bodies. The scope of such guidelines is to help operators perform the necessary interventions in case of Severe Accidents. As noted earlier, there are substantial differences between the description of mandatory actions given to operators in a Level 1 PSA context and that given during a Level 2 Severe Accident scenario. The SAMGs are informative in nature, describing the context in which the operator should act. The informative approach to management guidelines is justified by considering the following factors:
- The operators have received at least some training for reacting to Severe Accident scenarios
- The operators know in depth the plant response in standard conditions as well as the plant peculiarities and characteristics
- There is uncertainty on the feedback available to the operators and which alarms and signals are going to be available

- It could be necessary to operate systems outside their standard functions and objectives
- Some support from Emergency Response Team will be available.

When the SAM actions are modelled in a L2 PSA, usually the level of detail remains unvaried, as there is no task breakdown. Then, HRA analysts are called to evaluate a very general action, e.g. starting HP feed.

Figure 1 Severe Accident mitigation actions for Maanshan NPP

The Major Mitigation Actions in SAGs and SCGs

SAGs	Major Mitigation Actions	SCGs	Major Mitigation Actions
SAG-1 Inject into the SG	Inject into the SG by using auxiliary feedwater, condensate, and firewater pumps.	SCG-1 Reduction of radionuclide releases	Use containment sprays and fan cooler system, etc., to reduce the fission product release.
SAG-2 Depressurize the RCS	Open pressurizer PORVs and RPV head vent to depressurize the RCS.		
SAG-3 Inject into the RCS	Inject into the RCS by using high head safety injection, low head safety injection, and accumulators.	SCG-2 Containment depressurization	Use containment sprays, fan cooler system, and containment venting strategy to prevent containment overpressurization.
SAG-4 Inject into containment	Inject into containment by every possible path to cool RPV and allow recirculation.		
SAG-5 Reduce the fission product release	Use containment sprays and fan cooler system, etc., to reduce the fission product release.	SCG-3 Control hydrogen flammability	Stop containment heat sinks, operate H_2 recombiner, and resort to containment venting to prevent H_2 burn.
SAG-6 Control containment conditions	Use containment sprays and fan cooler system to control containment pressure and temperature.		
SAG-7 Reduce containment hydrogen	Reduce H_2 concentration by operating H_2 recombiner and igniter.	SCG-4 Containment vacuum control	Stop containment heat sinks and open pressurizer PORVs to pressurize containment.
SAG-8 Flood containment	Flood containment by every possible path to cool melt core debris.		

Currently, HRA practitioners mostly rely to expert judgement, using different frameworks and approaches to justify their evaluations. The external factors, pertaining to the environment, are probably similar to those considered in a Level 1 methodology. However, it is plausible that the external factors influence is going to be different due to the different operator mind-set in a Severe Accident, especially since the operators are expected to act according to their knowledge and deductions, and not simply abide by the procedures.

In some cases existing Level 1 methodologies have been adopted (e.g. use of THERP in Spanish PWRs [5]). If so, some degree of expert judgement is required, as Level 1 techniques do not provide descriptors fully applicable to a Level 2 scenario. The argument for doing so is that the systems claimed in a Level 2 scenario are not different from those used in Design Basis Accident (DBA) scenarios. In the Level 2 PSA for Beznau (Switzerland) power station [7] the author remarks that in the human error the diagnosis and the execution phase were assigned and the overall error probability was obtained by adding the two values. Due to scarcity of date some general HEP were assigned using THERP and ASEP.

Another source of uncertainty in a Level 2 framework which clearly influences the analyst's ability to evaluate the Human Error Probability is the dependency between Level 1 and Level 2 actions. The issue with modelling dependencies is due to difficulty in understanding how actions performed in a Level 1 time-frame influence those performed in a Level 2 time-frame. Also the dependency could be tied to the operators' mental state. For example, if an operator performs an action that does not achieve the expected outcome, he may become stressed and under more time pressure, resulting in an increased likelihood of error in any future performance. At the moment the authors are unaware of an explicit model that links the dependency between Level 1 and Level 2 operator actions.

POSSIBLE APPROACHES TO IMPROVE MODELS FOR HUMAN BEHAVIOUR IN L2 PSA

As introduced above, in a HRA context, the basic issue with modelling and quantifying human reliability in Level 2 PSA is thus one of uncertainty, first over what the operators will face in terms of exact sequence progression(s), plant state, instrumentation availability and reliability, etc., and second in terms of how the operators might react in extreme circumstances (faced with fatalities, fires, etc.). The first is uncertainty in the modelling, and (in theory) could be addressed by developing a number of specific and contextualised scenarios (using experts) to give a description of how one of these scenarios would look in real life. This is similar to what simulator training developers do today, albeit taking it dramatically forward. Nevertheless, this could result in some 'bounding scenarios' which are best engineering guesses about scenario progression and what the operators would see and have available, all in an unfolding timeline. These would be contextualised stories, similar to the CICAs used in the HRA technique MERMOS, but from them it would be possible to develop detailed task analyses and then apply HRA techniques to those task analyses.

For this to work, the HRA techniques must be able to address the second uncertainty problem, namely that of determining how people would react in such scenarios. HRA techniques from an early age considered 'stress' (e.g. THERP, 1983), and then later the concept of 'burden', which concerns the emotional stress associated with, e.g. fatalities, or having to make decisions that have grave and non-recoverable consequences. The effect of burden can be very real, and can even emerge in realistic simulations and emergency exercises, leading to (temporary) decision impairment and delay in action execution. This is well understood in military situations, where the counter-measure is a high degree of training and a very strongly disciplined command control hierarchy (the latter is not appropriate to civil nuclear installations).

The determination of the impact of burden on operator behaviour is difficult due to the lack of good psychological models in this area. The factor of stress can be used, but the derivation of the stress factor in most HRA techniques has not always considered burden in this way, instead focusing on time pressure and workload. A first and obvious solution is to consider utilising expert judgement (e.g. in the framework of a HRA technique such as ATHEANA or MERMOS), but then the question is one of whether we have experts who possess *substantive expertise* (i.e. they have experienced such burdensome scenarios themselves), because if not, then frankly it is not expertise we are eliciting but engineering judgment or even opinion, and it is likely that two different expert groups may differ significantly in their assessments. However, the use of a model to elicit expert judgement whatever the model may be, should allow to estimate uncertainties surrounding those estimates.

Human Performance Envelope approach

Another way to consider the issue is in terms of what is being called the *Human Performance Envelope* [8] in the aviation field. This new concept in human performance borrows from the engineering concept of a flight performance envelope for an aircraft: within the envelope (defined by a set of key parameters such as speed, thrust, angle, etc.) the aircraft performs as expected. Outside the envelope the aircraft will stall and fall out of the sky (if it is put into a deep stall, this is irrecoverable). This is illustrated below for air traffic control (ATC) operations.

Figure 2 Human Performance Envelope Model

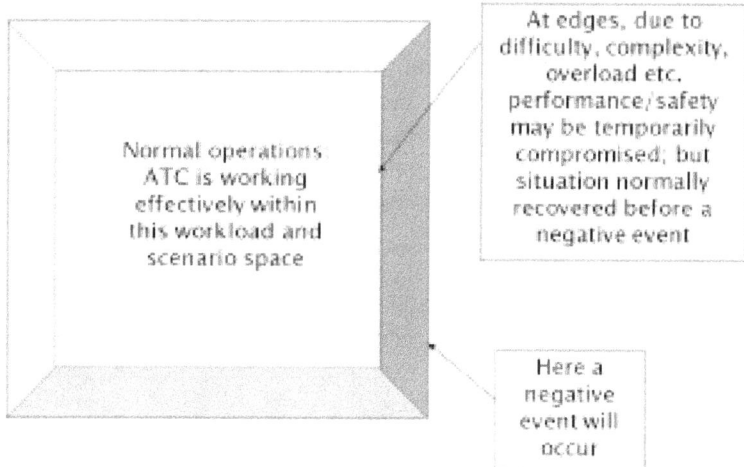

This concept can be applied to the nuclear power plant context, with the central 'plateau' area representing normal performance. The 'slopes' then represent off-normal conditions, as would be considered in Level 1 PSA. Here, behaviour is affected by conditions (including stress) but the operators are capable of dealing with these conditions, and HRA techniques can predict the outcome reasonably well.

Level 2 scenarios would either be on the lower edges of this slope or off it, at which point we are less certain of how operators would behave. Fukushima can certainly be considered to be such an extreme event. Yet the operators did try their best to recover from the situation, exhibiting a degree of heroism (i.e. personnel risking irradiation, as also seen in Chernobyl) and ingenuity (e.g. using car batteries to power up key instruments to better understand plant conditions). Fukushima was a very difficult and burdensome event, but did not result in the human equivalent of a 'stall'.

One advantage of the approach is that it can be used to develop behavioural 'markers' for operating crews and supervisors to help them understand when they are under-performing due to stress or burden, and help them get back to a better performance state. Concepts like the Human Performance Envelope therefore might help to address the safety of operator performance in Severe Accidents, even if it doesn't help with the actual quantification of that performance.

Using a performance envelope approach could be interesting, and increase the utility of the overall HRA approach, but the justification for it in front of regulatory bodies might require some explanations and more details, as it is a far cry from current accepted practice. Most likely it would only be countenanced as part of the qualitative analysis or functions provided to determine how to support operators in Severe Accidents.

Taking this qualitative perspective, the over-riding question becomes one of how to support operators in such events, since we will never be able to predict all the things that can happen in their various combinations and permutations. There is a need to focus on the collective *capability* of the shift team to respond, and understanding the principal factors that drive performance in Severe Accident situations. One study that addressed this, to an extent, is the DORRET approach.

The DORRET Approach

In the mid-90s, a nuclear R&D project called DORRET [9] (Determination of Operator Recovery Reliability over Extended Timescales) considered human performance in deteriorating situations over long timescales. To explain the context, at the time the UK had (and still has) mainly gas-cooled reactors rather than Light Water Reactors (LWRs). The accident dynamics of such reactors includes scenarios slower than those considered for the LWRs, e.g. up to 24 hours. The research also addressed not only reactors but reprocessing facilities, which also have some long timescale scenarios. No HRA technique at the time, other than expert judgement, addressed such scenarios (typically HRA was focused on scenarios ranging from 30 seconds up to 2-3 hours).

The DORRET Project gathered together a group of experts to consider human performance in such scenarios, which could include those that went outside the existing procedures and training system; hence the relevance of DORRET to Level 2 PSA. The DORRET team also analysed 78 relevant incidents (out of an initial dataset of >5400 incidents). Towards the end of the DORRET project a six-stage model of the operator-led recovery process was suggested:

1. Recognition of the need for action
2. Recognition of the recovery options
3. Nature of the complexity of the recovery options and their consequences
4. Decision making and support
5. Organisational capability to respond
6. Check/feedback on success of the task

For such events therefore, there needs to be continuous recognition of the need for action. This can be elaborated following Fukushima, since there was apparently a focus on the reactor area and less focus on the fuel cooling ponds; therefore recognition needs to extend to all major threat conditions (so there are no 'blind spots'). The second stage is recognition of recovery actions – because if the operators cannot think of anything to do, then they cannot act to mitigate on-going risks. The third stage is a recognition of the complexity of the recovery options, e.g. given that there could be fire/flooding/rubble/irradiated areas/communication or instrumentation difficulties, etc. that impede normal execution of actions. The fourth refers to the collective decision-making intelligence at the heart of the recovery effort. Most nuclear facilities that experience severe accidents will have either onsite or remote technical assistance (or both) within a couple of hours or less, and this is useful because it can help moderate decisions that may have to be made 'in the heat of the moment'. The organisational capability to respond is of course critical, and incorporates the training, selection and safety culture of personnel as well as the sufficiency of qualified and experienced personnel, whether during day or night shift, and the effective organisation of communications. The last stage of verifying task success links back to the first stage, closing the feedback loop.

From DORRET to NARA

The DORRET work was used to inform the NARA [10] (Nuclear Action Reliability Assessment) HRA technique, which has an 'Extended Timescale Factors' (ETFs) module, which also potentially has relevance to Level 2 PSA. The ETF module is based around five factors related to the DORRET recovery stages:

- **Information** – e.g. prioritised alarm system, diverse communication systems, diverse monitoring of key parameters/critical functions
- **Scenario characteristics** – amount of time available (segmented between 2 and 24 hours), environmental conditions local at the plant (e.g. fire, storm, rubble after explosion, etc.), confusion due to misleading indicators
- **Guidance** – quality of procedural guidance available (e.g. EOPs, SAMGs, etc.; shift changeover protocols (for > 6hrs)
- **Stress** – e.g. burden of coping with fatalities or operator concern about worsening the environment or causing major capital damage by extreme recovery measures
- **Teams** – the degree of team training in simulators and site incidents, support by technical support centre, etc.

At the moment the ETFs are not specifically for Level 2 PSA, but they could be developed for such a purpose. More generally, such work as in DORRET and NARA can be a starting point for deciding the most critical factors needed to quantify human reliability in Level 2 PSA scenarios, and also how to build a strong operating personnel capability in our nuclear facilities to respond to rare and extreme events.

Level 2 Event Training

Of all the factors considered, two seem to be dominant: information, and operator team preparedness (training/teamwork). Information available is primarily a design issue, so that as far as practicable there are good information systems that can highlight what is going wrong, and ensure that the operators can 'see the wood for the trees', and are diverse in nature so that misleading indications can be detected and compensated for by the operators.

Training needs to 'push the envelope' for operators, occasionally placing them in severe conditions that are not clear, and where normally available resources are compromised. Some nuclear power related organisations and institutions have indeed carried out such simulations and exercises, and operators usually find them challenging, but also rewarding, giving them the confidence that should such an event occur in reality, they stand more chance of 'keeping a cool head'.

CONCLUSIONS

Severe Accidents are rare but do happen, and we don't know where they might happen, given that they can be due to internal events (e.g. Chernobyl) or external events (e.g. Fukushima, triggered by a tsunami). Since operator action and resourcefulness is critical in such events, there is a dual need to prepare the operators as far as is reasonably practicable for such events via instrumentation, procedural guidance and training, and to assess the likely operator response via HRA in PSA in order to determine and mitigate vulnerabilities.

The main problem in developing a quantification approach for human action in a Level 2 PSA framework is linked to the uncertainty in how Severe Accidents develop. In answer to the question 'Can we quantify human reliability in Level 2 PSA?' the answer is a tentative 'yes': we can 'have a go' with some of the existing tools, and these can include both first and second generation tool s such as THERP and ASEP, NARA, ATHEANA and MERMOS. But we are largely at the stage of making educated guesses, since in all cases such tools have not been designed for Level 2 PSA, and/or rely heavily on expert judgment, either via analysts or subject matter experts, when in reality we have very few experts who really know what it is like in such scenarios.

What is missing is a valid psychological model of operator behaviour in such conditions, or at least a better understanding of the key factors and how they drive human performance in L2 scenarios. The DORRET approach, or a similar study, could be revised to inform the expert judgment process when using HRA techniques in L2 PSAs. More L2 real-time simulations where operating crews are pushed well beyond the normal scenario boundaries, as have been carried out in some NPPs, could be useful starting points to understand how operating crews react, and how behaviour alters when transitioning from a L1 scenario into a L2. Insights like this could also help in the development of better dependence modelling for L2 PSAs.

In conclusion, what is needed is a programme of work that seeks first to understand the limits of PSA modelling for L2, and then to understand human behaviour in such scenarios. From such understanding, factors can be extracted either to guide experts participating in or conducting HRAs for L2 PSAs, or to inform HRA techniques and models themselves. Given the inevitable limitation of modelling for L2 PSA, a related branch of work needs to focus on strengthening the operating crew capability, both via advanced simulator training and also by consideration of other Human Factors approaches (e.g. Human Performance Envelope) that may help operators continue to cope when everything appears to be failing around them.

REFERENCES

[1] IAEA (2010) Development and Application of Level 2 Probabilistic Safety Assessment for Nuclear Power Plants, Vienna, Austria

[2] IAEA (2009) "Severe Accident Management Programmes for nuclear power plants", Vienna Austria

[3] V. N. Dang, G. M. Schoen, B. Reer (2009) "Overview of the Modelling of Severe Accident Management in the Swiss Probabilistic Safety Analyses" ISAMM 2009, October 26 - 28, 2009 Böttstein, Switzerland

[4] Bye, A., Lois, E., Dang, V., Parry, G., Forester, J., Massaiu, S., Boring, R., Braarud, P., Broberg, H., Julius, J., Männistö, I. and Nelson, P. (2012) "International HRA Empirical Study – Phase 2 Report Results from Comparing HRA Method Predictions to Simulator Data from SGTR Scenarios." NUREG IA-0216 Vol 2, USNRC, Washington DC 20555.

[5] E. Raimond et al (2013) ASAMPSA2 "Best-Practices Guidelines For Level 2 PSA Development And Applications", April 2013

[6] M. L. Ang, N. E. Buttery (1997), "An approach to the application of subjective probabilities in level 2 PSAs" Reliability Engineering & System Safety, Vol 58, Pages 145–156, November 1997

[7] Martin Richner (2006) "Modelling of SAMG operator actions in Level 2 PSA" PSAM-8, May 2006, New Orleans, USA

[8] Edwards, T., Sharples, S., Wilson, J. R., Kirwan, B. (2010). "The need for a multifactorial human performance envelope model in air traffic control" *Presented at the HCI-Aero 2010 conference, 3rd-5th November, Cape Canaveral: USA*

[9] "Manual for the DORRET technique" Vectra Report No. 1005-215-TD01. Revision 1. October 1997.

[10] Kirwan, B., Gibson, H., Kennedy, R., Edmunds, J., Cooksley, G., and Umbers, I. (2004) "Nuclear Action Reliability Assessment (NARA): A data-based HRA tool. In Probabilistic Safety Assessment and Management" 2004, Spitzer, C., Schmocker, U., and Dang, V.N. (Eds.), London, Springer, pp. 1206 – 1211